T0092373

# The Power of Stars

Bryan E. Penprase

# The Power of Stars

Second Edition

Bryan E. Penprase
Division of Science
Yale-NUS College
Singapore, Singapore

ISBN 978-3-319-84943-0 ISBN 978-3-319-52597-6
DOI 10.1007/978-3-319-52597-6


Softcover reprint of the hardcover 1st edition 2017


Printed on acid-free paper

This Springer imprint is published by Springer Nature
The registered company is Springer International Publishing AG
The registered company address is: Gewerbestrasse 11, 6330 Cham, Switzerland

# Foreword to the Second Edition

This new edition of *The Power of Stars* benefits from a large number of people who have kindly offered their comments, donated photographs, and most importantly shared their impressions of the earlier draft as it has been read across the earth and used in many university courses as a textbook. Since the first edition, science and technology has also marched forward and is rapidly transforming our knowledge of the universe with new telescopes, dark matter experiments, and discoveries peering into the cosmic past and refining our understanding of our physical origins. Equally impressive progress in our study of the human past has enabled new X-ray techniques peering deep inside ancient artifacts and documents and new radars and lidar devices peering beneath the earth and penetrating jungle canopies. Since the publication of the first edition, there have been several major discoveries such as the first detection of gravitational radiation with LIGO, the measurement of cosmic background radiation with the Planck satellite, and ever-larger and more sophisticated telescopes and instruments studying supernovae, gamma ray bursts, and other mysterious sources in the night sky. These new tools have added immensely to our knowledge and have further increased the mystery and wonder presented by the night sky—that source of power within the title that evokes scientific study and that has inspired cultures throughout the centuries.

The new edition includes several improvements. Chapter 1 includes new sections on Native American sites such as Chimney Rock and Hovenweep with photographs from site visits, new information on Mercury and Venus transits, and new figures to help guide amateur observers in their viewing of planets. Chapter 2 includes new sections on Hawaiian and Polynesian navigation, finding guides for some of the main star clusters, nebulae, and galaxies in major constellations, new information on the Inuit skylore, and improved sections on Indian observational astronomy and observatories. Chapter 3 includes expanded sections on Aboriginal Australian creation, a new section on the ancient Hindu creation stories, and figures from trips to Angkor Wat, Cambodia, and Mamallapuram, India. Chapter 4 includes updates on the Chinese cosmologies and an extensive description of Hindu and Buddhist cosmologies. Chapter 5 includes new photographs and information from the Purple Mountain Observatory, Nanjing, China, and an expanded section on Chinese calendars and timekeeping. Chapter 6 includes an update on the Antikythera mechanism, including figures from reconstructions of the device based on new research, and updates on the rapidly developing world of atomic clocks—which have improved by more than a factor of 100 since the first edition. Chapter 7 includes the research on Stonehenge, photographs of the Hovenweep site at summer solstice, additional information on Islamic city orientations, a description on new research with lidar to uncover lost cities near Angkor Wat, and an extensive analysis of orientation of shrines and temples in Tamil Nadu based on site visits during 2014. Chapter 8 includes an improved section on Turrell's artistry and an updated look at the top skyscrapers—soon to be joined by ever-taller skyscrapers that attempt to defy the laws of physics and architecture in their reach for the stars. Chapter 9 includes updates on the latest research on dark energy and dark matter, including new images of dark matter from the Hubble Space telescopes, an update in the largest telescopes of the world, and a mention of the discovery of the Higgs boson and its connection to the early universe. Chapter 10 retains its philosophical tone but adds a summary of the "concordance cosmology" with updated results from new types of cosmic ray and neutrino detectors and new information on the discovery of gravitational radiation and its profound impact on our understanding of the universe.

In the new edition are a number of improvements that came from an extended stay in Asia at the new Yale-NUS College in Singapore. Being based at Yale-NUS College has helped the development of this new edition through innovative programs like Week 7, which allowed me to explore Tamil Nadu, India, with students and my anthropology colleague Bernard Bate and to explore China and its ancient and modern observatories with a team of students and Charlotte Evans, a long-term resident of Shanghai and expert in Chinese culture. Additional fieldwork within Asia included visits to Angkor Wat, Cambodia, and Borobudur, Indonesia, and collaborative research trips to Dubai, Taiwan, Kuala Lumpur, and many other cities hosting the world's highest skyscrapers. Visits to the Raman Research Institute (RRI) in Bangalore helped acquire new information on ancient Indian astronomy and cosmology. Collaborative visits to Swinburne University in Melbourne, Australia, allowed for additional resources on Aboriginal astronomy and culture at the Victoria Library and other cultural centers.

I would like to thank my hosts and collaborators on those visits—Lakshmi Sarapalli from RRI, B. S. Shylaja from the Jawaharlal Nehru Planetarium, Michael Murphy from Swinburne University, and Ray Norris from the CSIRO, Australia. I would also like to thank David Boyle from the Crow Canyon Archaeological Center for the chance to join in their summer solstice trip to Hovenweep and other important Native American sites in Southern Utah, Colorado, and New Mexico. Editorial comments came from Prof. Aparna Venkatesan from San Francisco University, who both "field-tested" the book within her class. I would also like to thank my departed colleague Bernard Bate for his eternal enthusiasm and joy in teaching and exploring the stars with me—as we attempted to impart "historical awe" to our Yale-NUS College students. I would also like to thank Gavin Flood for reviewing the sections on Hindu and Buddhist Cosmology.

Like science itself, the updates to this work will never reach a conclusion. And if the time since the first edition is any measure—the rate of advances of science and archaeology research will continue to utterly transform our understanding of our past. As we better understand the origins of the early universe, early civilizations, and the origins and nature of matter, we also better appreciate our own origins as citizens of a global human community—each responding to the power of stars in our own ways.

Bryan E. Penprase
Yale-NUS College, Division of Science
Singapore 138610, Singapore

# Contents

# Chapter 1

# Our Experience of the Sky and Star Knowledge

*If you will think of ourselves as coming out of the earth, rather than having been thrown in here from somewhere else, you see that we are the earth, we are the consciousness of the earth. These are the eyes of the Earth. And this is the voice of the earth.*

Joseph Campbell (Campbell and Moyers 1988)

We often hear that we live in a "global community" in which we are linked together across the Earth. The links we usually consider are materialistic—financial transactions, Internet services, trading of goods from distant continents. Yet we are also linked in a more profound way, as citizens of the Earth, surrounded by the planets of the solar system and the dazzling constellations of the night sky (Fig. 1.1). Before international trade and before the Internet, humans looked up at the night sky, and traded tales and theories of what it all means. Many of the tales endure in oral tradition, in mythology, and in more subtle forms.

**Fig. 1.1** The Earth, viewed from the Moon (image credit: NASA)

In these pages we will explore the many ways in which humans throughout time have responded to the sky. Sometimes they responded with fear, sometimes with great artistry and other times with great rationality. Each culture of the world for thousands of years viewed the same stars and planets we can observe on any clear night. And the power of the night sky, filled with shimmering stars, has left its mark on human civilization. Each culture has built systems to organize the stars into figures,


B.E. Penprase, *The Power of Stars*, DOI 10.1007/978-3-319-52597-6_1

and to tell tales using stars to embody the highest values of the civilizations. The power of stars has also moved humans to create giant celestially aligned structures—from the pyramids of Giza, to Stonehenge, to the large ceremonial kiva of the Anasazi.

The response of ancient people to the power of stars speaks to us across the centuries with a mysterious voice which resonates within us and somehow moves us as we recognize many familiar emotions from ancestors long past. The voices are stilled and, in many cases, canopies of jungle or vast deserts have enveloped their lands. Yet the ancient monuments speak of our common humanity and our common longing for knowledge of the meaning of our physical universe.

One of the most powerful elements of archaeoastronomy is the contrast between the still and vacant monuments, where once thronged a thriving empire, and the relentless rhythms of the cosmos, eternally completing cycles of planetary and stellar motions. These cycles often are measured in centuries, millennia, and even longer timescales. We may use this contrast to see in sharp focus our small place in the universe and share in the mysteries of how this universe works along with those who have tried to answer these questions in preceding centuries.

The story of archaeoastronomy is also an unveiling of the mind of the ancient astronomer through oral traditions, from often subtle and ambiguous clues left in rock, bone, and, in rare cases, writing. To understand these clues it is necessary to know the society, the land, the history, and the skies above the ancient land in question. We must seek to learn these lessons with humility, with compassion, and leave our Western viewpoints behind. We are about to explore a world in which shaman priests regularly summoned gods to and from the underworld, in which cycles of the universe threatened the entire society with destruction without the appeasement of human sacrifice, and in which gods and other spirits from the skies ruled the world by their will. We can expect to only have at best an incomplete vision from this process, but as we discover more about these long-lost ancestors we also learn much about ourselves.

What can we learn from the study of ancient astronomy and cosmology? We may dismiss (as do many scientists) these early cosmologies as “primitive” or “wrong.” Clearly if we assume our own science of astronomy and cosmology to be “modern” and “correct” we immediately share the common belief among nearly all cultures ancient and modern—that they are the best, most modern, and most sophisticated of all cultures before or since. In our lives we benefit from the perspective gained from centuries of effort from the great scientists who have assembled our scientific cosmology. This perspective allows us to easily see where the ancient cultures got it “wrong.” We can also imagine, however, that some future society will judge some aspects of our cosmology as quaint, with its expansion in four-dimensional space, and the formation of the universe in a “Big Bang” where all the present-day particles are created. As we explore the images, myths, astronomy, and cosmologies of the ancient world, it is important to realize that we “moderns” will someday too be ancients.

## The Human Experience of the Sky

The essential element of archaeoastronomy, which binds us to our ancestors, is available to all of us each night. A clear, dark, starry night sky brings out intense feelings in even the most sophisticated city dweller. These feelings vary among individuals, but for most they are intense and connect us to something sublime and transcendent. The night sky calls to us and over the years humans responded with their star tales, their monuments, their theories, and their observatories with gleaming polished glass and sophisticated instruments.

In early eras, the power of stars inspired tales of ancestors, their quest for food, the animals we lived with, and the delicate balance needed to maintain life. The stars taught us the way to live with each other and our environment. As our civilizations grew, the stars provided the basis for binding together hundreds, thousands, and now millions of us in commerce, in religion, in science, and also in war. This process began with setting the calendar, and each civilization from the world used the stars, the Moon, and the Sun to provide a tempo for the dance of its people as they planted crops, moved goods, and planned new ventures. The process continued as writing empowered people to record over many centuries patterns of events which previously were disconnected—eclipses, comets, and the motions of planets all seemed to repeat, giving great power to the ruler or astronomer who could predict the inner workings of the mysterious clockwork that seemed to run the universe. Our response to the sky continued and evolved from ritual sacrifice to a quiet contemplation of the mysteries of the sky in which the most sophisticated tools of our species were arrayed to wrest the secrets of stars. Counts of stars and catalogs gave way to spectra, which unlocked the secrets of what the stars were made of and how far their light has traveled.

With our knowledge of the stars we better understand our Earth, as we realize that the stars and the Earth are made of the same elements forged in the early universe and in the cauldron of earlier generations of stars. These elements have found refuge in the dirt and rock of this small outpost of life at the edge of a great galaxy. The stars call to us and exert a power on us, perhaps since we are formed from them as all the atoms in our bodies come from inside a long-lost star.

In archaeoastronomy we can combine our search for answers about the stars with a need to connect to the past—we can connect with the past lives of generations and civilizations long gone and even connect with the past of our universe, which each moment arrives to us in the light of ever more distant stars (Fig. 1.2).

**Fig. 1.2** The Eagle Nebula (figure credit: NASA)

## *Beginnings*

We can begin our study of the sky much as our ancestors did. Simply by looking up at the Sun, the Moon, planets, and stars, opening our minds, and by noticing subtle changes and remembering, we can become aware of the intricate clockwork of the skies. On any given day, the planets, the Sun, and the Moon form a pattern that changes. All that we need to see this pattern is time and a place to observe the sky. If the observations occur at the same time each day, the daily and annual stellar motions and planetary cycles begin to reveal themselves. In the following sections we will describe the basic daily and yearly motions of the Sun, the Moon, and planets, how these motions were observed and commemorated by civilizations around the world, and how you can observe and experience them yourself.

To the ancients, the sky was dominated by the "seven luminaries": the Sun, the Moon, Venus, Mercury, Mars, Jupiter, and Saturn. It is not a coincidence that our seven-day week is based on these objects and, in many languages, the day names correspond to these heavenly bodies. A quick look at day names in several languages shows this correlation in Table 1.1.

**Table 1.1** Names for the days of the week across the world

| English name | Spanish | French | Hindi | Luminary |
|---|---|---|---|---|
| Sunday | Domingo | Dimanche | Ravivar | Sun |
| Monday | Lunes | Lundi | Somvar | Moon |
| Tuesday | Martes | Mardi | Mangalvar | Mars |
| Wednesday | Miercoles | Mercredi | Budhvar | Mercury |
| Thursday | Jueves | Jeudi | Vrihaspatvar | Jupiter |
| Friday | Viernes | Vendredi | Sukravar | Venus |
| Saturday | Sabado | Samedi | Shanivar | Saturn |

Planetary associations with day names are nearly universal across cultures and nations. For example, day names in India correspond to the same celestial objects as the European day names. In Hindi, the planet Mars is *Mangal*, Mercury is *Budh*, Jupiter is *Vrihaspati*, Venus is *Sukra*, and Saturn is *Shani*, and the day names follow the planet names. The ordering of the day names is inherited from the ancient Babylonian astrologers, who gave each of the seven luminaries power over each of the 24 h of their day in a sequence based on the distance of the object according to their cosmology. For the Babylonians, our Earth was a slab of rock floating in a watery universe, with the planets themselves floating along rivers across the sky in the order (in distance above the Earth) of Moon, Mercury, Venus, the Sun, Mars, Jupiter, and Saturn, with the stars beyond. The later Greeks and medieval Christians adopted some aspects of this model, from which comes our sequence of day names (Fig. 1.3).

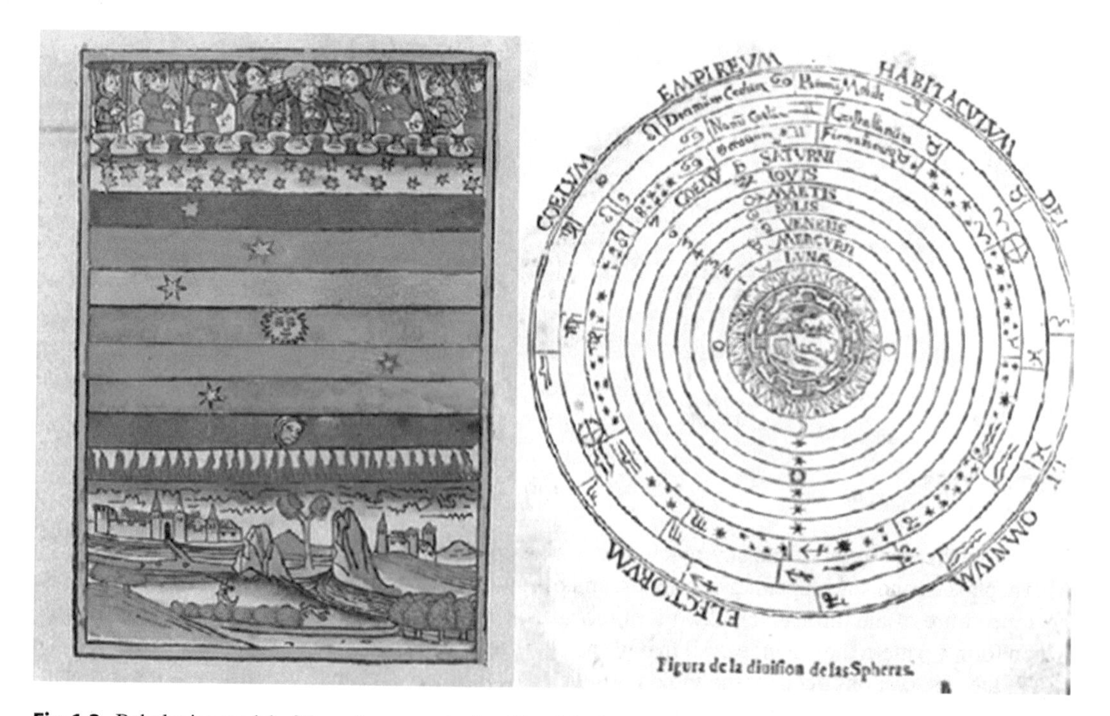


**Fig. 1.3** Babylonian model of the universe, as depicted in a Medieval Christian manuscript (*left*), and in a later exposition of the Ptolemaic cosmology (*right*). The *left image* comes from Konrad von Megenberg, Buch der Natur (1481), and the *right image* comes from Peter Apian, La Cosmographia (1551) (images courtesy of the Library of Congress and Claremont Colleges Special Collections)

The places of the seven luminaries above our floating Earth were determined based on the estimates of their changeability—the Moon varies most drastically, and so was closest to the Earth. Mercury and Venus, with their darting about through the sky on either side of the Sun, were deemed next, just below the Sun, while Jupiter and Saturn, with orbital periods of 12 and 26 years, respectively, were deemed farthest away and outside the Sun's layer.

Cycling through the seven luminaries for each of 24 h gives three complete cycles for 21 h and then an additional three objects to finish the day. In this way we arrive at our present sequence of day names, beginning with the Sun and ending with Saturn. This sequence of days and seven-day weeks is observed now by nearly all of the world's people.

The imbedding of astronomical information inside our seven-day week is an example of how ancient tradition exists among us, sometimes without us knowing it. The motions of the objects in our sky also continue with their clock-like regularity, also without our noticing. Our tour begins with a survey of these seven objects, sacred to nearly all ancient people, and the meaning of the Sun, Moon, and planets to people around the world. We will also describe the types of observations of these bodies, some subtle, and some obvious, that you can do to duplicate many of the observations performed by ancient cultures.

We will follow this same ordering of the Moon, the Sun, and planets, as the ancients used, as we describe the motions of the sky. You can see these motions yourself, with observations that mirror the practices of ancient astronomers. The observations are simple and require neither technology nor special training—only a good place to visit to sight the sky, imagination, and patience.

## *Moon*

The moon presents a metronome for the sky with which humans regulated their activities for centuries. Its dramatic waxing and waning give rise to our month and was the foundation for most ancient calendars and festivals. Ancient people associated the Moon with death and resurrection, with fertility, and with creation gods (Fig. 1.4).

We all have seen the Moon and its phases, but many are not aware that its shape, the meaning of the visible features on the surface, and the orientation in the sky are all a function of our culture and where we are located on the Earth. The "man on the Moon" is just one of many ways of looking at the Moon's disk.

The pattern of dark and light on the surface of the Moon, originating in lava flows of basaltic rock that erupted on the Moon's surface 3.8 billion years ago, provides a cultural Rorschach test. For many of the Native Americans the Moon's surface shows a rabbit, as evidenced by Zuni art, and from oral tradition of the pueblo people. For the Navajo, the First Man and First Woman emerged from the underworld into a dark and cold new world. They created the Sun from a slab of quartz crystal, decorated it with feathers, and attached the disk to the sky with darts of lightning. Then moon was made and decorated. But to move, the Sun and Moon required feathers to mark their path, and the Sun and Moon followed the path until the wind pushed some of the feathers into the face of the Moon, accounting both for its changing aspect and also its apparent drift on the sky (Williamson 1984, p. 44).

For the Chinese, the surface of the Moon shows both a rabbit and a frog, associated with the tale of Heng O, the princess married to the archer god, Shen Yi, who won an elixir of immortality from the gods. One day while Shen Yi was out, Heng O stole the pill, and when pursued by her angry husband, floated to the Moon to escape. Once on the Moon, she encountered a rabbit that was mixing medicines for the gods. When Heng O tried some of these medicines, she turned into a toad. To this day, the Autumn Moon Festival celebrates the fateful journey of Heng O, where she can be seen as a toad on the surface of the Moon, along with the rabbit (Krupp 1991, p. 72) (Fig. 1.5).

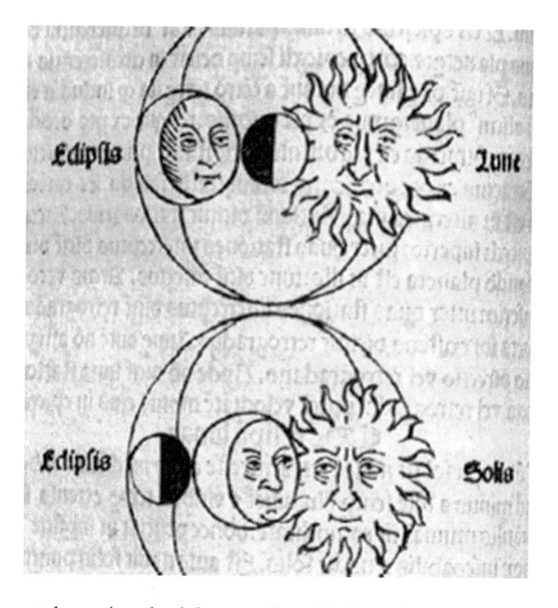


**Fig. 1.4** Illustration of the most dramatic celestial event viewable from the Earth—the lunar and solar eclipse—from Sacro Bosco, Johannes, Regiomantus (1482) (image courtesy of Claremont Colleges Special Collections)

**Fig. 1.5** For the Chinese, the dark Maria on the Moon reminded them of a rabbit cooking moon cakes. This figure shows the outlines of the rabbit and his pot over a photograph of the lunar surface (image from http://commons.wikimedia.org/wiki/File:Rabbit_in_the_moon_standing_by_pot.png)

The orientation of the Moon on the sky also varies with the season. In some parts of the world, the crescent moon points horizontally for much of the year, then tilts at the time of rainy season. In African tradition, the crescent moon is thought to hold moisture from the sky like a bowl.

When the crescent moon tilts relative to the horizon, the bowl is emptied and creates the rainy season (Aveni 1993). Interestingly, the Native American Chumash had the opposite belief, perhaps because of the different weather patterns in California compared to Africa. For the Chumash, a crescent moon which appeared vertical relative to the horizon meant the Moon was "empty" of celestial water, while a horizontal crescent moon indicated a rainy season, as if the Moon were filled to overflowing with water (Hudson and Underhay 1978, p. 76).

The division of the Moon in its phases has also been explained as a form of resurrection by the Egyptians, who associated these lunar phases with the dissection of the God Osiris by his rival Seth, into 28 parts, and his reassembly by his consort Isis (Krupp 1991, p. 66). The California Chumash and Yokut tribes considered the Moon's phase to be the flapping of the wings of the sky eagle, or condor, and the slow movement of the wings blocking the front of the Moon resulting in the cycle of lunar phases (Hudson and Underhay 1978, p. 80).

Regardless of how you consider the surface features or shape of the Moon, the degree of illumination is a very useful tool for telling the time. The Earth, the Moon, and the Sun form a huge triangle in space (Fig. 1.6). You can use your imagination and complete this triangle, to determine the location of the Sun from a view of the Moon and from the location of the Sun relative to earth you can determine the time. Simple rules to bear in mind—full moons rise at sunset, first quarter moon transits at sunset (and rises at noon), third quarter moons transit at sunrise (and rise at midnight). Once you see the Moon one night, you can also know that it will rise about 50 min later each night, causing it to move through the sky by about 12°, or approximately the width of your handheld at arm's length. While you are at it, you can measure the Moon's angular size—it is about the size of your pinky fingernail—only ½°.

**Fig. 1.6** NASA's Galileo spacecraft snapped this picture of the Earth and Moon together on its way to the planet Jupiter (image from http://grin.hq.nasa.gov/ABSTRACTS/GPN-2000-001437.html)

The Moon presents one easily tested observational challenge—is the Moon larger when it rises or when it transits? Many students in astronomy classes believe the Moon is larger at moonrise. You can test this—just use a small object, like a paperclip, and bend it to fit on both edges of the Moon as it rises. Wait a few hours and try again to see if this paperclip still fits on the Moon as it moves across the sky. The "earth moon illusion" is one that has been studied by psychologists and astronomers alike (Ross and Plug 2002) and you can see for yourself whether this illusion is real.

One important part of moon viewing is the date of the first crescent moon. A large number of civilizations base their calendar on the lunar month, and the first sighting of the crescent moon is essential for starting many of the most important festivals and celebrations of the culture. Ramadan is the most famous of these festivals. Traditionally it is the job of the imam to view the crescent moon and declare its first sighting to set the beginning of Ramadan. Many times due to atmospheric conditions or differing eyesight, disagreement will erupt in setting the date of this festival. Recently some Islamic countries have begun using technical means such as telescopes, and in some cases aircraft, for spotting the first crescent moon of the lunar month—an interesting mix of ancient lunar astronomy, and modern technology (Shaukat 2009).

In addition to locating the lunar month, both ancient and modern astronomers view the Moon as it moves across the sky and try to track its larger cycles of standstills and eclipses. Because the Moon's orbital plane is inclined 5° from the Earth/Sun orbital plane (known as the *ecliptic* plane), it will move 5° above and below the Sun. The Moon's orbit has its own 18.6-year cycle in which the orbital plane slowly precesses about the celestial sphere. The intersection of the Moon's orbital plane and the ecliptic plane is known as the "line of nodes." Each year the Moon will move a little in its progression around the plane of the Earth's orbit about the Sun, with a long 18.6-year period. While this long cycle continues, the Earth orbits and spins, and the path of the Moon sometimes crosses the line of nodes, giving rise to an "eclipse season." During an eclipse season, with each new moon it is possible to have a solar eclipse, and with each full moon a lunar eclipse is possible.

How can ancient people measure such long cycles? Usually the process involves the development of writing, recording eclipses, and observing recurrences of these eclipses. Ancient Chinese and Babylonian astronomers were excellent record keepers and from their centuries of eclipse observations it was well known that eclipses recur in approximately 18 year periods known as the "Saros cycle." With three of these Saros cycles, or approximately 54 years, eclipses can recur on the same part of the Earth.

But even without the invention of writing it is also possible to notice the long cycles of the lunar motion. The position of the Moon swings from one side of the Sun to the other, which at the extreme points give rise to what is known as a "lunar standstill." At the standstill, the Moon is at its highest or lowest point on the celestial sphere (about 5° above or below the ecliptic). When this happens the Moon rises in its extreme north or south positions, something recorded by ancient builders of Stonehenge and other Native American monuments such as Chaco Canyon. Many of the buildings in Chaco Canyon and other southwest sites such as Hovenweep monument appear to have walls and windows aligned with lunar standstill directions. One interesting example of a standstill observation from the Native Americans is the "Chimney rock" site in Colorado (Fig. 1.7). Here ancient people built a ceremonial kiva on a ridge overlooking a notch in two rock spires, which coincides with the direction of a lunar standstill. Why would they do this? The only answer is that the Native Americans were keenly aware of the cycles of the Moon's motion, and used the standstills to set multi-year calendar cycles (Fig. 1.8).

The Chimney Rock site is located high above an arid plain where the soil would not support productive agriculture. Nearby Taos Pueblo people associate the site with the War Twins of their mythology, giving some support to the site as a defensive location. Archeological evidence (including tree ring data from the kivas) provides exact dates for the construction and occupation of the site—between 700 and 850 AD for the first settlements and then an intensive period of construction between 900 and 1050 AD of the masonry structures on the site. By the end of the construction the site contained two kivas and 35 rooms, and was completed between 1076 and 1093 AD when the East Kiva and Great House were completed. These sophisticated structures included ventilation tunnels and shafts for viewing the sky. Evidence from excavation indicates that the tons of rock for the walls were carried over 1000 feet up the ridge. The purpose for these buildings was still unknown

**Fig. 1.7** Chimney rock, near Durango, Colorado, showing moonrise at lunar standstill, as viewed from the central kiva (photo by Heden Richardson; reprinted with permission)

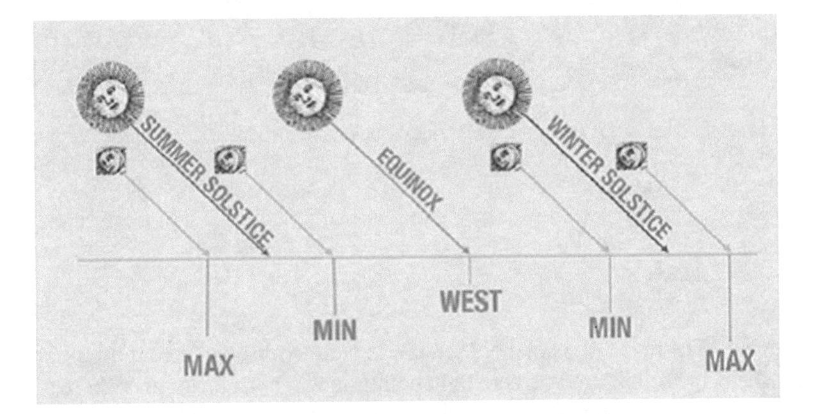


**Fig. 1.8** The location of solar and lunar extreme setting points on a horizon sweeps north and south with the seasons. At winter solstice a viewer looking west would see the Sun set to the south of west, and at summer solstice the Sun would appear to the north of west (*red lines*). The Moon's extreme or "standstill" points are on either side of these solstice positions, with opposite seasons; e.g., for lunar standstills in winter, the full moon would set in the NW, while in summer the full moon sets in the SW. The lunar standstill period roughly coincides with the 18.03-year "Saros cycle" useful for predicting eclipses and is thought to have played an important role in many ancient astronomical observatories (original figure by the author)

until archaeoastronomer J. McKim Malville, camping on the site, fortuitously witnessed the moonrise one evening from the Great Pueblo. To his amazement, the moon rose precisely between the two rock spires which towered above the site (Fig. 1.9). He noticed that this alignment, known as a "lunar standstill," is visible every 18.6 years and this interval coincided with the years of intense construction in 1056, 1075, and 1093 given by the tree ring data (Malville 2012). This intriguing interpretation suggests that the site was more than a defensive outpost, and was connected with Native American ceremonies observing long cycles of heavenly bodies.

**Fig. 1.9** The view facing toward the two spires at Chimney Rock from the Great Kiva, where the moon rises during its extreme "standstill" position. This site was one of many ancient observatories within the American Southwest (original figure by the author)

## *Mercury*

The mysterious planet Mercury is hard to observe from most urban locations. But in ancient times, where light pollution was not a factor and where people were very often about before dawn and after sunset, the planet Mercury was easily spotted at twilight. Mercury is an unusual planet physically, with a blasted hot surface of 900 °F basalt, a very eccentric orbit of 88 days, and a rotational period of only two-third of the year. Certainly an observer on Mercury would have a very hard time seeing the Earth in the glare of the Sun and considering too the fact that a day on Mercury lasts two-third of a Mercury year!

For us here on the Earth, Mercury can be spotted at dawn or at sunset, with a short period of maximum elongation, at which time Mercury is a full 18–24° from the glare of the Sun (depending on which side of the eccentric orbit is away from the Sun). To spot Mercury, look up its time of maximum elongation and whether it is visible at dawn or sunset. Find a clear site with an obstructed horizon and observe a yellowish star-like object that is close to the glare of the twilight sun.

While observing Mercury, you can contemplate on some aspects of Mercury's significance to our ancestors. Due to its darting motions on either side of the Sun, Mercury developed the name of "quicksilver" among alchemists, who associated properties of Mercury with the process of separating gold from lead. To the Greeks, Mercury was associated with the god Hermes and with knowledge, writing, records, and commerce. This latter association gives us the root for the word "mercantile," the business of Mercury. Egyptians knew Mercury as the "Seth of the evening twilight, a god in the morning twilight" (Neugebauer and Parker 1960, p. 181). Sumerian and Babylonian astrologers associated Mercury with the god *Nabu*, the God of wisdom and writing. The Chinese viewed Mercury as *Ch'en hsing* or the "hour star" which was associated with the primary element of water in the metaphysical system of Chou Yen (350–270 BC) (Kelley and Milone 2005, p. 318).

Mercury in our language is both a chemical element and a planet, which is a relic of the time when chemistry, physics, and astronomy were unified in the pre-scientific philosophies of alchemy and astrology. To alchemists, who include in their ranks such minds as the physicist Isaac Newton and the poet William Blake, Mercury was one of the primary elements, capable of initiating change, a fluid that was also linked to the intellect itself (Fig. 1.10). The planet Mercury is difficult to view, as it is usually lost in the glare of the morning or evening sun. One dramatic way to view Mercury is during a "transit" when the planet crosses through the disk of the sun. Our ancient ancestors did not have access to telescopes, but modern viewers can catch a site of Mercury during its transit, which due to Mercury's relatively short orbital period, happens on a regular cycle. The transit shows the planet in silhouette, and dramatically reveals the tiny size of Mercury compared to the Sun (Fig. 1.11). Table 1.2 provides dates for Mercury transits over the next century, which offers the rare chance to catch mysterious Mercury in the act of transiting between its morning and evening visibility. Since Mercury is associated with swiftness, these events are relatively quick—lasting about 5 h, and are separated on average by 7 years.

**Fig. 1.10** An image of the planet Mercury, from the work by Sentinus Jacobus, "*Poeticum Astronomicon*" (1482), which shows Mercury's symbolic association with speed (the chariot towed by birds and athletes) and with healing and medicines (the scepter with two snakes) (image courtesy of Claremont Colleges special collections)

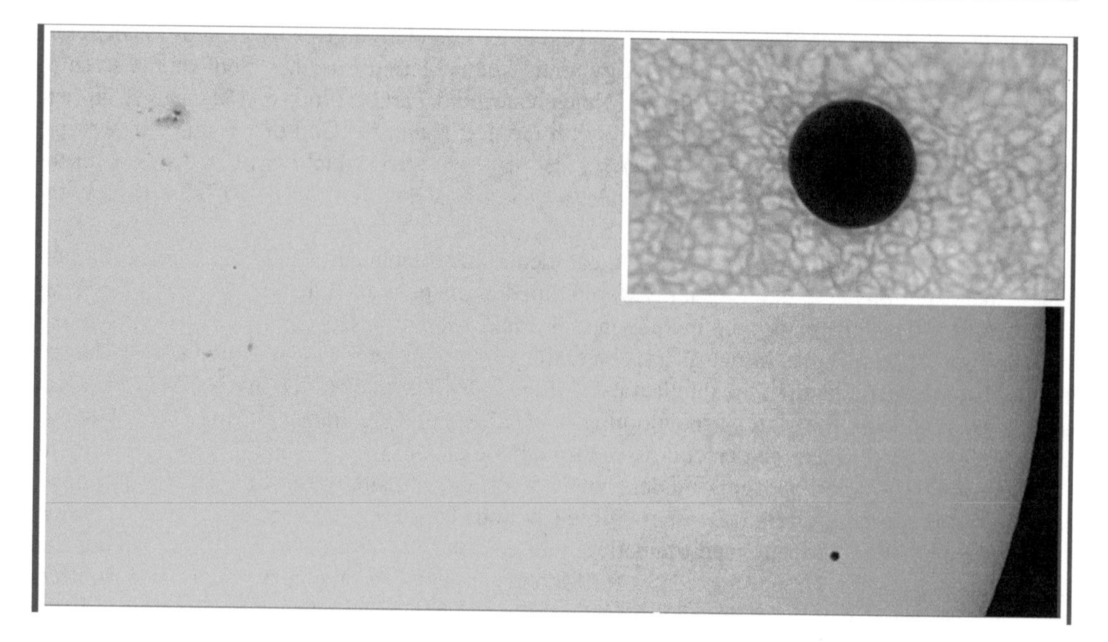

**Fig. 1.11** Mercury caught in the act of its transit across the sun during May 9, 2016. Notice the tiny size of Mercury (*black dot in lower left*) compared to the sun and to a sunspot (*in the upper left* of the image). The inset box shows a close-up of Mercury against the mottled surface of the sun taken with the Big Bear Solar Observatory. (Figure from Jay Pasachoff and Glenn Schneider with permission; inset image from BBSO/NJIT; credit to Jay Pasachoff, Glenn Schneider, Dale Gary, Bin Chen)

**Table 1.2** Dates of Mercury transits visible from Earth during the next two centuries

| Year | Date | Time (UT) |
|---|---|---|
| 2019 | Nov_11 | 15:20 |
| 2032 | Nov_13 | 8:54 |
| 2039 | Nov_07 | 8:46 |
| 2049 | May_07 | 14:24 |
| 2052 | Nov_09 | 2:29 |
| 2062 | May_10 | 21:36 |
| 2065 | Nov_11 | 20:06 |
| 2078 | Nov_14 | 13:41 |
| 2085 | Nov_07 | 13:34 |
| 2095 | May_08 | 21:05 |
| 2098 | Nov_10 | 7:16 |
| 2108 | May_12 | 4:16 |
| 2111 | Nov_14 | 0:53 |
| 2124 | Nov_15 | 18:28 |
| 2131 | Nov_09 | 18:22 |
| 2141 | May_10 | 3:43 |
| 2144 | Nov_11 | 12:02 |
| 2154 | May_13 | 10:58 |
| 2157 | Nov_14 | 5:40 |
| 2170 | Nov_16 | 23:15 |
| 2174 | May_08 | 3:26 |
| 2177 | Nov_09 | 23:09 |
| 2187 | May_11 | 10:24 |
| 2190 | Nov_12 | 16:48 |

### *Venus*

The next planet in the line of ecliptic and in ancient cosmologies is Venus. Venus appears for a longer period in both evening and morning sky than Mercury and disappears completely for a time while it crosses the Sun (Williamson 1984). The entire period of each apparition is approximately 260 days, which was the basis of the central timekeeping period of the Aztec and Maya. For these people Venus presented a more dramatic cycle than any other in the sky. Its dramatic re-emergence in dawn or evening and its path across the sky and plunge toward the horizon portended many things to Mayan astronomers, who charted each of the five different Venus trajectories in their books and created tables predicting the path of Venus with a precision that still provides accurate forecasts within 1 day, even after 1500 years (Fig. 1.12).

The cycle of Venus for the Maya was associated with the *tzolkin*, the sacred calendar, in which groups of 13 × 20 days were kept, in combinations of number and day names, each with a special omen. The 260 days seemed to correlate well with the gestation period of humans, the period of growth for corn, and with other natural cycles of the Mayan environment.

For the Sumerians and Babylonians, Venus was associated with the goddess Inanna/Ishtar, respectively. Ishtar was the goddess of love, sex appeal, and war, an insatiable goddess of passion. The re-emergence from twilight of Venus was celebrated with a precession through the Ishtar gate in which a parade of lavishly decorated "floats" would recreate the movement of the planets through the heavens. The Chinese saw Venus as *Tai po* or the "grand white" and associated the planet with the primordial element of metal (Fig. 1.13).

The early Greeks associated Venus with Phosphoros, the light bearer. Certainly when Venus is at its brightest it appears much as a phosphor flare in the sky, the brightest object in the sky other than the Sun and Moon! The brightest time for Venus is just after maximum elongation, when it comes toward the Earth as it passes us on the inside. A separate name for Venus applied for the morning apparition of Venus (Phosphoros) and the evening apparition (Herperos) (Kelley and Milone 2005, p. 37). Later, the Greeks associated Venus with Aphrodite, goddess of love, perhaps from its changeable nature, or perhaps since Venus shines in the romantic early evening (Fig. 1.14). For the Egyptians, Venus was associated with the god Osiris, lord of everything, along with the Moon and the constellation Orion.

Perhaps one of the most striking interpretations of Venus comes from the Pawnee Native Americans, as the Morning Star was central to an elaborate ritual of human sacrifice. The Pawnee would capture a girl, often from a neighboring tribe, and paint half of her body red to symbolize the dawn and half black for the night. She would be carefully dressed with symbolic clothes and a headdress and led to a scaffold, itself designed to embody symbolic woods and colors. The victim would then be sacrificed by an arrow through the heart. The ritual practice, ended in 1838, was an attempt to summon the power of Venus and the skies to renew the Earth (Williamson 1984, p. 220).

The planet Venus is an interior planet and moves from one side of the Sun to the other in a period of about 260 days, and as it does so changes in size and phase due to the changing distance and angle with the Sun. Venus is easy to view and you can mark its motion from 1 week to the next as it reaches its "maximum elongation" (the greatest distance from the sun—about 47°) to the time when it disappears in the evening sky and remerges in the pre-dawn sky. But the real treat is to view Venus in the telescope—just as Galileo did 400 years ago. The planet Venus reveals its crescent or gibbous shape with modest magnification, confirming for your own eyes the giant celestial racetrack of our solar system—with Venus on an inside track to pass the Earth each year (Figs. 1.11 and 1.12).

On rare occasions, as Venus crosses in front of earth, it undergoes a "transit" and moves across the disk of the sun. This phenomenon comes in pairs—separated on average by 8 years—and these pairs of transits themselves separated by an average interval of 121 years. Ancient people were unable to see this remarkable event, but in the 18th century the explorer James Cook was sent to Tahiti for the

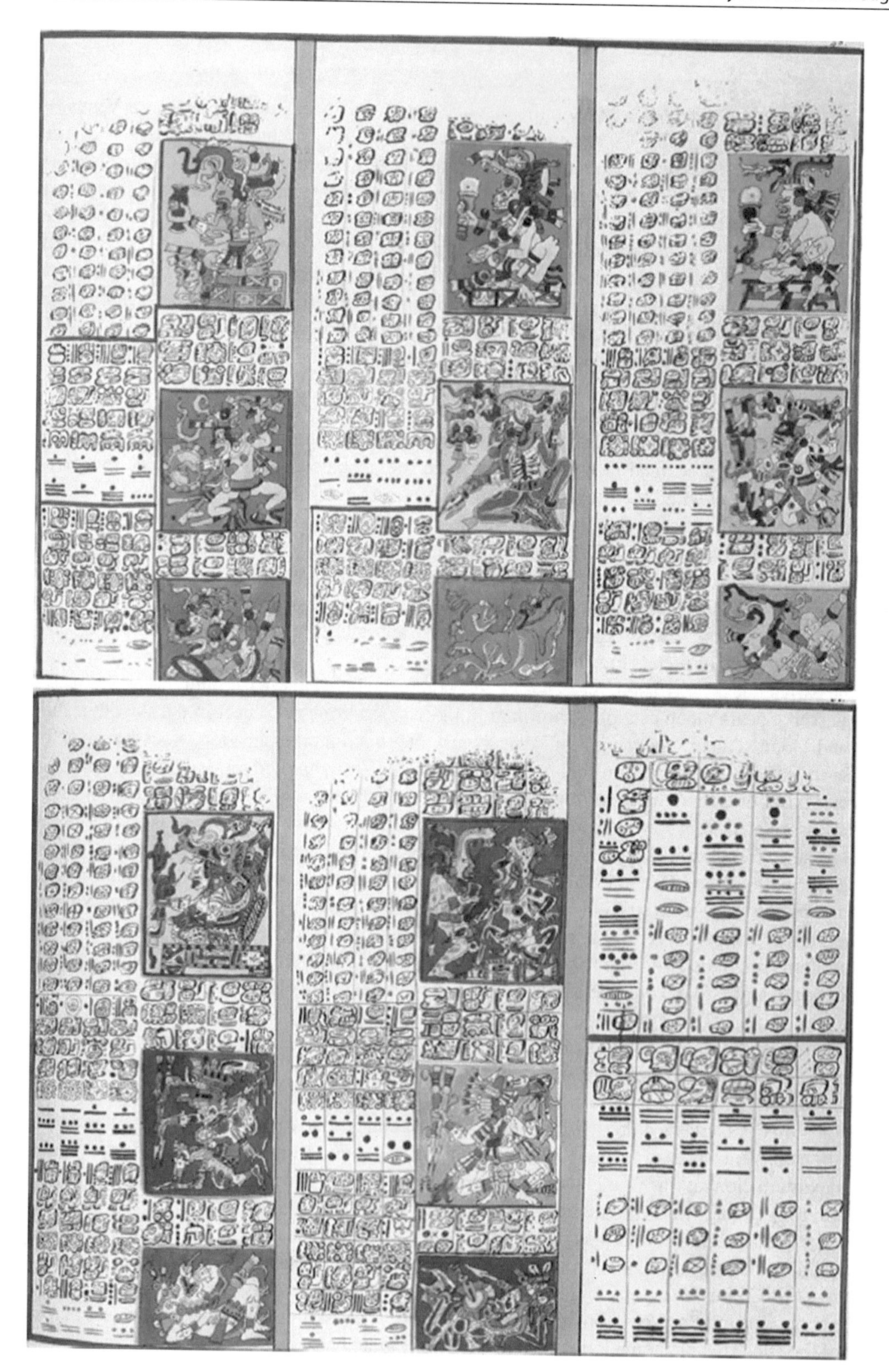

**Fig. 1.12** The Dresden Codex Venus pages, showing an ancient Mayan almanac of Venus that dates from the Classic Mayan period, some 1500 years ago. The five panels represent the five tracks of Venus in the sky and dates of appearance and disappearance are given, that are still accurate to within a day (image from the facsimile edition of the Dresden Codex within the Claremont Colleges Special Collections library)

**Fig. 1.13** The phases of Venus as seen from the Earth. Galileo was the first to see the phases of Venus and sketched this impression by hand for the work *Sidereus Nuncius* (1610) (image from the Galileo project at http://www.astronomy2009.org/static/archives/images/screen/galileo_12.jpg)

**Fig. 1.14** Image of Venus, with associations of Cupid, and the goddess of love, as portrayed in Sentinus Jacobis, *Poeticum Astronomicon* (1482) (image courtesy of Claremont Colleges Special Collections)

purpose of observing the Venus eclipse of 1769. Cook's journey at the time was like an interplanetary space voyage, and the observations of Venus that Cook and ship astronomer Charles Green were hoping to take could indeed unlock the distance scale of the planets.

In the eighteenth century, the Venus transit was the critical "missing link" in the distance scale, and observations of small angular shift of the planet Venus as viewed from different locations on Earth were needed to measure the size of the solar system. Cook and Green observed the transit diligently and recorded the event from their location in Tahiti. Cook's images were drawn by hand (Fig. 1.15), and the

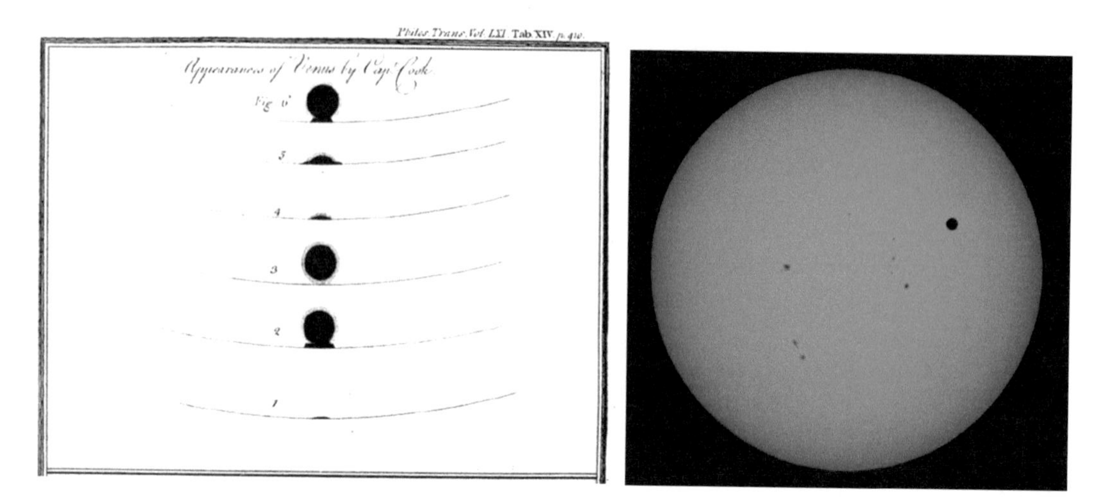


**Fig. 1.15** The transit of Venus, as viewed by James Cook and Charles Green in Tahiti on June 3, 1769. The drawings were published in 1771 and while not accurate enough to measure the distance to Venus, inspired astronomers to persevere. In 1874 photographic images were able to give astronomers the first accurate measurements of the solar system. The photograph at right was taken during the last Venus transit of June 6, 2012, and shows the planet Venus (*upper right*) in silhouette along with several sunspots (*center*). The next Venus transit is in 2117, so there is plenty of time to prepare for the pair of transits in 2117 and 2125! (figure credits—http://star.arm.ac.uk/history/transit.html (*left*), and https://upload.wikimedia.org/wikipedia/commons/e/e4/Venus_transit_2.jpg (*right*))

timing of the transit was recorded carefully with the ship's clock. While Cook and Green's data were not precise enough to give an accurate measure of the distance to Venus, they gave the astronomers of the time an approximate estimate for the distance of Venus. Astronomy had to wait until the next transit in 1874, when more accurate clocks and astrophotography enabled astronomers to calculate the first accurate measurements for the astronomical unit, the fundamental distance scale of the solar system. A complete listing of Venus transits—for the next 1000 years—is given in Table 1.3.

**Table 1.3** Dates of Venus transits visible from Earth during the next 1000 years

| Year | Date | Time (UT) |
|---|---|---|
| 2012 | Jun_06 | 1:29 |
| 2117 | Dec_11 | 2:48 |
| 2125 | Dec_08 | 16:01 |
| 2247 | Jun_11 | 11:33 |
| 2255 | Jun_09 | 4:38 |
| 2360 | Dec_13 | 1:44 |
| 2368 | Dec_10 | 14:45 |
| 2490 | Jun_12 | 14:17 |
| 2498 | Jun_10 | 7:25 |
| 2603 | Dec_16 | 0:13 |
| 2611 | Dec_13 | 13:34 |
| 2733 | Jun_15 | 17:18 |
| 2741 | Jun_13 | 10:17 |
| 2846 | Dec_16 | 23:11 |
| 2854 | Dec_14 | 12:19 |
| 2976 | Jun_16 | 19:44 |
| 2984 | Jun_14 | 12:49 |
| 3089 | Dec_18 | 21:31 |

## Sun

The Sun is of course the most brilliant of celestial bodies, and yet for many the basic movements of the Sun are not well understood. Many people will be surprised to know that the Sun does *not* rise in the east and set in the west, except for 2 days each year. The Sun instead moves about on the horizon in a range of angles, known as azimuths, which vary with the seasons. For those in the Northern Hemisphere, the Sun will sweep an arc along the horizon through the year, starting at a southernmost location near winter solstice and continuing toward the northern "standstill" points until the summer solstice. Depending on one's latitude, these standstill angles can range from 23.5° north and south of east, to as large as 40 or 50° in northern Europe (Fig. 1.16).

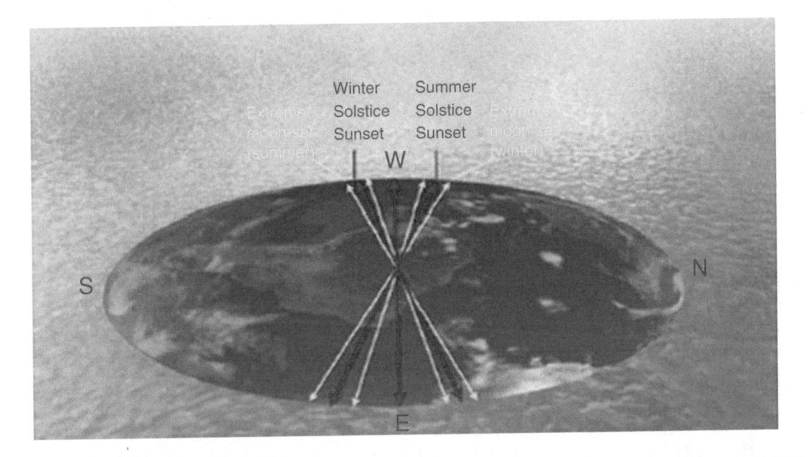


**Fig. 1.16** Illustration of solar and lunar standstill points as viewed toward the west on a hypothetical disk-shaped Earth near the equator. The Sun swings across the horizon on an annual cycle, while the Moon cycles through extreme positions or "standstills" in an 18.61-year cycle. For any observer on the Earth, these directions point to the horizon toward unique directions that depend on the observer's latitude (figure by the author)

The trajectory of the Sun in the sky each day is an arc which meets the horizon somewhere between standstill points in the eastern half of the sky and "transits" when it reaches its highest point, which is the invisible line crossing from due south to due north known as the meridian. From the meridian transit came the acronyms AM and PM, for ante meridian and post-meridian, for morning and afternoon.

The height of the Sun on the meridian of course will vary through the year as well, reaching its extreme values on the solstices. The maximum height of the Sun at the tropics (locations on the Earth at latitudes of 23.5N and 23.5S) is directly overhead and for those of us at latitudes beyond the tropics, the Sun will approach but never reach directly overhead or zenith. If we consider the Sun's position on the celestial sphere, it moves through the year between celestial latitudes (known as declinations) of +23.5 and −23.5 for the summer and winter solstices, respectively. A simple calculation will reveal that for your location's latitude, the height or *altitude* of the Sun (a) above the horizon will be located at the angle a given by:

$$a = \text{declination} - \text{latitude} + 90$$

where a = 90 corresponds to directly overhead. The declination of the Sun can be found from considering the season—the declination will be −23.5 at winter solstice, +23.5 at summer solstice, and 0° at the equinoxes. It is easy to use this information to help design a sundial or to use the changing height of the Sun from different locations as a navigation tool (Fig. 1.17).

**Fig. 1.17** The height of the Sun and stars above the horizon, or their altitudes, depends on the observer's latitude and the season. This figure from an early astronomy work allows the reader to dial in their latitude and observes a side view of the celestial sphere from their location on the Earth (image from Peter Apian, La Cosmographia (1551), Claremont Colleges special collections)

The azimuths and altitudes of the Sun are not the entire story of the Sun's motion across the sky. The exact location of the Sun relative to the meridian at noon will lag and lead a bit east and west, as it moves north and south in a 47° arc from winter to summer. The exact shape of the Sun on the meridian at noon describes a figure eight shape known as the *analemma*, which includes the combined effects of the tilt of the Earth's spin and the rate of travel in the orbit (Fig. 1.20). From the tilt comes the variation in the solar height with the year and from our elliptical orbit about the Sun, the variable times of the Sun's rising and transiting.

Both the changes of the Sun's azimuth and elevation were well known to ancient people. One of the principle axes of many ancient monuments is the line toward the northern "standstill point" which will enable ancient people to look at the Sun at the solstices within a sacred space. From Britain's Stonehenge to Casa Rinconada in the Anasazi Chaco Canyon monument, a wide variety of ancient structures include the standstill to commemorate the solar year. Cardinal directions for many Native American tribes referred not to north, south, east, and west, but instead to the four directions on the horizon for the standstill points of sunrise and sunset, in the NE, SE, SW, and NW (Fig. 1.18).

It is easy to verify the motion of the Sun for yourself. One approach is simply to watch the sunrise or sunset from a fixed point over several weeks. You can notice the changing location of the Sun on the horizon by this kind of repeat observation. In doing this experiment, you are recreating for yourself the "horizon calendar" which was very important for many Native Americans—they would identify each hill and mesa on their local horizon with dates of their year, important festivals, and times to plant or harvest crops (Fig. 1.19).

Perhaps a window in your house can also serve as an observatory—just watch the location of the Sun at dawn or sunset and you too can see the celestial machinery at work in your own house! Many Native American buildings include windows that face key horizon directions where the Sun and Moon can indicate annual seasons or even longer cycles. One example is the Casa Grande in Arizona, in which a central room is illuminated twice a year (on the equinox) by two small portholes that admit light through the building. This tradition is also found in Western buildings. Many cathedrals in Europe make use of the motion of the Sun through the year to include indoor calendars, which mark the position of the Sun's light through a window, and locate the day of the year from a scale inscribed on the floor.

**Fig. 1.18** Diagram of Stonehenge, showing the summer solstice sunrise positions over the heel stone (*upper right*). Later authors have suggested the possibility of alignments between the trilithons and lunar standstills, perhaps as part of observing the 18.6-year cycle of the Moon between eclipses (image from Nordisk familjebok (1918) at http://commons.wikimedia.org/wiki/File:Stonehenge_vid_midsommar_1700_f_Kr,_Nordisk_familjebok.png)

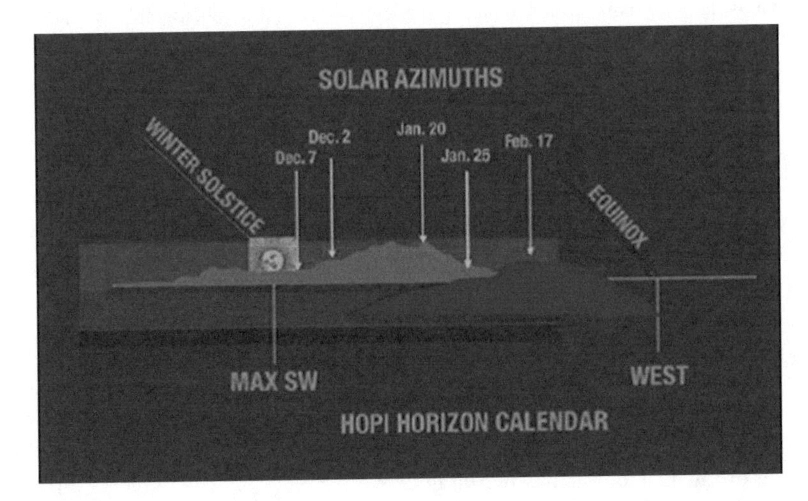


**Fig. 1.19** Schematic of the Hopi Horizon calendar, showing the azimuth location of the Sun over familiar landmarks throughout the year, and which was used by the people of Walpi village to set the agricultural and ritual calendar (figure by the author)

We can make observations on any day of the week. You can verify on any given day whether the Sun transits at noon—just look at shadows at noon and see whether they point north or south. The length of the shadow indicates the season and departures of the shadow from clock time sense the variable speed of the Earth in its orbit. Many have also succeeding in taking pictures of the Sun each day at the same time to photograph the analemma (Fig. 1.20). These pictures, from multiple exposures of the Sun moving across the celestial sphere, provide dramatic proof of the tilt and orbit of the Earth!

**Fig. 1.20** A photograph of the analemma at sunrise with the Tholos (360–350 BC) at Ancient Delphi. The photograph was taken on successive days at the same time and shows the apparent shift in the position of the Sun each day due to the seasons known as the analemma, with summer at top and winter below (photograph by Anthony Ayiomamitis; reprinted with permission)

## Sun Lore

The Sun was a source of great spiritual meaning and power to nearly all people and figured prominently in nearly every account of the creation and structure of the universe. We will cover some aspects of the Sun's significance to ancient cultures in additional detail in Chap. 3, but we can survey some of the diverse lore and meaning of the Sun to ancient people from around the world.

A listing of names for the Sun from around the world is a who's who of the celestial pantheons of ancient culture. Table 1.4 gives a concise listing of the names of the Sun from many cultures and highlights the primacy of the Sun in the hierarchy of gods and goddesses to ancient people.

Many cultures embody the Sun in gods or goddesses with connections to the sky god or with powers to create the universe, music, or the arts. Both *Apollo* of the Greeks and *Lugh* of the Celts possessed the power of music and arts. Many Sun gods were creators of the universe, including the God *Sin* of the Haida, *Oloron* of the Yoruba, and *Atum Re* of the Egyptians. In some cases the Sun is the child of the creator god, as in the case of *Dahzbog* of the Slavs, and *Shamash* of the Babylonians. For alchemists, the Sun was associated with the most precious element of all, gold, which never tarnishes and shines with a mysterious luster.

In some cases, tales of the Sun provide insight into the climate of a culture. For example, the Inuit or Eskimo people talk of the Sun and moon in terms of a sister and brother chasing each other holding some burning moss in the weak, smoldering light of a burnt-out oil lamp. After a while the two chased each other and their fires began to glow brighter as they ran faster and faster. As they chased each other they rose to the sky and the sister *Siqiniq* became the Sun while her brother *Aningaat* became the Moon and still smolders for her (MacDonald et al. 1998, p. 98). In the Arctic latitudes, the Sun would disappear for long periods, which lowers the land into a period known in Igloolik as *Tauvikjuaq*—"the great darkness." In true sibling rivalry, the Sun and Moon would compete with

**Table 1.4** Partial listing of Sun names from cultures around the world (O'Hara 1997)

| | | |
|---|---|---|
| Atum Ra/Amon Re | Egypt—Middle Kingdom | Monad—Lord of Everything. Center of Sun Cult |
| Huitzilopochtli | Aztec | Hummingbird God central in sacrifice rituals |
| Odin | Norse | Sun is "Odin's Eye" while the God Odin rules the heavens |
| Apollo | Greek | Inventor of music, passes through sky in a chariot |
| Awonawilona | Zuni | Creator God |
| Oloron | Yoruba—Nigeria | Creator God—sky and Sun |
| Gitche Manitou | Arapaho | "Great Spirit" manifest as the Sun |
| Sin | Haida—British Columbia | Creator of universe |
| Agni | Hindu | God of solar fire; important for Hindu wedding and associated with love |
| Mitras | India/south Europe | God of the Sun |
| Shamash | Babylon | God of wisdom, children include Misharu (law) and Kitsu (justice) |
| Dahzbog | Russia/eastern Europe | Son of sky god |
| Lugh | Celtic/Ireland | God of all arts, gold clad |
| Hae-Sun | Korea | Sun maiden, turned into Sun |
| Unelanuhi/Aaghu Gugu | Cherokee | Sun goddess |

each other to emerge first from the darkness. "When the sun returns before the first new moon of the year it was said that the moon had been defeated and that the spring and the summer would be warm" (MacDonald et al. 1998, p. 111).

For the Chumash Native Americans of California, the Sun is a forbidding old man who travels across the sky on a cord carrying a torch. The Sun is part of a team with the Moon that plays the game *peon* against Sky Coyote and Eagle for fate of people down below on the Earth. If Sun and Moon win, the people will have a year of drought, and famine and disease will ravage the land. If Sky Coyote's team wins, rain will be abundant and food will be abundant. The Sun was needed for life, but was not one to be crossed. He was an irritable widower, whose only clothing was a headband where he might place Chumash children plucked from this earth, which would feed his two daughters later when arrived home from his daily journey (Hudson and Underhay 1978; Williamson 1984). The dangerous nature of Sun reflects the nature of the southern California climate, with its modest amount of rainfall and propensity toward drought.

For the Aboriginal Australian people, who live in a very sunny climate, tales of the Sun and Moon play a central part in the culture. The Australian culture dates back for more than 15,000 years, and has diverse lore about the Sun and Moon. In most regions of Australia, the Aboriginal people saw the Sun as a woman. She prepared herself for her journey across the sky by covering herself with red ochre, which sprinkled onto the sky as dawn's colors. The Sun not only brings the beautiful red and yellow colors of sunset and sunrise from her ochre powder, but also provides the life-giving light and warmth of the world.

We can think of the Sun in many terms, as a source of life, energy, and warmth—all practical benefits of the Sun's light. But we should also remember the powerful symbolism of the Sun as it appeared to many cultures, as it appears to us today in its journey around the celestial sphere.

## Mars

♂

Mars is our first of the outer planets in our survey and its motion is the most dramatic and rapid of all the outer planets as it moves across the sky. Mars and the Earth "race" each other around the Sun in the same counterclockwise direction when viewed from above the North Pole of the Sun. Mars has an orbital period around the Sun of about 2 years, and so the Earth is able to "lap" Mars in just more than one Earth year. Due to the proximity between the Earth and Mars, the change in the brightness and size of Mars is the most extreme of all the outer planets, and it reaches its maximum brightness when the Earth is passing Mars in its orbit (Fig. 1.21).

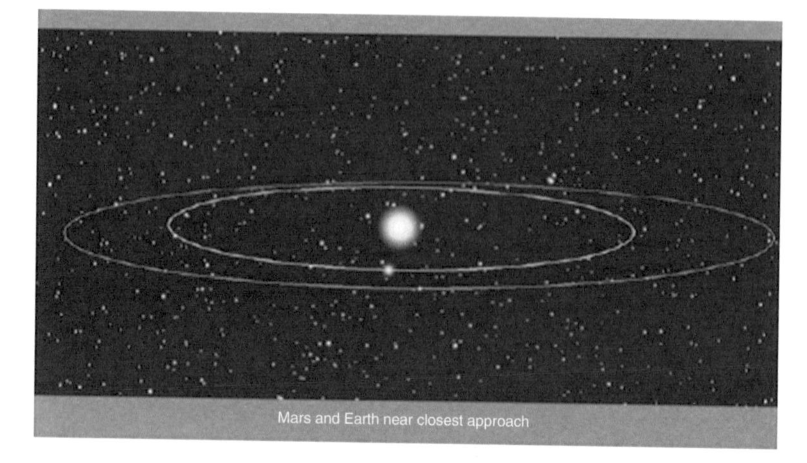


**Fig. 1.21** Mars and the Earth approach each other approximately every 18 months. This figure shows their relative orbits and proximity of the two planets when Mars is at opposition (figure by the author)

You can easily find the times at which the Earth "passes" Mars—these times happen approximately every 18 months. During the time of Earth's approach, Mars turns backward in its apparent motion about the sky and comes to a stop as the Earth makes its inside move, and then slips behind as the Earth takes the lead in its inside track around the Sun. The path of Mars in the sky during this period describes a large loop known as a "retrograde loop" due to the apparent backward or retrograde motion of Mars during the time the Earth is passing the planet (Fig. 1.22). While at closest approach or opposition, Mars appears at its brightest, crossing the meridian at midnight, and with larger angular size and brightness than at any other time.

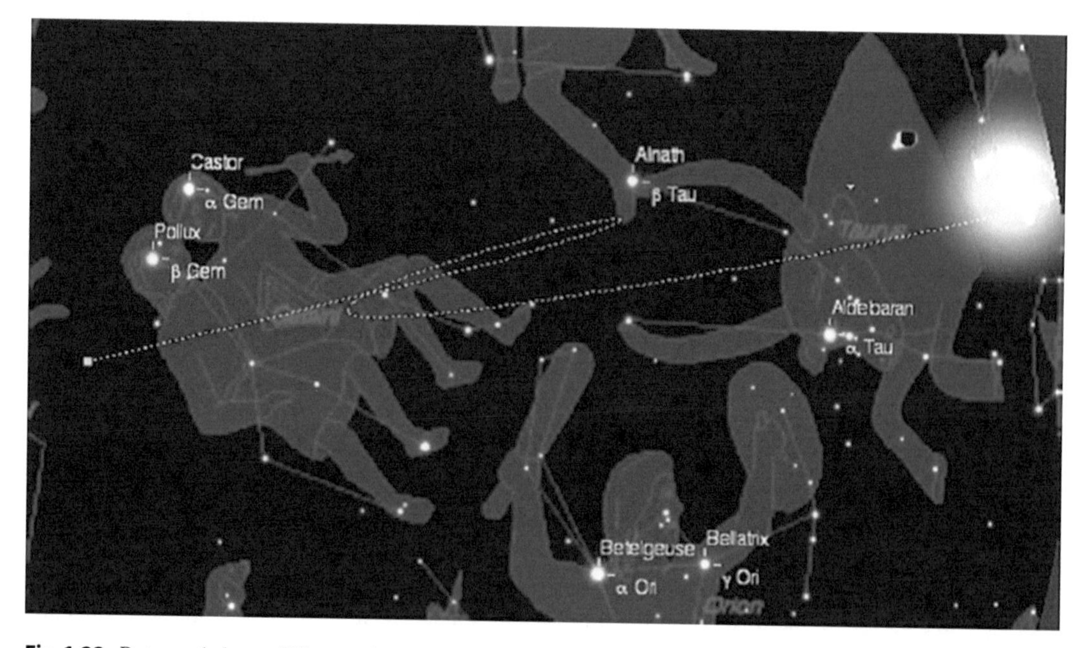


**Fig. 1.22** Retrograde loop of Mars against the stars, as generated by the Voyager IV computer program (figure by the author)

It is easy to estimate from basic observations where Mars and the Earth are in their orbits. If Mars appears in the pre-dawn sky, the Earth is making its move and approaching Mars in its race around the Sun. If instead, Mars appears to rise at sunset, it is close to opposition. Finally, if Mars is visible in the early evening post-dusk hours, it is past opposition and the Earth has passed it. The sequence can be observed over many months—Mars will shift in its position on the sky, and its rising times will get earlier each day as the Earth and Mars race around the Sun. You can watch this race unfold by just noting the location of Mars as it rises over several nights.

For ancient astronomers, the dramatic retrograde loops of Mars made this planet very important for predicting the future. Its rapid march across the sky could also be one reason why this planet was assigned its war-like properties by the Greeks and Romans. The Babylonians associated Mars with the god *Nergal*, who is both the god of plague and war. In Greece, the planet Mars was known as *Pyroeis* or "fiery star." The Egyptians considered Mars to be the "Horus of the Horizon" and spoke of Mars as "He who travels backward" in reference to the dramatic retrograde loops of the planet Mars. For the Chinese, Mars was *Ying Huo* or "fitful glitterer" and was associated with the primordial element of fire, appropriate for the red planet. As with the planet Mercury, Mars had important properties for early alchemists, who believed that Mars could bring fire, and "the warrior's lance" to "separate the pure from the impure and thus to renew the elixir" (S. Trismosin, quoted in Roob 1997). Mars was associated by alchemist with the element of iron, and the color red, which were deemed appropriate for the war-like planet.

While not exactly ancient, the most famous encounter with the planet Mars from the last century is well known to all. Beginning with a misunderstanding of the features on Mars described in Italian as *canali* by Italian astronomer Giovanni Schiaparelli (1835–1910), Percival Lowell embarked on a quest to learn about the civilization on Mars. Using his newly built observatory in Arizona, constructed on Mars Hill using his considerable personal fortune, Lowell was convinced he could see evidence of aqueducts and roads on the surface of Mars and published three books, *Mars (1895)*, *Mars and Its Canals (1906)*, and *Mars as the Abode of Life (1908)*. From these speculations came inspiration for both science fiction of H.G. Wells and many other popular culture ideas of Mars. Perhaps our current cultural infatuation with alien life was inspired by the misunderstanding of the nature of the planet Mars (Fig. 1.23).

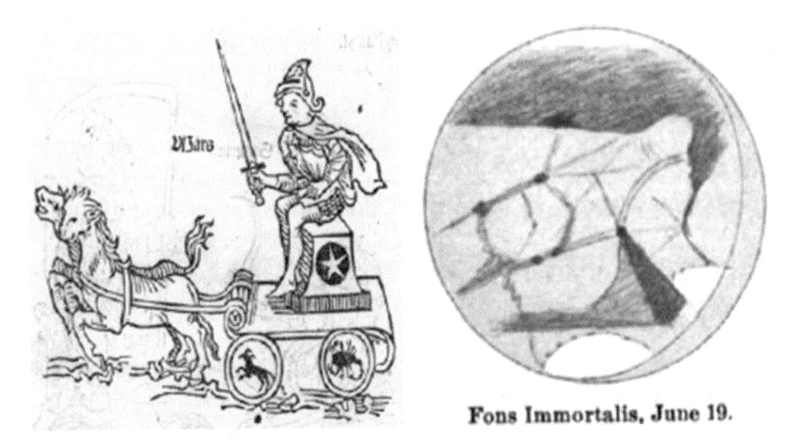


**Fig. 1.23** (*Left*) Image of Mars, emphasizing its association with warfare, as portrayed in Sentinus Jacobis, *Poeticum Astronomicon*, from 1482 (image courtesy of Claremont Colleges Special Collections). (*Right*) A more modern and equally imaginative conception of Mars, as being filled with canals and a "Fountain of Youth" as drawn by Percival Lowell (Lowell 1911)

### *Jupiter*

To view Jupiter through a small telescope or even binoculars brings one into the modern age of astronomy. Jupiter's stripes become visible with small telescopes, and its group of moons visible along the plane of the stripes. Galileo first viewed this wonder in 1610 and named the moons the "Medician stars" after his patron Cosimo de Medici. To ancient people, however, Jupiter was known mostly for its stately motion across the sky over the months in a more measured and gradual pace than Mars, but with the same basic type of motion. The retrograde loops shift 1/12 of the way across the sky with each year, highlighting a different zodiacal constellation each year, in a 12-year cycle. The slower speed of Jupiter comes from its greater distance from the Sun, which Kepler showed results in a longer orbital period. But the numerology of the 12-year cycle, and the grace with which Jupiter loops through the zodiac, impressed ancient people with the power of this object, and perhaps gave it the place as the ruler of the solar system (Fig. 1.24).

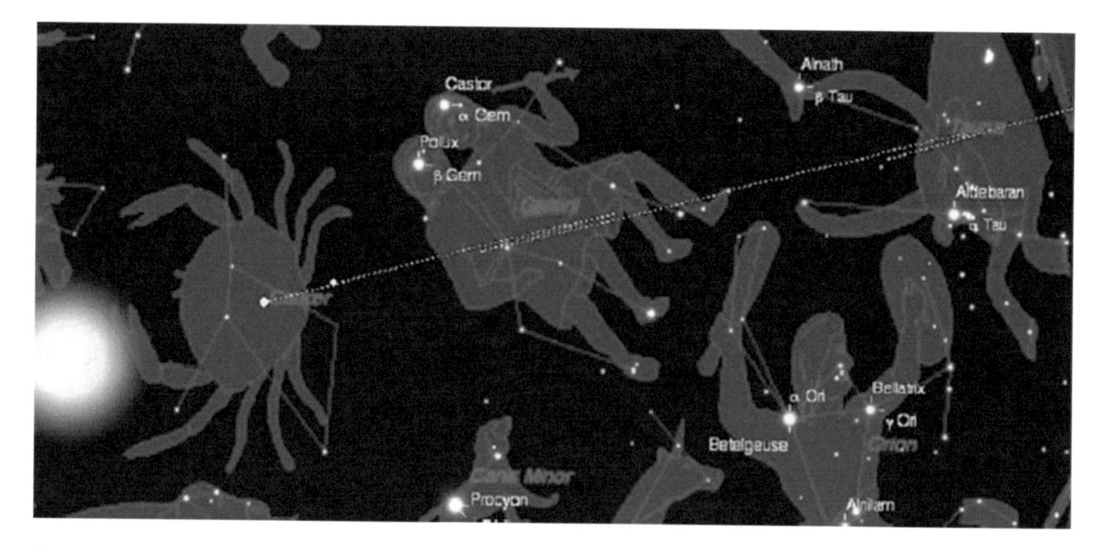


**Fig. 1.24** Jupiter retrograde loop against the stars generated with the Voyager IV program. Note that since Jupiter is more distant than Mars, its retrograde loops are smaller, due to the smaller parallax of Jupiter (figure by the author)

You can observe Jupiter's location in the sky and immediately determine where the Earth and Jupiter are in their respective trips around the Sun. Just as for the planet Mars, Jupiter will appear first in the pre-dawn sky and will rise closer to midnight each day over several months as the Earth approaches closer to Jupiter. When Jupiter transits at midnight it is at opposition, its closest point to the Earth, and the midpoint of its retrograde loop.

Jupiter for the Babylonians was associated with the god *Marduk*, the ruler of the sky, and the one that slays the primordial sea monster *Tiamat* in their creation story. According to legend, Marduk split Tiamat into two parts, which became the Earth and sky and then he set the planets into motion. Marduk as the ruler of the sky was watched carefully by Babylonian astrologer/priests for omens that reflected upon the earthly ruler. The Greeks and Romans continued the royal association between the planet and ruler of the Gods—Zeus and Jupiter. For the Greeks, Jupiter also was the lord of atmospheric phenomena such as rain, storms, lighting, and thunder (Krupp 1991, p. 176).

Egyptians referred to Jupiter as "Horus who Bounds the Two Lands" or "Horus who Illuminates the Two Lands" (Krupp 1991, p. 180), referring to the role of Jupiter in uniting the Upper and Lower Egypt, and thereby giving Jupiter some of the properties of the Pharaoh. Our day name of "Thursday" inherits this same divine rulership—the Norse God Thor was something of a superhero, ruling warily over the gods, giants, and monsters in the Norse pantheon:

> "Thor is the foremost of the gods… he is the strongest of all the gods and men. He has three valuable properties. The first is the hammer Miollnir which the frost-monsters and cliff giants recognize the moment it is raised on high. A second splendid thing he owns, his belt of strength. A third thing he has which is of the greatest value: his iron gloves."
>
> (Snorri and Byock 2005, p. 41)

For the Chinese, Jupiter was known as *sui hsing*, or the "year star," and was associated with the element of wood. Many of us are acquainted with this Chinese astrology from Chinese restaurants, or from Chinese New Year, where the "Chinese zodiac" gives rise to named years, such as "Year of the Tiger." Few realize that this system was originally dictated by Jupiter's orbit. The position of Jupiter in the sky moves slowly each year and travels completely around the sky over a long cycle of 12 years. This cycle served the Chinese astronomer much as an hour hand on the celestial sphere and allowed them to mark the passing years. Each of the years in the cycle, by long tradition, was associated with 1 of 12 animals, giving rise to the "Chinese zodiac." A famous bronze disk, reproduced below, commemorates the Chinese folk astrology and includes rings of the four primordial animals, hexagrams from the I Ching, the 12 year animals associated with Jupiter's 12 year path around the sky, and the 28 animals associated with the lunar mansions or *hsiu*, which are the constellations in which the Moon resides in its journey around the sky (Fig. 1.25).

**Fig. 1.25** Chinese bronze zodiac mirror disk, from the Sui Dynasty (AD 581–618), showing a ring of the 12 animals of the zodiac (*outer ring*) as well as four dragons symbolizing the four cardinal directions (*image source*: http://commons.wikimedia.org/wiki/File:Zodiaco_Chino.jpg)

It is also possible to learn the Chinese year name from the stars near Jupiter. Each of our modern constellations is listed in Table 1.5, along with the Chinese equivalent and the years at which Jupiter appears in the constellations. By taking multiples of 12, you can also work back to your birth year and determine the animal of your birth year.

**Table 1.5** Chinese new year names and locations of the planet Jupiter, which determine the 12 year Chinese zodiac

| Chinese year name | Jupiter location | Year |
|---|---|---|
| Rat | Sagittarius | 2008 |
| Ox | Capricornus | 2009 |
| Tiger | Aquarius | 2010 |
| Rabbit | Pisces | 2011 |
| Dragon | Aries | 2012 |
| Snake | Taurus | 2013 |
| Horse | Gemini | 2014 |
| Sheep | Cancer | 2015 |
| Monkey | Leo | 2016 |
| Rooster | Virgo | 2017 |
| Dog | Libra | 2018 |
| Pig | Ophiuchus/Scorpius | 2019 |

Galileo Galilei was the first to observe Jupiter's moons and their motions with a telescope, and his excitement at these telescopic observations was recorded in a letter to the Doge of Venice in 1609. In Galileo's words:

> Most Serene Prince Galileo Galilei most humbly prostrates himself before Your Highness, watching carefully, and with all spirit of willingness, not only to satisfy what concerns the reading of mathematics in the study of Padua, but to write of having decided to present to Your Highness a telescope that will be a great help in maritime and land enterprises. I assure you I shall keep this new invention a great secret and show it only to Your Highness. The telescope was made for the most accurate study of distances. This telescope has the advantage of discovering the ships of the enemy two hours before they can be seen with the natural vision and to distinguish the number and quality of the ships and to judge their strength and be ready to chase them, to fight them, or to flee from them; or, in the open country to see all details and to distinguish every movement and preparation.
>
> Letter to Leonardo Donato, Doge of Venice (Galileo 1609).

Below these words, Galileo diagrammed the motions of the four satellites, which were the first ever satellites observed around another planet (Fig. 1.26). The observations of Jupiter by Galileo played a key role in establishing the Sun as the center of the universe, and in so doing helped restructure the European conception of the universe (Fig. 1.27).

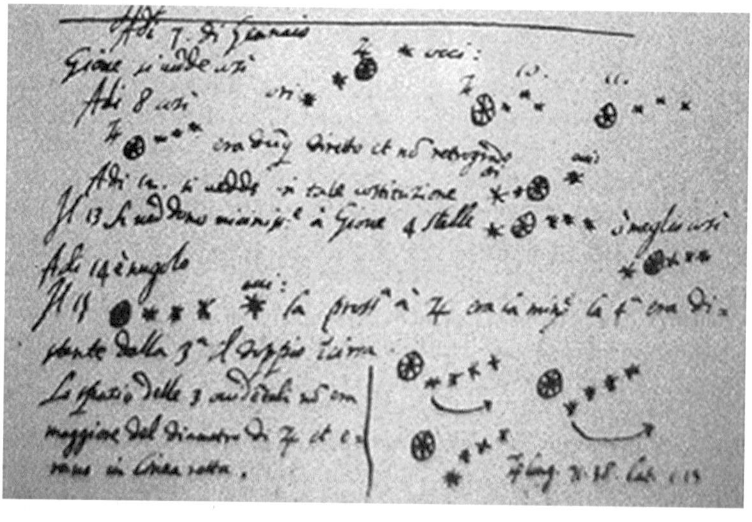

**Fig. 1.26** Galileo's sketch of Jupiter and its moons, then called the Medician stars. Galileo sketched these notes in his own hand and included with them a description of their positions and motions (image from the Galileo project at http://www.lib.umich.edu/spec-coll/largegal1.html)

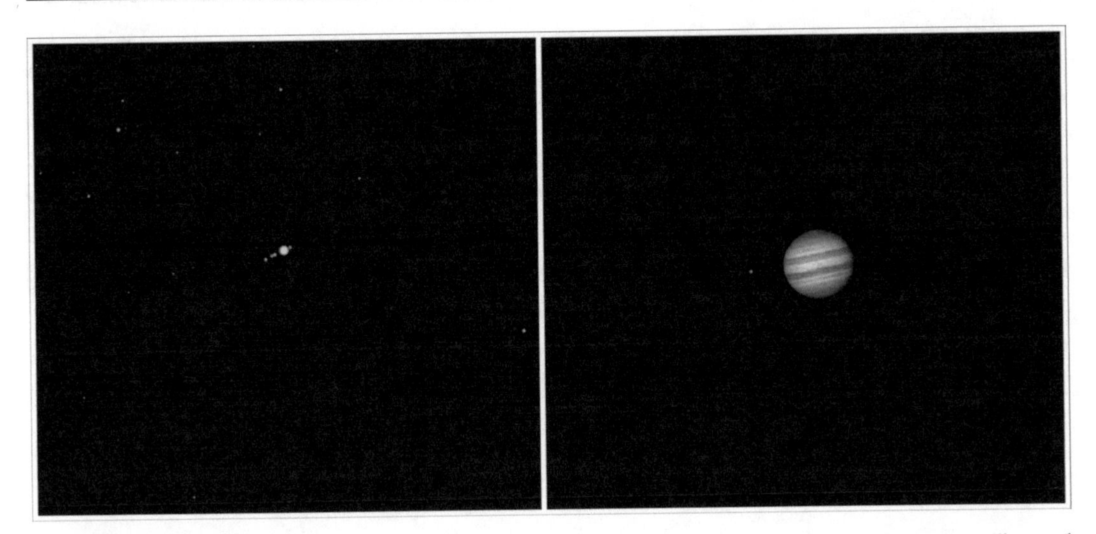

**Fig. 1.27** The planet Jupiter as viewed through a small telescope or binoculars (*left*) shows the four Galilean satellites, and can be recognized as a planet due to its larger angular size compared with surrounding stars. A slightly larger telescope will reveal the bands of clouds on its surface (*right*), which modern astronomers have shown are decks of high clouds of white ammonia ice and lower orange/brown clouds that include ammonium hydrosulfide

## *Saturn*

♄

The outermost planet known to the ancients is Saturn. Its slow speed of motion and small retrograde loops are dramatically different from Mars and Jupiter. Taking over 29 years to orbit the Sun, Saturn's orbit is slow and subtle to observe. With a small telescope, Saturn is a beautiful sight, and its rings and moons often shock new observers with their brilliance. You can also see the effects of the Earth's motion as Saturn's rings are seen edge on in some years, and then appear at an angle in other years. Saturn, like the Earth, is tilted from its orbital plane. This effect gives Saturn seasons, with storm cycles visible with the Hubble Space telescope. With a more modest telescope, or even binoculars, it is possible to see Saturn's rings at different angles depending on its inclination, which can range from edge-on to about 30°.

The ancient people had only their eyes to view Saturn and were only aware of its pale and faint yellow glow, its slow movement about the sky, and its smaller retrograde loops, which we know now is just a result of its greater distance from the Earth and from the Sun (Fig. 1.28).

The ancient Sumerians associated Saturn with the God *Ninurta*, who was the god of war and hunting. In cuneiform the name given to Saturn is "*Genna*" which means "small" or "tiny," in reference to the dim light of Saturn. For the Egyptians, Saturn was considered "Horus the Bull of the Sky," perhaps in reference to the plodding motion of Saturn across the celestial sphere (Krupp 1991, p. 180). Early alchemists associated Saturn with the element lead, one of the densest and heaviest known elements. Some alchemists credit Saturn and lead with bringing reflection and wisdom or as "a great, wise and understanding lord, the begetter of silent contemplation" and a "keeper and discoverer of mysteries" (Roob 1997, p. 189). Others saw Saturn as a bringer of poverty and dissolution. As described by one alchemist "Within lead there dwells a shameless demon who drives men to madness" (Roob 1997, p. 196) (Fig. 1.29).

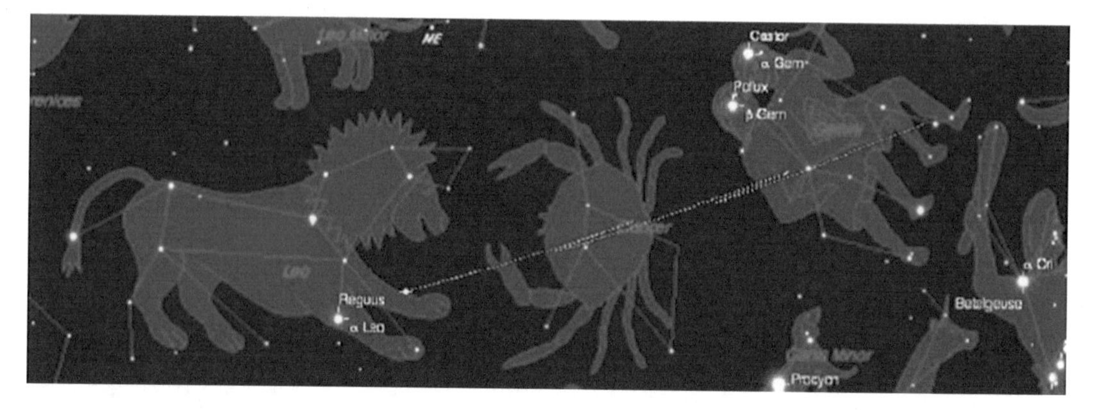


**Fig. 1.28** Two of Saturn's retrograde loops against the stars, which are smaller due to Saturn's great distance (figure by the author)

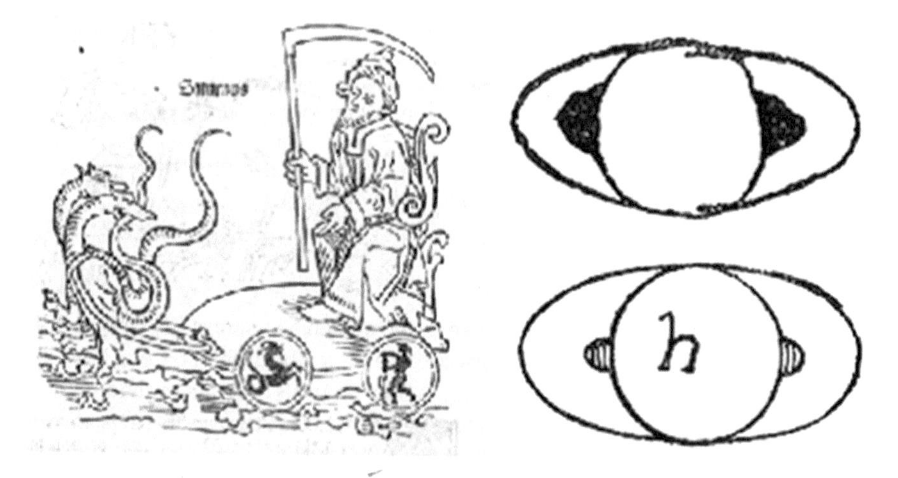

**Fig. 1.29** (*Left*) Image of Saturn showing symbols associated with death, as portrayed in Sentinus Jacobis, *Poeticum Astronomicon* of 1482 (image courtesy of Claremont Colleges Special Collections). (*Right*) Sketch of Saturn from 1616 by Galileo, recording his impressions of the first telescopic views of Saturn and its rings (image from the Galileo Project http://galileo.rice.edu/sci/observations/saturn.html)

The Chinese assigned to Saturn the name *T'ien Hsing*, which translates to either "filler" or "the Exorcist, or Quelling Star" (Krupp 1991, p. 182). The Chinese associated Saturn with the cardinal direction of center, and the element of the Earth (Kelley and Milone 2005, p. 328). Indian folk tales associate Saturn with *Prajaptai,* "Lord of the Animals," and father to the Lunar Mansions or Nakshatras.

The mysterious distant light of Saturn has inspired ancient people in many different ways and continues to inspire us today. With the planet Saturn we have reached the outer limits of our ancient solar system and now consider the outermost sphere beyond—the stars (Fig. 1.30).

## *The Stars*

The stars were the outermost sphere of the Greeks and Chinese, and many other cultures literally believed a shell of stars surrounded the Earth and planets (Fig. 1.31). The exact nature of stars was unknown, and very imaginative speculations exist across the ages explained the glowing light of the stars. Ancient Egyptians told two stories to describe stars. One is that the sky is the underside of a celestial cow, whose hoofs rest on the four corners of the Earth, with the stars hanging from her belly.

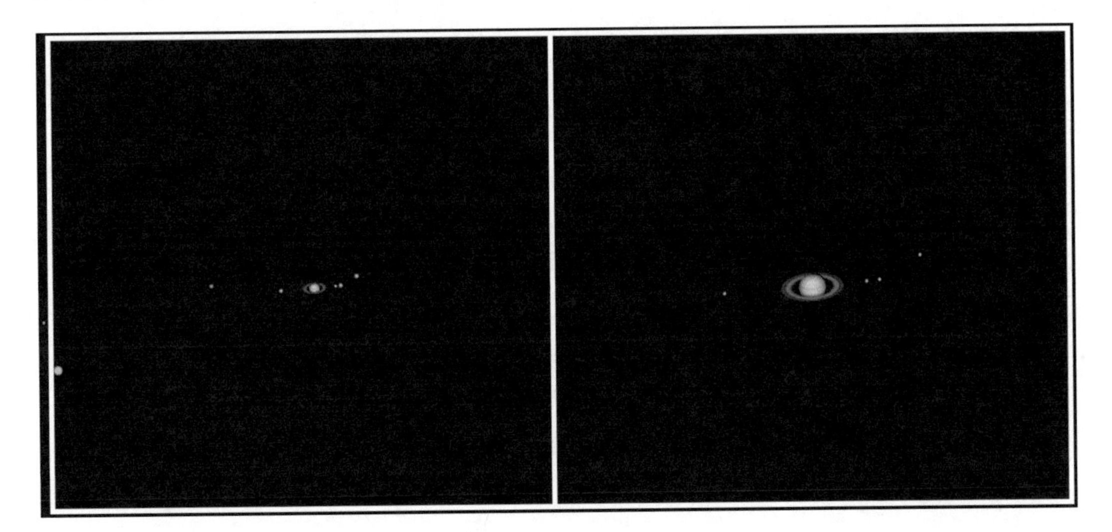

**Fig. 1.30** The planet Saturn, as viewed through a small telescope or binoculars (*left*), shows a hint of its ring system and its family of moons. A view from a larger telescope (*right*) reveals the bands of clouds on the planet surface, as well as the structures in the rings such as the Cassini Division, caused by a gap in the ice particles that comprise the rings

**Fig. 1.31** An early cosmological illustration of stars from Camille Flammarion, L' Atmosphere (Paris, 1888)

The second more widely known Egyptian conception is of stars being decorations on the body of Nut, the sky goddess. Egyptians later developed a set of decan stars for setting their solar calendar, by observing the locations of stars at sunset and sunrise and looking for the first appearance of a star in the dawn sky, known as its heliacal rising, which would happen on the same date each year. The Egyptian calendar needed this help, as their year would drift significantly without the invention of the leap year. As we will see in later chapters, the Egyptians also worked very hard to provide the departed Pharaoh with the tools needed to take his place among the stars in the afterlife, using both the pyramids and other funerary rites to assure a successful journey.

Several Native American tales discuss the origin and nature of the stars themselves. One creation tale from the Cheyenne describes the world of the creator god Maheo as dark and silent until he started the universe as a fire and clapped his hands together within the embers of that fire, scattering the embers which became stars through the dark universe (Tsonakwa and Evenson 1987).

Another famous tale from the Navajo describes Black God carefully placing the stars in the sky by taking crystals from his pouch, to make a beautiful arrangement. While he rested, Coyote saw the pouch and thinking it was something good to eat, ran off with the pouch, spilling stars across the sky. This tale explains both the random nature of the stars and the howling of coyotes at night, who howl to apologize for crossing Black God (Miller 1997, p. 188).

Some interesting and very original thoughts about the stars come from the Aboriginal Australian people. Some Aboriginal people thought the stars were campfires of the sky people, while others thought they were nautilus shells in the sky, while others imagined them to be departed ancestors. Perhaps the most interesting tale, however, involves stars as slivers cut by a giant:

> Beyond the horizon, where no-one has ever been, there is a beautiful land with grassy valleys and tree-covered hills… The inhabitants of that land are moons.. they have no arms or legs but they can move quickly across the grass by rolling over and over… Outside the valley there lives a giant. He catches the wandering moon, and with his flint knife, cuts a slice from it each night, until after many nights there is nothing left but a number of shining slivers. The giant cuts them up very finely and throws them all over the sky. They are timid little creatures, the cut-up moons that become stars.
>
> (Johnson 1998, p. 20)

For other civilizations, the observations of stars were part of a ritual in which the gods would reveal their will in the interplay of star and planet over the course of centuries. Motivated by this cosmic research effort, early Babylonian astronomers charted the stars and divided them into 3 groups of 12, which include a band near the celestial equator, and two bands to the north and south of the central band. Key gods controlled 3 groups of 12 stars, arranged in bands across the sky. Anu the sky god, and father of Ishtar, controlled the central group of stars, while Ea, "lord of the Apsu" or freshwaters controlled the northern group. Enlil, Anu's son, and owner of the "Tablet of Destinies" in which men's fates were recorded ruled over the southern group of stars. In the tablets of Mul Apin from 687 BC the dates of the heliacal risings of each of these 36 key stars were recorded, and each of these stars was watched carefully for omens of the future, beginning the practices of Western astrology.

The Chinese also studied the stars for omens of the future, with large bureaus of professional astronomers who created daily readings of each of 28 sectors in the sky known as *hsiu* or lunar mansions. The Chinese set to work mapping the stars with unprecedented accuracy, with continuous records of eclipses, comets, meteor showers, and supernovae from the time of unification (280 BCE) until the time of European contact in the sixteenth century. The night sky was like a living organism, to which the astronomer was a doctor, carefully examining and looking for symptoms of malady. The "mandate of heaven" required the emperor to rule with justice and wisdom, and study of the stars helped assure this mandate. The heavens were like a "hen's egg" with the Earth in the center like a "yolk" and the stars "supported by vapors" (Needham and Ronan 1980).

Early Greeks also considered the nature of stars and came up with diverse and creative explanations. Heraclitus (500 BC) considered the stars to be "bowls turned with the hollow sides toward us, in which bright exhalations are collected and form flames" (Hetherington 1993a,b, p. 58). Xenophanes (570–475 BC) stated that "the sun comes into being each day from little pieces of fire that are collected.." and that "there are innumerable suns and moons, and all things are made of earth." One pre-Socratic view of stars that survived among the Greek and later Roman public is that the stars are a celestial fire, which we view through holes in an outer shell in the universe. This view was proposed by Parmenides (510 BC) and expanded upon by later thinkers. From the pre-Socratic ideas we have our word "empyrean" which refers to this outermost realm of fire that generates starlight.

Later Greek thought began to imagine the stars to be made of cooler stuff, known to Aristotle as ether, and arranged in patterns on a fixed outermost crystalline sphere, that spins once daily around the Earth, and the enclosed clockwork of the solar system, which including all the spheres for the planets and the Moon, is a series of 53 nested spheres (Toulmin and Goodfield 1965).

The stars indeed do seem to move as if fixed in an invisible sphere rotating about the Earth. Each day the stars rise and set and those stars near the celestial pole (now marked by the bright star Polaris) wheel around endlessly in circles centered on Polaris. Modern astronomers still use the "celestial sphere" as a method of marking positions of stars and consider positions of stars in terms of their latitudes and longitudes on the sphere, which are given the names of declination and right ascension. The spinning Earth provides the rotation of the celestial sphere, which we see as the daily or "diurnal" motion of the stars.

The basic rising and setting of stars is easy to observe, but requires one to be patient and observe the locations of stars against landmarks on the horizon over a period of several hours. If one is outdoors and examines the same stars on successive days, the close observer can see an interesting mystery. The sphere of stars not only rotates once a day, but also experiences a slight shift or extra rotation of 1° toward the west each day.

The motion of stars relative to the Sun is such that in a year the celestial sphere makes one extra turn (resulting from the Earth's orbit around the Sun), and therefore different constellations are visible from month to month. This can even be seen in a few days—each day a star will rise 4 min earlier or shift about 1° west each day. The result is different stars are visible each day at dawn, during their "heliacal risings" and other stars disappear in the glare of the Sun at sunset during their "heliacal settings." This more gradual or annual motion of the stars is the basis for tables of heliacal risings of stars constructed by the Greeks, Chinese, and Babylonian astronomers.

You can then fix your location in the year by observing which stars are visible near dawn or just after sunset. If you have a look from one night to the next these stars will change. You can see the shift of constellations from night to night over several weeks. Or with a simple digital camera and a tripod, it is easy to see the stars moving overhead. Just set your camera to its maximum ISO setting of 1600 or use "fast" film of 800 or 1600. Then with a camera held open on a tripod for 30 s, a minute, or several minutes, the entire stream of stars can be seen to spin overhead, leaving beautiful star trails in your picture (Fig. 1.32).

**Fig. 1.32** Star trails over Mauna Kea, Hawaii. The north celestial pole is visible as the central "hub" of rotation of the sky. This picture was taken over the course of about an hour, as evidenced by the length of the star trails in the figure, which are moving through their diurnal or daily rotation (image reprinted with permission courtesy of Gemini Observatory/AURA; http://antwrp.gsfc.nasa.gov/apod/ap051220.html)

In making the observation of the annual motion, you are recreating the basic stellar astronomy of ancient people, who told tales and mapped these motions both for the intellectual satisfaction and for the practical needs of setting their calendar right during the year. We will discuss in the next chapter some of the tales and maps that were constructed of stars in the different cultures of the world.

## References


Aveni, A.F. 1993. *Ancient astronomers*. Washington, DC: Smithsonian Books.

Galileo, G. (1609). Letter to Leonardo Donato, Doge of Venice. From http://www.lib.umich.edu/spec-coll/largegal1.html

Hudson, T., and E. Underhay. 1978. *Crystals in the sky: An intellectual odyssey involving Chumash astronomy, cosmology, and rock art*. Socorro, NM: Ballena Press.

Johnson, D. 1998. *Night skies of aboriginal Australia: A noctuary*. Sydney, NSW, Oceania: University of Sydney.

Kelley, D.H., and E.F. Milone. 2005. *Exploring ancient skies: An encyclopedic survey of archaeoastronomy*. New York: Springer.

Krupp, E.C. 1991. *Beyond the blue horizon: myths and legends of the sun, moon, stars, and planets*. New York, NY: HarperCollins.

Lowell, P. 1911. *Mars and its canals*. London: The Macmillan.

MacDonald, J., et al. 1998. *The Arctic sky: Inuit astronomy, star lore, and legend*. Toronto, ON: Royal Ontario Museum/ Nunavut Research Institute.

Malville, J.K. 2012. *A guide to prehistoric astronomy in the southwest*. Boulder: 3D Press.

Miller, D.S. 1997. *Stars of the first people: Native American star myths and constellations*. Boulder, CO: Pruett.

Needham, J., and C.A. Ronan. 1980. *The shorter science and civilization in China: An abridgement of Joseph Needham's original text*. New York: Cambridge University Press.

Neugebauer, O., and R.A. Parker. 1960. *Egyptian astronomical texts*. Providence, RI: Published for Brown University Press, by L. Humphries, London.

O'Hara, G. 1997. *Sun lore: Myths and folklore from around the world*. St. Paul, MN: Llewellyn.

Roob, A. 1997. *Alchemy & mysticism: The hermetic museum*. New York: Taschen.

Ross, H.E., and C. Plug. 2002. *The mystery of the moon illusion: Exploring size perception*. New York: Oxford University Press.

Shaukat, K. 2009. Moonsighting. From http://moonsighting.com/faq_ms.html.

Snorri, S., and J.L. Byock. 2005. *The prose Edda: Norse mythology*. London: Penguin.

Toulmin, S.E., and J. Goodfield. 1965. *The fabric of the heavens: The development of astronomy and dynamics*. New York: Harper.

Tsonakwa, G. and D. Evenson. 1987. Echoes of the night, Soundings of Planet. http//:www.amayon.com/Echoes-Night-Dean-Evenson/dp/B000017QT

Williamson, R.A. 1984. *Living the sky: The cosmos of the American Indian*. Boston: Houghton Mifflin.

# Chapter 2
# A World of Constellations in the Night Sky

*The arch of sky and mightiness of storms*
*Have moved the spirit within me,*
*Till I am carried away*
*Trembling with joy.*

Uvavnuk, Inuit shaman woman (Lionberger 2007)

All men have stars, but they are not the same things for different people. For some, who are travelers, the stars are guides. For others they are no more than little lights in the sky. For others, who are scholars, they are problems… But all these stars are silent.

Antoine de Saint-Exupéry, The Little Prince

Look up in any modern city and you may only see a few or a dozen stars. Travel to the country, and the sky is filled with stars, of a wide variety of colors and brightness, arrayed across the sky with a mysterious splendor as it has for thousands of years. Our reaction to this sky is primal and universal. The night sky can bring us strong feelings—A feeling of "falling from a great height" as Carl Sagan describes in the opening of his series Cosmos:

Our contemplations of the cosmos stir us. There is a tingling in the spine, a catch in the voice; a faint sensation, as if a distant memory of falling from a great height. We know we are approaching the grandest of mysteries. The size and age of the cosmos are beyond ordinary human understanding. Lost somewhere between immensity and eternity is our tiny planetary home, the Earth.

Carl Sagan, Cosmos.

The vertigo that arises from the contemplation the heavens is reminiscent of how Emily Dickinson describes how she describes how she knows what is true poetry:

If I read a book and it makes my whole body so cold no fire can warm me I know that is poetry. If I feel physically as if the top of my head were taken off, I know that is poetry. These are the only way I know it. Is there any other way?

Emily Dickinson from a letter from 1870

Many generations have ordered the realm of the night sky with symbols, names, and lessons that mirror the values and experience of our earthly world below. In this chapter we will explore some of the ways in which cultures from around the world have given order and meaning to the formless feelings invoked by the night sky.


B.E. Penprase, *The Power of Stars*, DOI 10.1007/978-3-319-52597-6_2

## A Tour of Constellations North and South

The celestial sphere is like an inverted globe, in which the latitudes and longitudes of the Earth lay below and with the celestial sphere a map of the sky looking up away from the Earth. Instead of countries, continents, and cities, our celestial globe is marked by groups of stars known as constellations. The celestial sphere also features dense clusters of stars such as the Pleiades and visible cities of billions of stars such as our Milky Way and nearby galaxies such as the Magellanic Clouds and Andromeda.

As we can map the Earth, we can map the sky. The latitudes of the sky are known as declination, while the longitudes of the sky are known as right ascension. And as we can explore the Earth, we can explore the night sky with telescopes and with our imagination. In the following sections we will explore portions of the "celestial sphere" with some of the main stars labeled, with the constellation names (and figures) assigned by the European civilization of our times (Fig. 2.1).

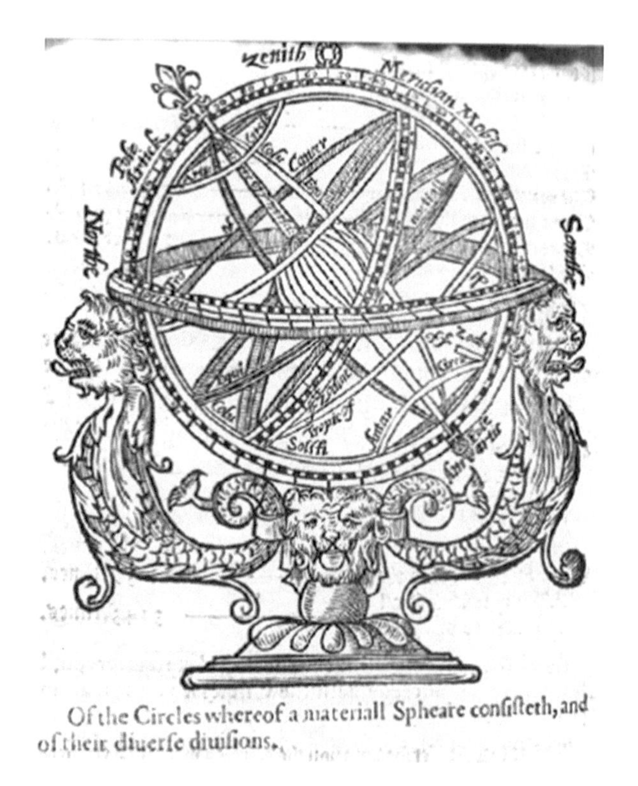


**Fig. 2.1** Early illustration of the celestial sphere, as centered on the Earth, with a view of the zodiac, and projections of the other lines of latitude on the Earth onto the sky (From Blundevil, Blundeville (1594); image courtesy of Claremont Special Collections)

Just as the Earth can be arranged in different countries, with unique names for mountains, rivers, and hills that change from each culture, the landmarks of the sky also have names that have changed through the years. The "countries" on the globe of the sky are the constellations, groupings of stars that fill a region about 15° across. Western astronomical tradition has identified 88 constellations, largely derived from the Greek and Roman traditions, and includes hundreds of named stars, which carry in their names the heritage of astronomy from Egypt, Babylon, Persia, the Arab World, India, and Greece.

Our survey will travel from north to south and explore these constellations as they were described by the inhabitants of Europe, Asia, North America, South America, Australia, and the Pacific Islands.

## Arctic Skies and the Northern Circumpolar Stars

The north of the Earth is home to the civilizations of the Inuit, the Vikings, and many Native American groups. The top of the celestial sphere includes Polaris, the pole star for the past thousand years or so, as well as Ursa Major, our "Big Dipper." Additional stars in the northern region include the constellations of Cassiopeia and Cepheus, and some of the stars of Cygnus, and less well-known constellations such as Draco (the Dragon) and Camelopardalis. Figures 2.2 and 2.3 show a modern projection of the northern celestial sphere, with constellations from the Western European tradition labeled.

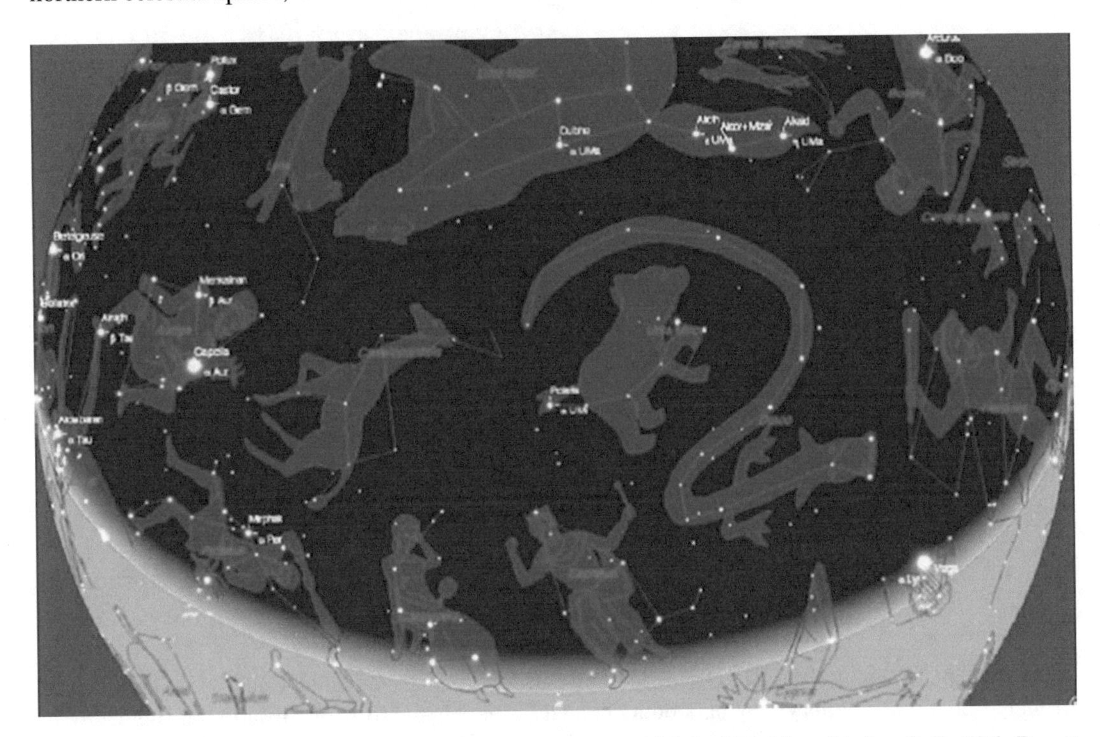

**Fig. 2.2** The north circumpolar sky, with the Great Bear at *top* (in which the "Big Dipper" is found), the Little Bear at *center*, the North Star (Polaris) at its tail. Cassiopeia and Cepheus are visible at the *bottom* (figure by the author)

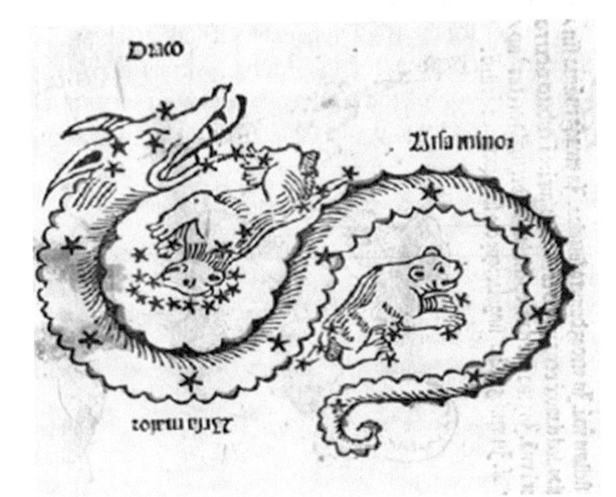


**Fig. 2.3** The northern circumpolar stars, from Sentinus Jacobis, *Poeticum Astronomicon* of 1482 (image courtesy of Claremont Colleges Special Collections)

## *Ursa Major or "The Big Dipper"*

Since the northern sky appears to be "circumpolar" to many of the Northern Hemisphere cultures, a great deal of attention is paid to these stars, as they seem to defy death in their endless circling of the celestial pole, and are visible throughout the night and also throughout the year. We will start by considering just the constellation of Ursa Major, or the Big Dipper, as it was viewed from cultures around the world (Fig. 2.4).

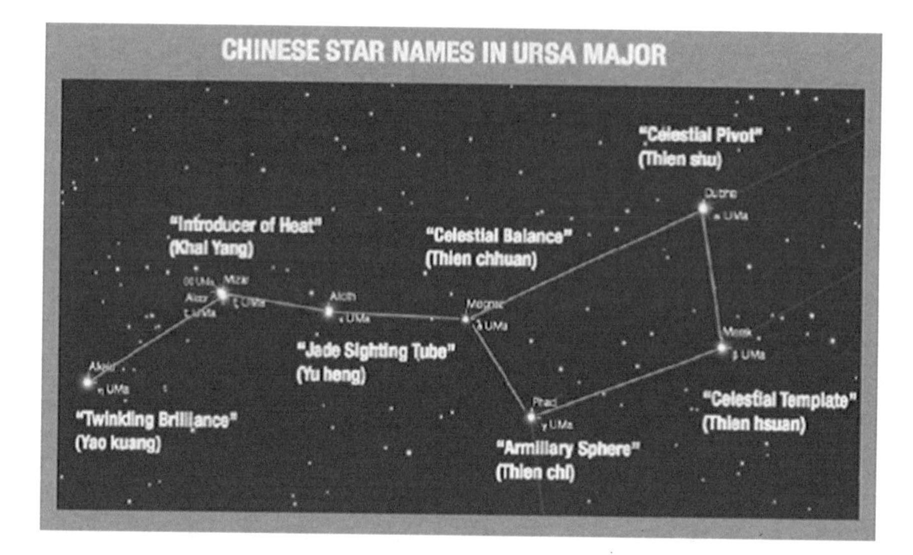


**Fig. 2.4** Big Dipper or Ursa Major, with Translations of Chinese star names labeled (figure by the author; based on information in Needham and Ronan 1980)

In Western mythology, the Big Dipper is known as Ursa Major, the Great Bear, based on the myth of Jupiter hiding one of his young paramours named Callisto in the form of a bear to hide her from the wrath of his wife Juno. Her young son Arcas was hunting in the woods one day after Callisto had been changed into a bear, and accidentally shot Callisto after she rushed toward him with maternal joy. Jupiter placed them in the sky as Ursa Major and Ursa Minor in pity for their loss.

The most widely known names for the Big Dipper stars listed in Table 2.1 are Arabic names, which are a tribute to the excellent observations of the Arabic astronomers, who extended the work of Ptolemy with new theories of motion of the planets, and new observatories built between 829 and 1440 AD across the Arab world. Additional star names from the Chinese culture are also shown in

**Table 2.1** Chinese star names for the Big Dipper region (Needham and Ronan 1980, p. 93)

| Star | Name | Chinese name | Folk name | Function in empire |
|---|---|---|---|---|
| α | Dubhe | Thien Shu ("celestial pivot") | Wen Chang's Throne | God of literature and writing |
| β | Merak | Thien hsuan ("celestial template") | Wen Chang's Throne | "." |
| γ | Phecda | Thien chi ("armillary sphere") | Wen Chang's Throne | "." |
| δ | Megrez | Thien chuan ("celestial balance" or "heavenly authority") | Wen's Throne/K'uei | Minister of literary affairs |
| ε | Alioth | Yu heng ("jade sighting tube") | Chu-i | Minister who looks after student welfare |
| ξ | Mizar | Khai Yang ("introducer of heat") | Chin-Chia | Minister who searches for talented young people |
| η | Benetnash | Yao kuang ("twinkling brilliance") | Kuan-ti | God of wars and war prevention |

Table 2.1—both in their scientific astronomical tradition and from their "folk" astronomical tradition. The circumpolar region was the home of the emperor for the Chinese. Each of the stars of the Big Dipper was associated with portions of the emperor's celestial bureaucracy from which he received "the mandate of heaven." The various stars in the Big Dipper are listed below with their Chinese names, along with their translations. Chinese names for these stars include the technical name used by the professional astronomers of the time, and the folk names, which are from mythological ministers of the Chinese empire. The Big Dipper was divided into two groups of stars—the "bowl" or "box" known as *Khuei* or "the chiefs" and the "handle" known as the *Piao* or "the Spoon."

In the Chinese "folk" system, which could date back to 2000 BC, the box of the stars form a throne occupied by the god of literature and writing known as *Wen Chang*, and the handle is a group of petitioners receiving an audience with the exalted Wen. The petitioners include *K'uei*, the "minister of literary affairs of the world," *Chu-i*, "minister who looks after the welfare of students," *Chin-chia*, "minister who searches for talented young people for the government," and *Kuan-ti*, a god who managed and prevented war. In the "technical" system, the star names refer to many of the instruments used by the bureaus of Chinese astronomers and set the basis for dividing the sky into the 28 lunar mansions or *hsius*. Each of the Ursa Major stars was associated with a function of the empire, and "keyed" to a sector of sky that stretched from north to south much like segments of an orange. Groups of professional Chinese astronomers scanned the skies looking for correlations between "guest stars," "sweepers," and other transient celestial events and the corresponding ministry, which was linked from the circumpolar stars to each sector of the sky.

The Egyptian cosmology, with its emphasis on eternal life, placed great importance to the Big Dipper. The figure for the Big Dipper to Egyptians was a bull's leg, which in many murals is shown in the company of a hippopotamus, an alligator, and several other creatures (Fig. 2.5). The bull's leg

**Fig. 2.5** The northern circumpolar sky from Ancient Egypt, as depicted in the sky ceiling of the temple of Dendera, built in 54 BC. This figure is a sketch of the ceiling, which is centered on the North Star, and features a large hippopotamus and bull's Leg (*center*) for the region of the Big Dipper (figure from Library of Congress, "Heavens" online exhibit, http://www.loc.gov/exhibits/world/heavens.html)

shape was used in one of the main implements of the mummification rituals, known as the *Meskhetiu*, used in the "opening of the mouth" ceremony, which helped place the departed Pharaoh into the eternally living stars of the celestial pole by helping him speak the correct spells from the *Book of the Dead*. It is even thought by some that shafts built into the great pyramids of Giza look out into the north celestial pole, which at that time was marked by the faint star Thuban, a minor star in the constellation Draco (Krupp 1983).

The same group of stars in Ursa Major has been seen as the wagon by the Babylonians, a wagon with a team of oxen or horses in parts of Northern Europe, a plough in northwest Europe, a skunk by the Sioux, a camel by North Africans, a group of boys turned into geese by the Chumash tribe of California, and even as a shark by groups of people from the East Indies (Staal 1988, p. 137).

One interesting take on the Ursa Major is a tale from the Micmac tribe of Nova Scotia, which like many other Native Americans sees the Big Dipper as a Bear pursued by hunters. The "bowl" is the bear, and in our story the hunters include Robin, Chickadee, MooseBird, Pigeon, Blue Jay, and Owl, who form a long line of stars that go beyond the handle of the Big Dipper. The story begins with a hunt, and the birds pursue the bear across the land starting in spring when the birds and bear both emerge hungry from a long winter. As the chase goes on, the various birds fall back to form the line in the sky. In autumn, the bear rears back and Robin hits the bear with an arrow and it falls on its back, whereupon Robin jumps on the bear to begin devouring its prey. Robin tries to shake off the blood, and in the process scatters the red color to all the trees of the Earth. This is why Robin has a red breast, and why the trees turn red each autumn, according to the story. The bear cycles endlessly in the sky, its upside down form crosses the sky in winter, and then is "reborn" in Spring, ready to repeat the cosmic hunt (Miller 1997, p. 39). This story is particularly beautiful in the way it highlights both celestial and earthly cycles, with the changing of the seasons and the rotation of the night sky in perfect synchronicity.

The following section will summarize the sky lore of the northern circumpolar sky in depth for three different northern cultures, European, Chinese, and Inuit, to give a sampling of some of the diverse ways of seeing the same part of the sky from different cultures.

## Northern Circumpolar Sky from Around the World

### *European Circumpolar Constellations*

#### European Constellation Lore and Stars

Most people will start to learn constellation lore with the most widely told tales of the Greek and Roman skies. The tales of the Greeks and Romans were incorporated into the popular culture in most European cultures, even though many of the star names come to us from Arabic sources. A common theme within the star tales of the Greeks and Romans is divine lust, envy, and vengeance. Some examples include the tales of Ursa Major (with Zeus' lover being turned into a bear) and the tale of Cassiopeia (with her punishment from Neptune to be placed above the sea). In this section we briefly summarize some of the principal northern stars and constellations, and star lore, as they are traditionally known in the western European cultures (Fig. 2.6).

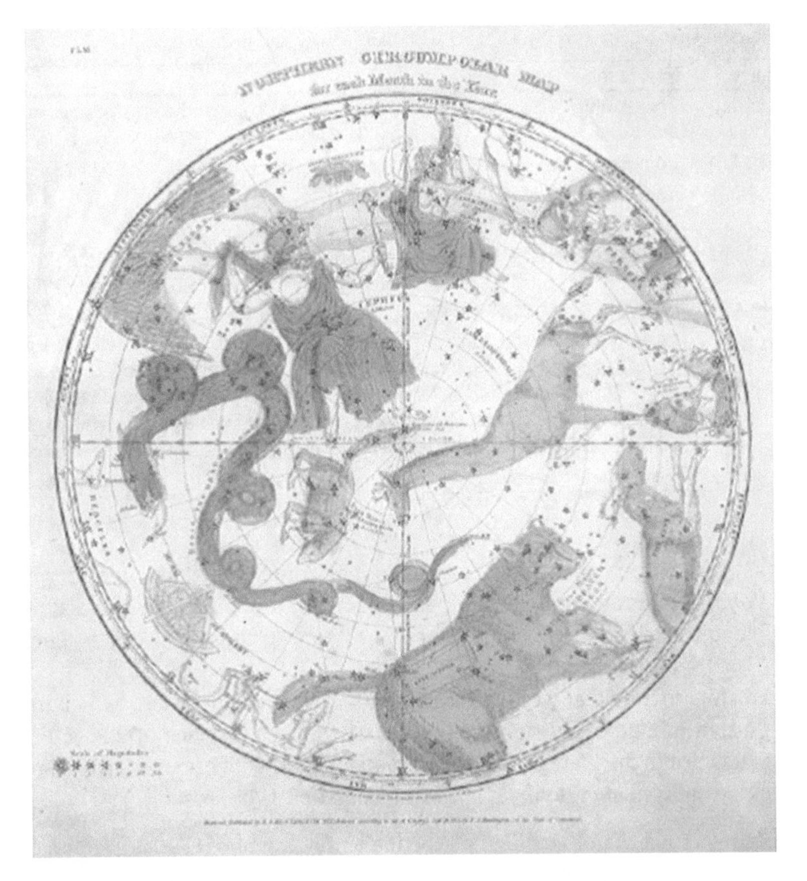

**Fig. 2.6** Circumpolar sky from the Burritt sky atlas (Burritt 1833), showing figures for Ursa Major, Ursa Minor, Draco, Camelopardalis, Cassiopeia, and Cepheus. Names of the principal stars and sky lore for each of these constellations from the European/Greek/Roman cultures are described concisely in the text. Accompanying supplementary content includes an interactive digital sky map for each culture (image courtesy of the Claremont Colleges Special Collections)

## Ursa Major and Ursa Minor

The Ursa Major constellation star lore has already been discussed, and some of the major stars within Ursa Major and Ursa Minor are listed below. The Greek letters refer to the order of brightness of the star, beginning with α for the brightest star. This tradition began with the earliest star atlases of Bayer (1603), Hevelius (1690), and Flamsteed (1729).

The Arabic names for the principal stars of Ursa Major are still in use today, as is the case for many of our western European constellations. The Arabic names are mostly based on anatomical parts of the bear (Allen 1963, p. 438). The listing of the star names and meanings is given in Table 2.2 and shows both the logical and some improbable origins of our modern star names.

**Table 2.2** Key stars in Ursa Major (the Big Dipper) with their names and meanings

| Name | Letter | Translation |
|---|---|---|
| Dubhe | ($\alpha$ UMaj) | From Arabic "*Thahr al Dubb al Akbar*" or the "back of the greater bear," a yellowish binary star |
| Merak | ($\beta$ UMaj) | From Arabic "*Al Marakk*" or the "loin of the bear" |
| Phaed | ($\gamma$ UMaj) | From Arabic "*Al Falidh*" or "the thigh" |
| Megrez | ($\delta$ UMaj) | From Arabic "*Al Maghrez*" or the "root of the tail" |
| Alioth | ($\varepsilon$ UMaj) | The name is perhaps corrupted from Arabic "*Al Hawar,*" "white of the eye," or "intensely bright", or perhaps "*Alyat,*" improbably translated to "the tail of eastern sheep" |
| Mizar | ($\xi$ UMaj) | From Arabic "*Mi'Zar*" or "Girdle." A binary star which is split by 14 separations and also is paired with the star Alcor, which is itself a famous star, barely visible with naked eye but easily picked up with a telescope. Alcor is named from Arabic for either *Al Kawwar,* "the faint one," or *Al Sadak*, "the test" as Alcor was used as a vision test by ancient civilizations. The combination of Alcor and Mizar is a beautiful sight through even a very small telescope and is easily found as the bend in the handle of the Big Dipper |
| Benetnash | ($\eta$ UMaj) | This name is from the Arabic "*Ka'id Banat al Na'ash*" or "the chief of the mourners" |

Our discussion of circumpolar star names would not be complete without considering the pole star, Polaris. While Polaris seems an intuitive and logical name for a star near the celestial pole, the Greeks called the star Phoenice, and some accounts refer to the name "Cynosura" from the Greek name "dog-tailed one."

It is also interesting to note that the star Polaris is itself a set of three stars, which includes a bright Cepheid variable star, orbited by a bright yellow dwarf star at a distance of 224 billion miles, and a faint third star that is lost in the glare of the brightest star we see with our eyes. So even within the pole star is a complex astrophysical system, full of mystery (Allen 1963, p. 444).

**Sky Lore Summary** Ursa Major is thought to be the embodiment of Callisto, Jupiter's earthly lover, who was hidden from Juno as a bear. Ursa Minor is thought to be the embodiment in the sky of Arcas, the son of Callisto, who accidentally shot his mother while hunting and was placed in the sky in a gesture of compassion by Jupiter.

## Draco: The Dragon

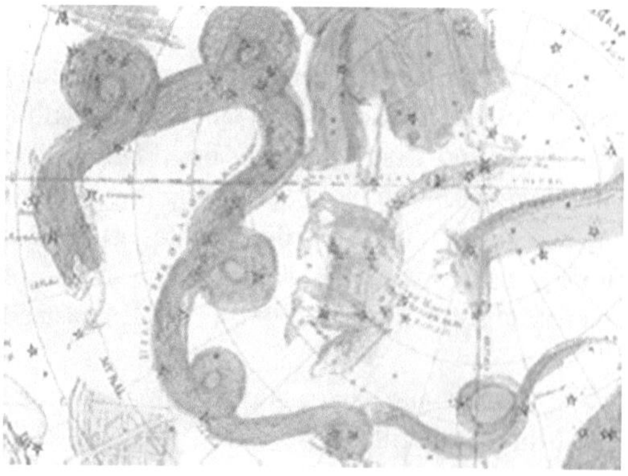

The constellation of Draco wraps around and between Ursa Major and Ursa Minor and is a faint sinuous line of stars surrounding the celestial pole and separates Ursa Major (the "Big Dipper") and Ursa Minor (the "Little Dipper").

**Principal Stars** *Thuban ($\alpha$ Draconis).* The name Thuban also comes from Arabic and is the title for the entire constellation of Draco. This star was close to the north celestial pole during the time of Egyptians, and it is alleged that the Great Pyramid of Giza is aligned with a shaft pointed toward Thuban's "upper culmination" or highest point in the sky during 2500 BC.

**Sky Lore Summary** Draco is the name for "The Dragon" and the constellation has been connected to several mythological dragons, including the Dragon of Ladon, who guarded the land of the Hesperides, and the other a dragon who sided with the Titans in an epic clash against the gods of Mt. Olympus. The dragon in this story was placed in the sky after Minerva (Athena) grabbed its tail and swung it up into the sky.

## Cassiopeia: The Seated Queen

The constellation Cassiopeia forms a distinctive "W" or "M" in the circumpolar sky (depending on the season) and is easily spotted by the bright group of stars that make up the main part of the constellation.

**Principal Stars** *Schedar (α Cass).* This name perhaps comes from Arabic *Al Sadr*, "the breast" which is appropriate for its location within the constellation of Cassiopeia, the Seated Queen of Ethiopia.

**Sky Lore Summary** Cassiopeia is the legendary queen of Ethiopia, famed for both her beauty and vanity. When she insulted Neptune by claiming her beauty surpassed his Nereids, Neptune placed her in the sky in a place where her hair could never touch the sea, preventing her from ever washing her prized locks.

## Cepheus: The King of Ethiopia

Cepheus is a faint trapezoid in the sky, vaguely in the shape of a house. None of the stars are particularly distinctive.

**Principal Stars** Alderamin (α Ceph). The name comes from the Arabic Al Dhira al Yhamin, "the Right Arm" (Allen 1963, p. 157). It is also interesting to note that this star will become the new pole star in AD 7500, due to celestial precession.

It is also important to note that the star δ Cephei, located at the base of Cepheus, is the original Cepheid Variable star and changes in brightness by about a factor of 2 over a period of 5.4 days. If you can find this star and observe it visually over several days you can see it fade out. This star was one of the Cepheid variables that enabled astronomers like Edwin Hubble and Alan Sandage to measure the distance to the nearby galaxies in the 1920s and 1960s, and from these observations came evidence for the Big Bang, which predicted that the galaxies would appear to be moving apart from each other due to cosmic expansion.

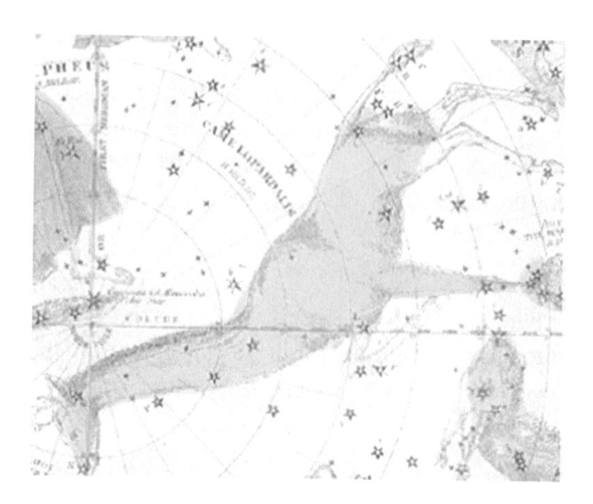

### Camelopardalis: The Giraffe

Camelopardalis is a very faint and weakly defined group of stars in the circumpolar sky. It is improbably named Camelopardalis, for the "camel leopard" or "spotted camel" which were names for the giraffe in Europe before the seventeenth century. The figure first appeared in one of the star atlases of this time and is not a constellation with significant star lore or bright stars. It does, however, complete our celestial menagerie of the northern circumpolar sky from the western European tradition.

## *The Circumpolar Sky from Ancient China*

### Chinese Constellation Lore and Stars

Chinese astronomy worked on two parallel levels. The folk tradition, with its rich lore and mythological characters, provides a colorful picture of ancient Chinese values, traditions, and moral lessons. At the same time, the professional astronomers had the most extensive observatories and astronomical bureaus of the ancient world, and patiently assembled impeccable records of novae, comets, meteor showers, and other celestial phenomena for over 1000 years. Both levels will be considered here with an emphasis on the technical aspects of the Chinese astronomical system.

The Chinese folk tradition is based on thousands of years of belief and oral tradition, mixed in with some mystical Taoist philosophy and some Confucian moral lessons. The bronze planisphere disk, shown earlier, shows a set of the primary images of Chinese folk astronomy. The four directions, colors, and animals are on the inside. Each direction provides an association between a color and an animal that includes north (black tortoise), east (azure dragon), south (red bird), and west (white tiger). In the Chinese metaphysics, a fifth direction is also recognized as the center, which is given the color yellow, and associated with the Earth. The addition of 12 animals associated with the 12 year Jupiter cycle and a division of these 12 years into four "trines" give a rich basis for astrological forecasting. Even more detail can be found from the hour of a person's birth, which will fall into 1 of 12 2 h intervals, each ruled by a "secret animal."

Chinese folk tradition is more than astrology, however, and embodies a system of values based on thousands of years of culture. One of the most famous tales involves the constellations of Lyra and Aquila, two parts of the "Summer Triangle," easily visible from China during the early evenings in the summer months. In the tale, Lyra is the weaving princess, *Tchi-niu*, and Aquila is the cowherd *Kien-niou*, who gaze at each other across the barrier of the Milky Way, which in this tale is a river:

> The Princess lives with her father the Sun God, and falls in love with the cowherd, and convinces her father to bless the marriage. The two were so smitten with love that they forgot their duties, and the royal herd became scattered all over the sky, while the Princess forgot about her weaving. The King was forced to banish Kien-niou to the opposite side of the Milky Way. The Gods were saddened by this separation and arranged a bridge of magpies to be built one day each year, on the seventh day of the seventh month. On this day the two were reunited, and the land is filled with warm rain, which is the tears of joy of the reunited lovers.
>
> (Staal 1988, p. 186)

The tale of Tchi-niu and Kien-niou brings together meteorology, astronomy, and a lesson about work, love, and family all in one story, which can be remembered on any clear summer night.

Chinese professional astronomers divided the sky into several zones. The central region was known as the "purple forbidden enclosure" and included the circumpolar stars, which were associated with

**Table 2.3** Chinese constellation names for the 28 lunar mansions and the corresponding Western star names (from Needham and Ronan 1980)

| Lunar mansion | Ancient constellation name | Principal star | Correlated phenomenon with circumpolar key star |
|---|---|---|---|
| Chio | Horn | α Vir (Spica) | Lower transit of Polaris |
| Khang | Neck | κ Vir | Transit of $\alpha$ and $\beta$ Centaurus |
| Ti | Root | α Libr | |
| Fang | Room | ρ Sco | Culmination of $\alpha$ Dra (Thuban) "pivot of the right" |
| Hsin | Heart | σ Sco | |
| Wei | Tail | μ Sco | |
| Chi | Winnowing basket | γ Sag | Lower transit of $\kappa$ Dra and upward position of tail of Great Bear |
| Nan Tou | Southern Dipper | ϕ Sag | Culmination of $\iota$ Dra "celestial unity" |
| Niu/Chhien Niu | Oxherd boy | β Cap | Lower transits of $\alpha$ and $\beta$ U Majoris or "pivot of heaven" and "star of the emperor" (edge of Big Dipper); also coincides with culmination of a Lyra, the "weaving girl" |
| Nui/Hsu Nu | Girl serving maid | ε Aqr | |
| Hsu | Emptiness | β Aqr | Lower transit of $\gamma$ Ursa Majoris "celestial armillary" |
| Wei | Rooftop | α Aqu | Lower transit of $\delta$ and $\varepsilon$ Ursa Majoris "celestial balance" and "celestial sighting tube" |
| Shih | House/encampment | α Peg (Markab) | Lower transit of $\varepsilon$ Ursa Majoris the "celestial sighting tube" |
| Tung Pi | Eastern wall | γ Peg | Culmination of β Ursa Majoris "celestial template" |
| Khuei | Legs | η And | Culmination of 3233 Ursa Majoris |
| Lou | Bond | β Ari | Lower transit of h Ursa Majoris "twinkling brilliance" |
| Wei | Stomach | 41 Ari | |
| Mao | Graph of star group (Pleiades) | η Tau | Lower transit of α Dra (Thuban) "pivot of the right" |
| Pi | Net (Hyades) | ε Tau | |
| Tsui | Turtle | λ Ori | Culmination of κ Dra |
| Shen | Graph of 3 stars | ξ Ori | Downward position of tail of Great Bear |
| Ching | Eastern Well | μ Gem | Culmination of α and β Ursa Majoris |
| Kuei | Ghosts | θ Can | |
| Liu | Willow | δ Hyd | |
| Hsing | Seven stars | α Hyd | |
| Chang | Extended net | μ Hyd | |
| I | Wings | α Cra | Lower transit of γ Ursa Majoris "celestial armillary" |
| Chen | Chariot | γ Corvi | Lower transit of β Ursa Majoris "celestial template" |

the emperor. Two outer rings of stars were at lower declinations on the sky. Stars were grouped into constellations with smaller numbers of stars than Western charts, since the constellations each fit into 1 of 28 separate radial sectors of the sky. As mentioned earlier, each of the 28 sectors, known as lunar mansions or *hsiu*, was keyed to one of the circumpolar stars, and teams of astronomers watched carefully for changes in the sky to diagnose the health of the empire. Within these 28 *hsiu* are principal stars, grouped into small constellations, which are summarized in Table 2.3 with their Chinese names and properties. Each of these *hsiu* is highlighted in our interactive Suchow planisphere in the supplementary content which comes with the book (Fig. 2.7).

The principal stars of the Chinese hsiu system are summarized in Table 2.3, with their Chinese and translated names (from Needham and Ronan 1980). Typically the first sighting of the main stars in each of the 28 *hsiu* is correlated with an event in the circumpolar region. These events include "culminations" which is when a circumpolar star reaches its greatest height in the sky or "lower transits" which is when the circumpolar star circles around to its lowest height in the sky and crosses the meridian in the north.

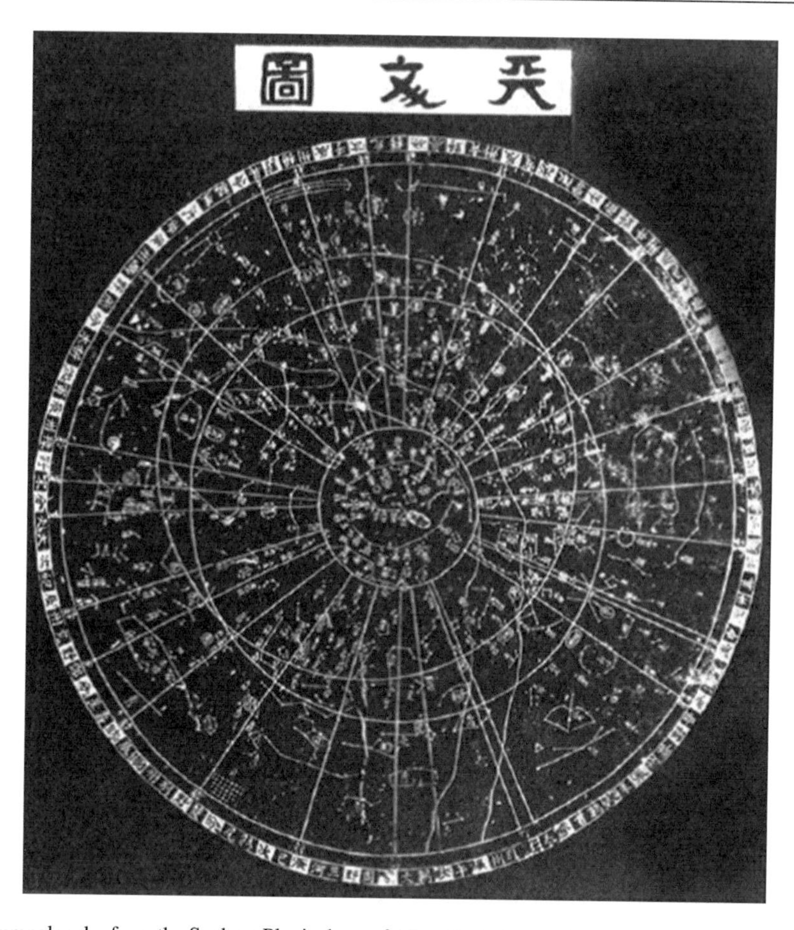

**Fig. 2.7** Circumpolar sky from the Suchow Planisphere of AD 1193. This planisphere was carved in stone in AD 1247 based on an earlier design created by the imperial tutor Huang Shang as a gift for his young student, the future emperor Ning Tsung. Central in this map of the sky are the circumpolar stars, seen in the *small inner circle*. These stars were associated with the emperor and were keyed to the 28 sectors of the sky, which are shown with *radial lines* (photograph of replica planisphere at Claremont Colleges; the supplementary content includes an interactive digital version of the Suchow planisphere, in which the constellations and main features of the diagram can be explored in detail)

## Star Maps for World Constellations: Northern Hemisphere

In the sections below, we will provide maps of the sky which feature constellation figures from a variety of world cultures—Chinese, Navajo, Inuit, and several others. As a way of getting oriented, within the star maps it is possible to locate familiar "landmarks" such as the Big Dipper, and other groups of stars such as the Summer Triangle and Winter Hexagon. Figure 2.8 provides a view of the Northern hemisphere sky in the same projection used throughout our tour of constellations. In Fig. 2.8 we have indicated the location of the Winter Hexagon (which groups together six of the western constellations) and the Summer Triangle (formed by bright stars in Cygnus, Aquila and Lyra), as well as the Ecliptic—the path of the sun through the sky.

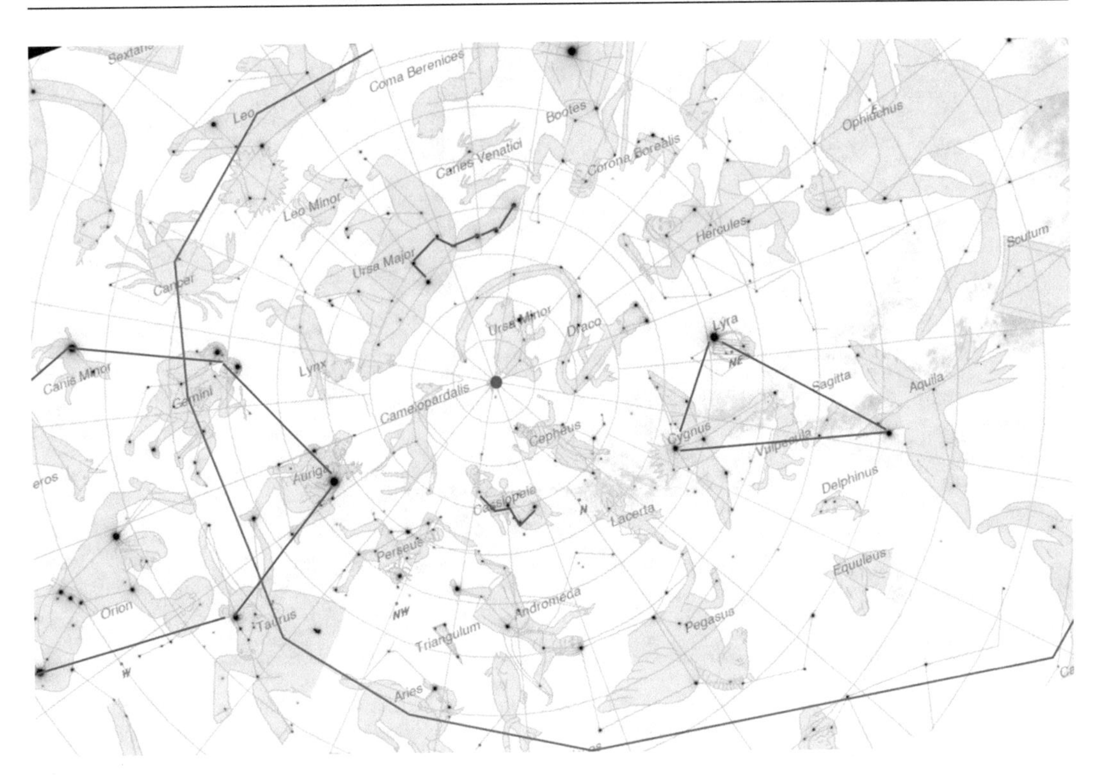


**Fig. 2.8** Legend for the Northern hemisphere star maps. The familiar groups of constellations are highlighted in *Blue*—the Winter Hexagon (*left*) and the Summer Triangle (*right*). The center of the image is the North Celestial Pole, and above this is the Big Dipper (highlighted in *blue*), and below it Cassiopeia (shaped like a "W" and highlighted in *blue*). The zodiacal constellations, which form the backdrop for the apparent motions of the planets and sun is shown with a *red line*

## Northern Constellations from Around the World

### *Chinese Constellations and Asterisms*

Most of the small asterisms of the Chinese *hsiu* are very different from Western constellations. However, some similar groupings include the two groups of stars in Pegasus, known as *Shih* and *Tung Pi*, which include the sides of the "Great Square" but as separate groupings. Also the constellation of *Shen* is very close to what Western astronomers would call Orion and can be seen at the 4 o'clock position of the Suchow planisphere. Like most cultures, the Chinese also recognized both the Pleiades and Hyades clusters in our constellation of Taurus and assigned them separate lunar mansions of *Mao* and *Pi*.

It is interesting to note that both levels of Chinese astronomy, folk and professional, are united in the lunar mansion named *Chien Niu*, which translates to "Oxherd Boy." The location of this constellation is on an opposite side of the Milky Way to the star α Lyr, which we know as Vega, and the Chinese as *Chih nu*, or "weaving girl." When the lunar mansion of the "Oxherd" appears in the twilight, the "weaving girl" reaches its greatest height in the sky, connecting the same two characters of Chinese astronomy folk tradition in the intricate hsiu system used by the imperial astronomers of China (Fig. 2.9).

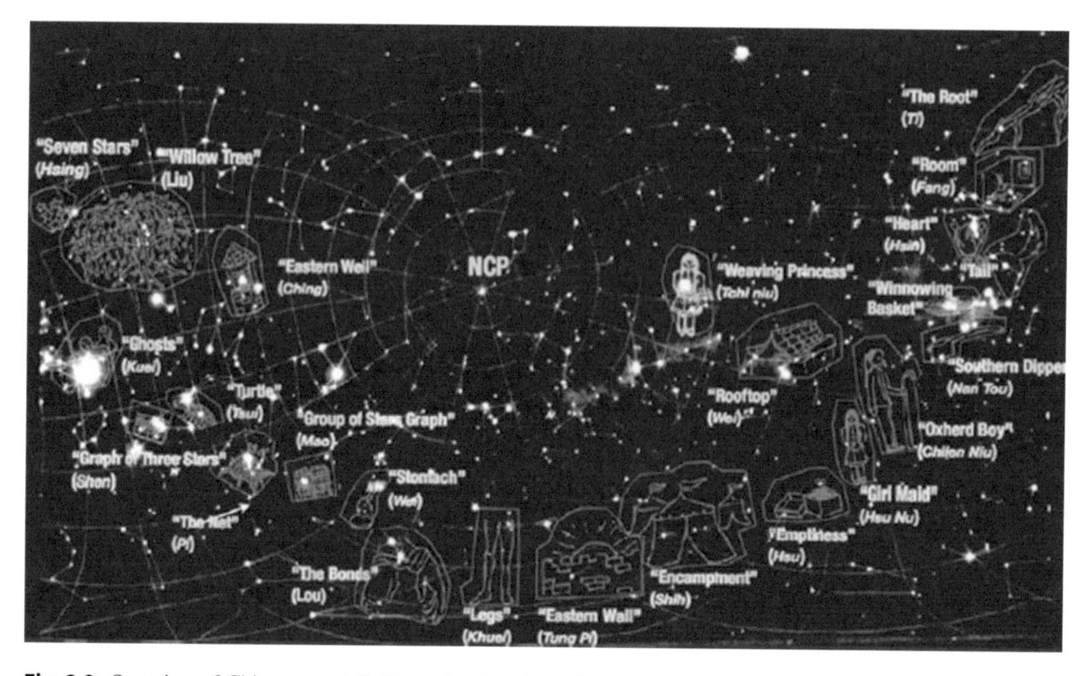


**Fig. 2.9** Overview of Chinese constellations, showing the main constellations within Chinese star lore and the hsiu system (figure by the author; artwork by Kim Aldinger. An interactive version of all the star maps is included in the supplementary content for this title)

## Northern Circumpolar Sky from Around the World

### *The Arctic Inuit Sky*

#### Inuit Constellation Lore and Stars

The Inuit approach to sky lore was intimately connected with the harsh conditions of the Arctic and the struggle to survive and hunt in the intense cold and dark of the region (Fig. 2.10). The canopy of night sets in for the long arctic winter and at Arctic latitudes the stars wheel around the edges of the horizon in circles that graze the edge of the horizon. The Sun and Moon are seasonal visitors to the region, like many of the animals in the Arctic. The Sun can be completely below the horizon for weeks or months at a time and will only make a brief appearance above the horizon in late fall and early spring. However, during summer the days are endless and the Sun wheels above giving 24 h of light for months.

The celestial sphere in Inuit is known as *Qilak*. To the Inuit, the *Qilak* is the home for souls of departed people, as well as the Sun and the Moon. The Qilak circles above the Earth, or *Sila*, and can be reached by certain shamans, who are able to travel between the Earth and sky and perhaps to worlds beyond the sky (MacDonald et al. 1998). Many diverse traditions exist among the various tribes of Greenland, Northern Canada, and the Arctic coast of Alaska. All of them include accounts of multiple worlds and use stars, weather, and folk tales to help in hunting, navigation, and teaching the young about the workings of the universe. In nearly all accounts, the special behavior of the few non-circumpolar stars is of special interest. Seasonal appearances of the Western constellations of Orion and Taurus (known to the Inuit as *Ullaktut* and *Sakiattiak*, respectively) mark the winter. The first visibility of the stars of the Western constellations of Aquila and Lyra (known to the Inuit as *Aagjuuk* and *Kingulliq*, respectively) signals the passage of the Winter solstice and the beginning increasing light for the region. Also in the absence of the Sun, the appearance of these stars on the horizon can signal

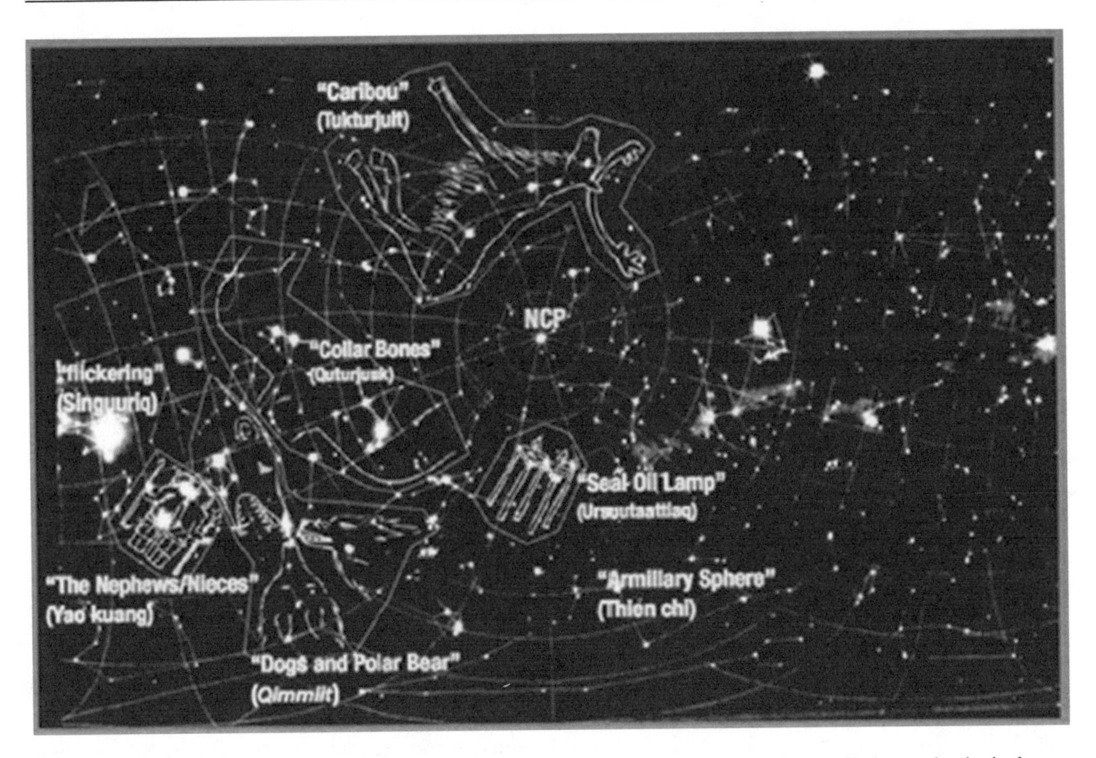


**Fig. 2.10** Stars and constellations from the Inuit sky, with the Inuit names for the constellation and principal stars labeled. The Inuit sky lore necessarily focused on the circumpolar region, although the seasonal appearance of our constellations of Orion and Taurus (visible in the *bottom of the figure*) played an important role (based on information from MacDonald et al. 1998, p. 39) (figure by the author; artwork by Kim Aldinger)

the coming of dawn and help keep time for the Inuit. In the dark days of winter without sunrise, the stars signal the time for villagers to wake up, for children and hunters to begin their days, and for the village to start the routines of the day.

The Western constellation of Taurus, and its bright star Aldebaran, is known as *Nanurjuk*, which means "like, or having the spirit of a polar bear" (MacDonald et al. 1998, p. 57) (Fig. 2.11). One tale tells of how *Nanurjuk* is a polar bear being held at bay by a pack of dogs, which are the rest of the stars of the Hyades. In some versions of the story, *Nanurjuk* is already dead, and its red color is the blood of the polar bear. The remaining stars of Taurus are hunters and their dogs with their fires, getting ready to share the meat. Other versions of the tale have the Pleiades serving as a group of dogs or hunters, with the brightest star of the cluster serving as the bear.

The nearby constellation of Orion, known to the Inuit as *Ullaktut*, is also associated with hunting. The name *Ullaktut* translates to "runners," referring to three heroic hunters running through the sky, as seen as the belt stars of Orion. Interestingly, the star Polaris was a minor one for the Inuit, perhaps because it was so high in the sky it was difficult to use for navigation. The Inuit recognized the entire grouping of the Big Dipper, which they called *Tukturjuit* or "caribou." Many of the stars were used as hour hands on the night sky to indicate hours of the night, or as calendar stars to help determine the date in fall, winter, or spring.

Several of the Inuit constellations are partial groupings of the Western constellations (Pituaq/Cassiopeia and Kingulliq/Orion, for example), while others combine multiple Western constellations (Quturjuuk = Gemini + Auriga). Table 2.4 summarizes some of the Inuit stellar groupings and associations, which provide insight into the values of the Inuit people, who used the power of these stars to help them survive the Arctic winter.

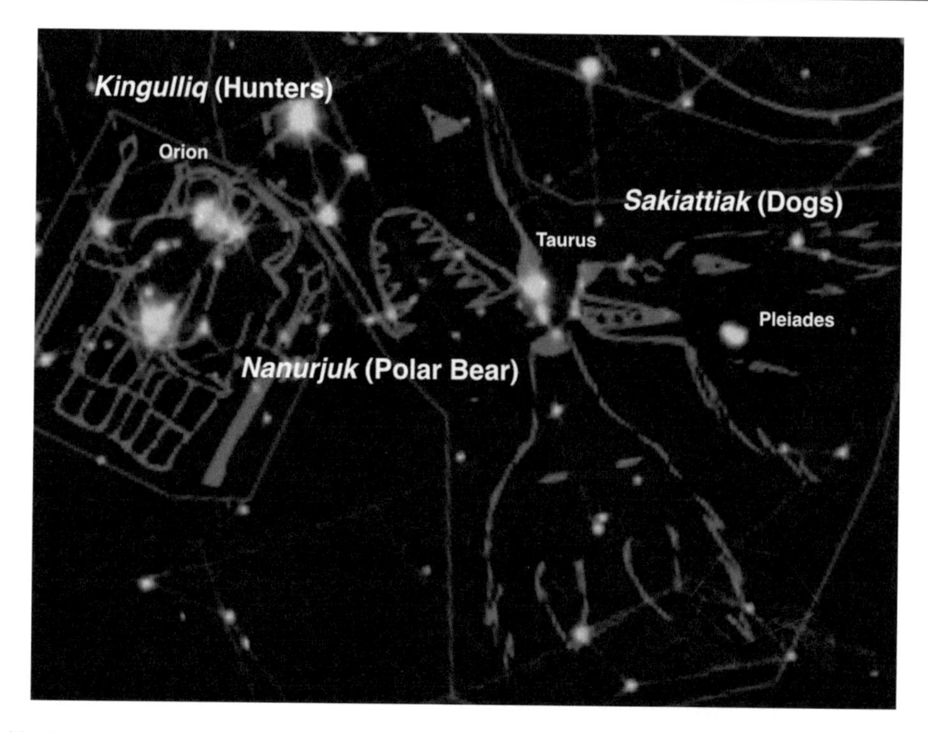


**Fig. 2.11** A star chart of the region near the constellation Taurus, showing sky figures from the Inuit culture. These figures include *Kingulliq*, corresponding to part of the Western Constellation of Orion, and the stars near Taurus that include Aldebaran and the Hyades cluster *center* and the Pleiades on the *right*. These two clusters are known to the Inuit as *Nanurjuk* (Hyades) and *Sakiattiak* (Pleiades) and are thought to be a polar bears and hunting dogs, respectively (figure by the author, using the Starry Night program)

**Table 2.4** Principal Inuit stars and constellations with their meanings (adapted from MacDonald et al. 1998, p. 43)

| Inuit name | European constellation | Principal stars of the Inuit | Explanation |
|---|---|---|---|
| Aagjuuk | Aquila | Altair, Tarazed | First visible after winter solstice, used to mark daytime and coming of spring |
| Akuttujuuk | Orion | Betelgeuse, Bellatrix (top two stars or "shoulders" of Orion) | Indicator of winter coming on; seen in the early evening as fall begins |
| Kingulliq | Lyra | Vega | Second star visible during spring, in some legends is "a brother of the Sun" |
| Kingulliq (second) | Orion | Rigel (right "foot" of Orion) | Name means "the one behind"; legend has it as a hunter trailing the three belt stars from dropping his glove |
| Nanurjuk | Taurus | Aldebaran | Name means "like or having spirit of polar bear." Many tales with this star as polar bear being hunted by other stars nearby |
| Nuutuittuq | Ursa Minor | Polaris | Name means "never moves" and is a minor star and only used for navigation for the southernmost Inuit people |
| Pituaq | Cassiopeia | Scadar, Caph (western edge of the "W") | Name means "lamp-stand" |
| Qimmiit | Taurus | Hyades | Name means "dogs" and in one legend the stars are the dogs who has cornered a polar bear (Aldebaran) |
| Quturjuuk | Gemini + Auriga | Pollux, Castor, Capella, Menkalinan | Two pairs of stars which means "collar bones" and used as an hour hand to mark the hours of the night |
| Sakiattiak | Taurus | Pleiades | Name means "breast bone" and is sometimes said to be dogs or hunters in the sky |

(continued)

**Table 2.4** (continued)

| Inuit name | European constellation | Principal stars of the Inuit | Explanation |
|---|---|---|---|
| Sikuliarsiu-juittuq | Canis Minor | Procyon | This long name means "the one who never goes onto the newly formed sea-ice" and refers to a large man who went hunting |
| Singuuriq | Canis Major | Sirius | Name means "flickering" probably from the fact that this star would be very low on the horizon at the northern latitudes of the Inuit |
| Sivulliik | Bootes | Artcturus, Muphrid | A pair of stars with a name that means "the first ones" referring to the fact that this star was visible during the beginning and mid-winter |
| Tukturjuit | Ursa Major | Dubhe, Merak, Phecda, Megrez, etc. | This grouping is the same as our Big Dipper and in Inuit means "caribou" |
| Ullaktut | Orion | Alnitak Alnilam Mintaka (belt stars) | The three stars of Orion's belt, with a name that means "the runners" after three hunters running after their prey |
| Ursuutaat-tiaq | Cassiopeia | Scadar, Caph, Cih, etc. (same as Western Cassiopeia) | Name means "seal-skin oil or blubber container" |
| Qangia-mariit | Orion Nebula | M42 | Name means "nephews" or "nieces" and the nebula was thought to be a group of children |
| Aviguti | Milky Way | | Name means "divider" or separator. In one account the Milky Way is the track made by Raven's snowshoe when he walked across the sky creating the inhabitants of the Earth |

## The Equatorial Sky

In the center of the celestial sphere lies the Zodiac—the path of the Sun and planets through the sky—and all of the background zodiacal constellations. For the Babylonian astrologer–priest, the conjunction of the Sun and star provided a compelling source of power, and it is this belief that underlies our most commonly known system of "astrology." The zodiac marks the plane of the solar system and is tilted from the equator of the celestial sphere by 23.5°. From the equator of the celestial sphere comes the "equatorial sky" which includes constellations halfway between north and south poles. The zodiacal and equatorial constellations are seasonal and lost in the glare of the Sun for the time when they are not visible.

Some of the most distinctive constellations in this central part of the celestial sphere include the Pleiades, Orion and Taurus, Sirius, Scorpius, and the "Summer Triangle" which includes the stars Vega, Altair, and Deneb (Fig. 2.12). Hundreds of star tales exist for each of these constellations—even the lore for the Pleiades alone could fill an entire book!

We will focus on the star lore of three cultures in depth as "case studies" of how different cultures can approach the same stars to contrast with the Greek and European star tales. The following section describes the star lore and constellations of the Greeks and Europeans, and of three "equatorial cultures"—the Hawaiian, expert navigators of the open oceans; the Chumash, an ancient Native American culture from Southern California; and the Navajo, a well-known North American tribe which used star lore as part of their healing ceremonies. All three of these cultures offer completely original views of the same stars and yet all three share the same human impulse to explain and order the power of the stars above their lands.

**Fig. 2.12** The equatorial constellations of Pisces, Aquarius, Capricorus, Sagittarius and Scorpius, which all are on the "zodiac" or ecliptic plane. The Sun in this diagram is in Capricorn, the basis of the Babylonian system of astrological birth signs, and the location of the intersection of the celestial equator and ecliptic in ancient times (figure by the author)

## The Equatorial Sky from Around the World

### *Equatorial Constellations from the European Tradition*

#### European Equatorial Constellation Lore and Stars

The sky in the European tradition is a swirling mass of Greek gods, goddesses, monsters, and animals (Fig. 2.13). This cosmic zoo was part of every child's education long ago, as people spent many hours under the night sky. Unlike many of the traditional cultures we will consider, the Western sky lore is often not passed on in either oral or written tradition in modern times. Glimmers of astronomical knowledge come to many from the astrological page of a newspaper, where reports from the different zodiacal signs are presented. These same constellation names are used in modern astrophysics research, yet few of the scientists are aware of the long tradition of myth and legend behind the names of stars and constellations.

We will briefly review each of the main equatorial constellations below, first with the European system of star names and constellations in order of the seasons in which they are visible. Each of the four seasons offers a different "cast" of constellations and figures as the Earth and Sun change their positions.

#### The Autumn Stars of the Zodiac and Equator

Our tour of the European sky begins at the "Prime Meridian" of the celestial sphere, the vernal equinox. When the Sun is at the autumnal equinox in fall, stars opposite to this point (near the vernal equinox) appear overhead at night. These are the autumnal constellations, which include Andromeda, Perseus, Pisces, Aquarius, Pegasus, and Aries, along with Cassiopeia and Cepheus (discussed earlier). The Vernal equinox in fall crosses the meridian at midnight. Currently the Vernal equinox is located in the constellation Pisces, but due to the Earth's wobbling spin axis, the equinox slowly precesses in a circle around the sky in a 26,000-year cycle, causing all the constellations to shift over thousands of years (Fig. 2.14).

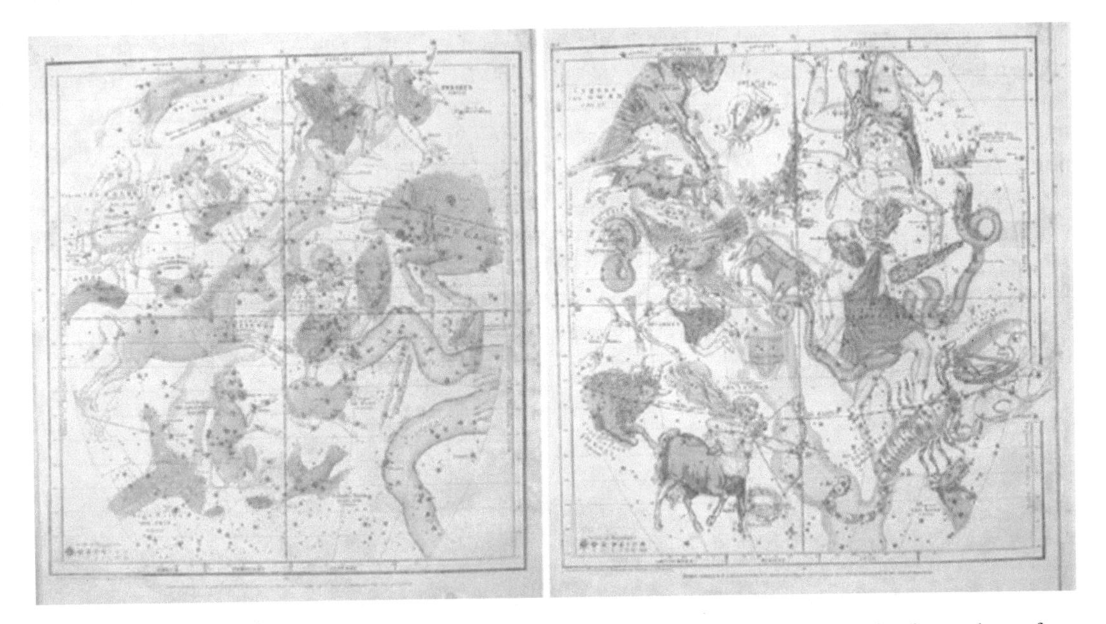

**Fig. 2.13** Constellations from winter (*left*) and summer (*right*) with Greek and Roman constellation figures drawn, from the 1833 celestial atlas of Burritt (1833). Many of the easily recognized zodiacal figures are visible, along with the path of the Milky Way and the celestial equator (image courtesy of Claremont Colleges special collections)

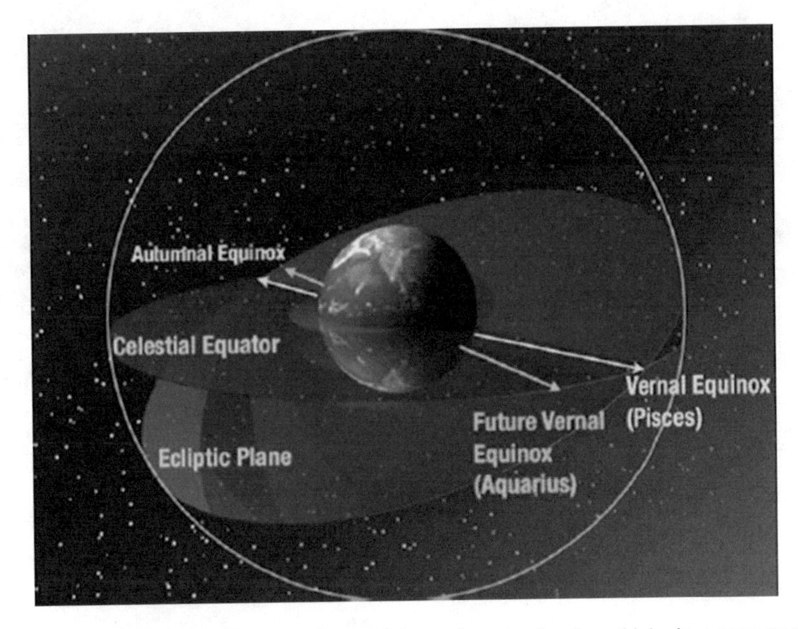


**Fig. 2.14** Illustration of celestial sphere and location of the equinox on the sky, which gives name to the "age" of an era, such as the "age of Aquarius" (figure by the author)

As a result of celestial precession, all of the zodiacal "birth signs" commonly reported in astrology columns are several thousand years out of date. Your actual zodiacal birth sign, defined as the location of the Sun when you were born, is the next later sign due to the precession of the equinox! (i.e., Virgos are really Libras, etc.). In addition, the equinox slowly drifts around the sky in a 26,000 year cycle, shifting from its present location in the constellation of Pisces to the constellation in Aquarius within 600 years, giving rise to the expression "the dawning of the age of Aquarius."

This precession was discovered independently by several cultures. The ancient Greek astronomer Hipparchus first noted the shift of star positions by 130 BC, while the ancient Chinese astronomer Yu Xi in 320 AD estimated that the Earth's axis shifted by 1° in 50 years. More accurate measurements by Zu Chongzhi enabled the Chinese astronomers to correct their charts to account for this effect, which by modern astronomical measurements is 1° in 71.6 years. Indian astronomers wrote about the precession of the equinox in their text Suryasiddhanta, from 1400 AD, and reported a value of 54″ per year, which provides 1° of motion in 66 years, very close to the modern value (Rao 2014). The Indian astronomers measured this shift using their giant instruments at Jantar Mantar in Jaipur, which included giant quadrants and sextants (some as high as 90 ft) which could measure the height and timing of the sun's transit and the positions of stars above the horizon to less than an arc minute (Rajawat 1990) (Fig. 2.15).

**Fig. 2.15** The tower of the Jaipur Astronomical Observatory, constructed in 1720. This instrument is known as the Samrat Yantra or "King of the Instruments" and consists of a 90 foot tower, and flanked by observation piers. The Jaipur observatory instruments could measure the transit of the sun accurately to within seconds, and also chart the positions of stars on the celestial sphere to within an arc minute, enabling Indian astronomers to detect the shift of the stars and the equinox over the course of centuries (image courtesy of Sheila Pinkel)

## The Autumn Constellations

### Andromeda: The Princess

Andromeda was the daughter of Cassiopeia, who was chained to a rock by a vengeful Neptune in punishment for her mother's bragging about her beauty surpassing the Nereids. Perseus arrived on his flying horse Pegasus to rescue Andromeda. Andromeda is also the name given to the nearest major galaxy to the Earth, a large spiral galaxy 2.2 million light years distant known to astronomers as M31. On a clear and dark night, you can spot the Andromeda galaxy off to the side of Pegasus without a telescope, making it the most distant naked eye object in the sky (Figs. 2.16 and 2.17).

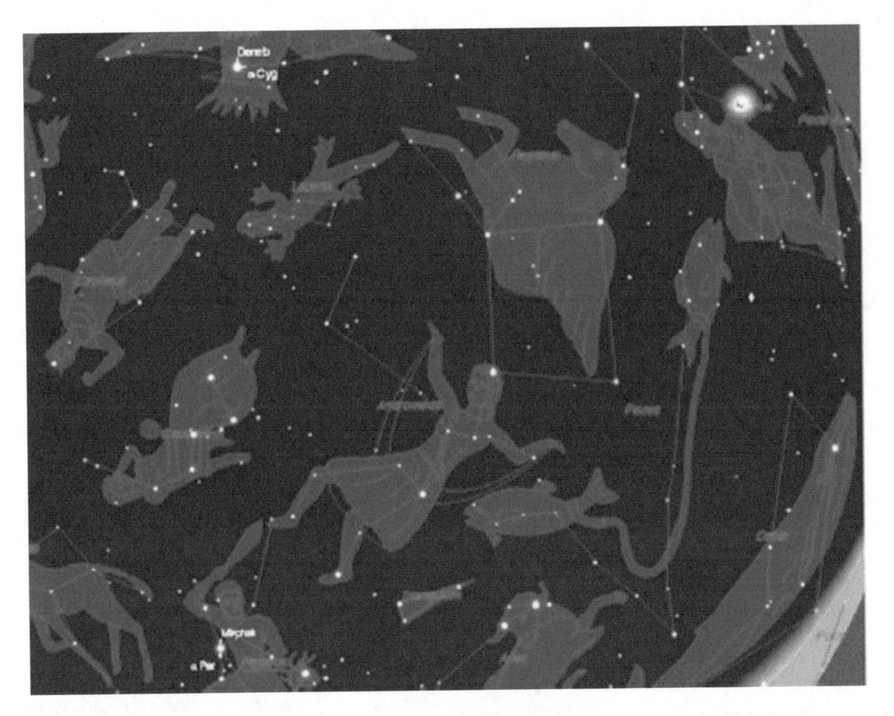

**Fig. 2.16** The autumn constellations, featuring the constellations of Perseus, Andromeda, Pegasus, and Cassiopeia, as viewed from the Northern Hemisphere (figure by the author, using the Voyager IV planetarium program)

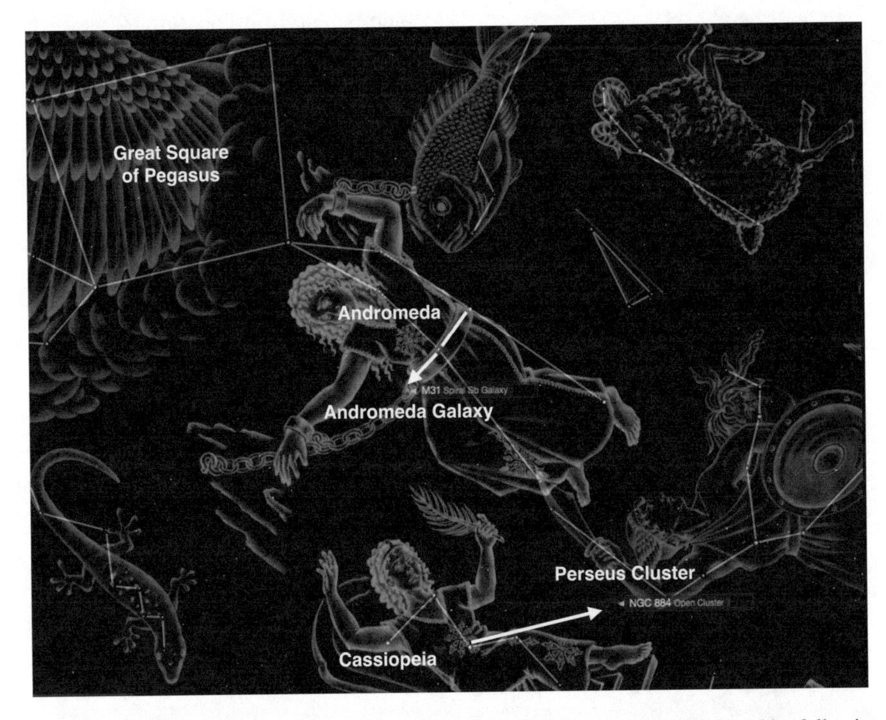


**Fig. 2.17** Both the Andromeda Galaxy and the Double Cluster in Perseus can be easily found by following key landmarks in the sky. To find the Andromeda Galaxy, it is best to look for the Great Square of Pegasus (*upper left*), and follow it downwards until you see a pair of stars which are in the middle of Andromeda's body in the figure. Sweeping to the *left from this point* reveals the galaxy in a pair of binoculars, or even with the naked eye in a dark site. The Perseus Double Cluster is easily found by following the "W" shape of the Cassiopeia constellation, where it is about 2/3 of the distance toward Perseus (figure by the author, using the Starry Night planetarium program)

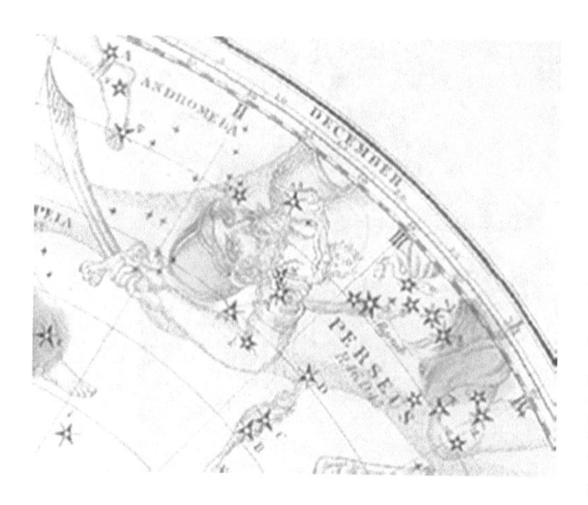


## Perseus: The Hero

Perseus is the son of Jupiter and the princess Danae. He is famous for many adventures, including the freeing of Andromeda, slaying the Medusa, and for using Medusa's head to turn the sea monster Cetus into stone. Perseus appears just to the left of Andromeda and a star cluster within Perseus in legend is the head of Medusa (Figs. 2.18 and 2.19).

**Principal Stars and Objects** α Persei (Mirfak), β Persei (Algol)—an eclipsing binary and the winking eye Medusa's head, the Perseus Double Cluster—NGC 869 and 884.

## Pegasus: The Flying Horse

Pegasus was a legendary winged horse formed from the blood of Medusa and the foam of the sea. The constellation Pegasus forms the "great square" which is visible in fall nights and which is helpful in locating many of the fall constellations and the Andromeda galaxy.

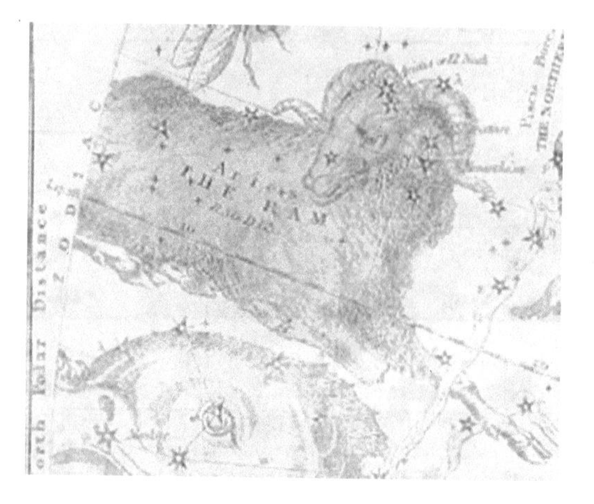


## Aries: The Ram

Aries is one of the 12 zodiacal constellations and also the location in the sky in ancient times of the "Vernal Equinox"—the location where the plane of the Sun's apparent motion and the celestial equator intersect.

**Fig. 2.18** The Andromeda Galaxy (M31) as it would appear in a pair of binoculars (*left*), and through a small telescope (*right*). It is the most distant object visible with the naked eye, at over 2.2 million light years away. The size of the galaxy is comparable to the full moon, and so it is easy to find, and in a dark site will be quite spectacular to view. The presence of light pollution will prevent viewing the larger disk of the galaxy, and cause M31 to appear as a round smudge of light, which is the view of the center or nucleus of the galaxy

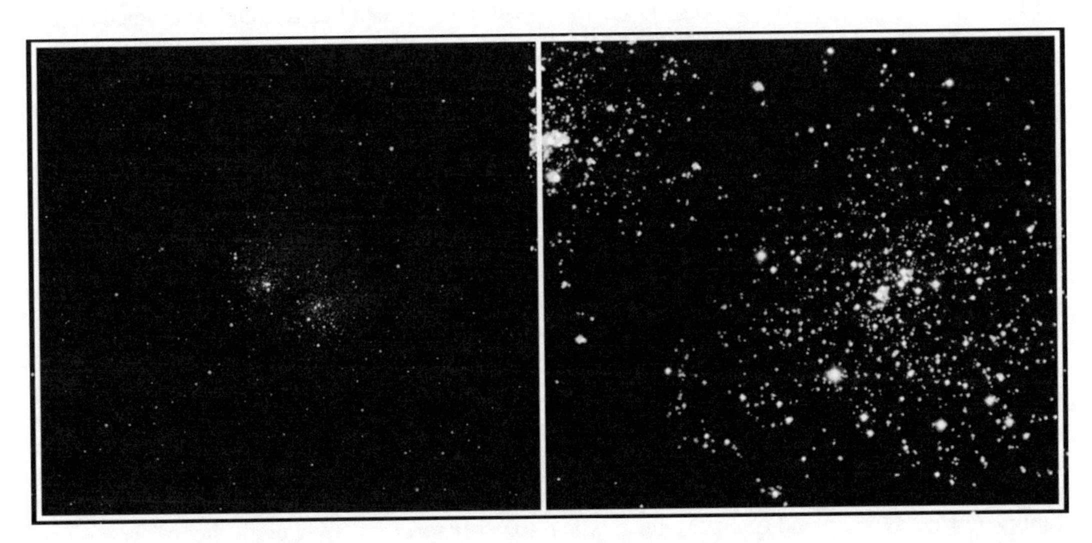

**Fig. 2.19** The Double Cluster in Perseus (NGC 869 and NGC 884) is over 7000 light years away, and provides a spectacular view through binoculars (*left*) and through a small telescope (*right*). The large size of the object makes it easy to find in the fall sky, and rewards the viewer with literally hundreds of visible stars. Many fainter stars exist within the clusters, with over 4000 stars in NGC 869, and over 700 stars in NGC 884, according to the SIMBAD database (http://simbad.u-strasbg.fr/simbad/)

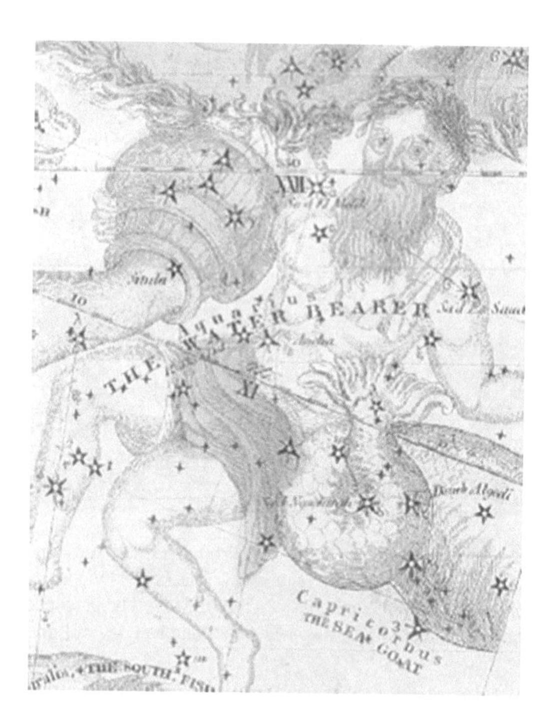


### Aquarius: The Water Bearer

Aquarius is a constellation with a very ancient association with water. The ancient Egyptians observed that the spring rainy system came when the Sun was in this part of the sky. Aquarius was also cupbearer to the gods and appears toward the new location for the vernal equinox of 2600, giving rise to the expression "the age of Aquarius."

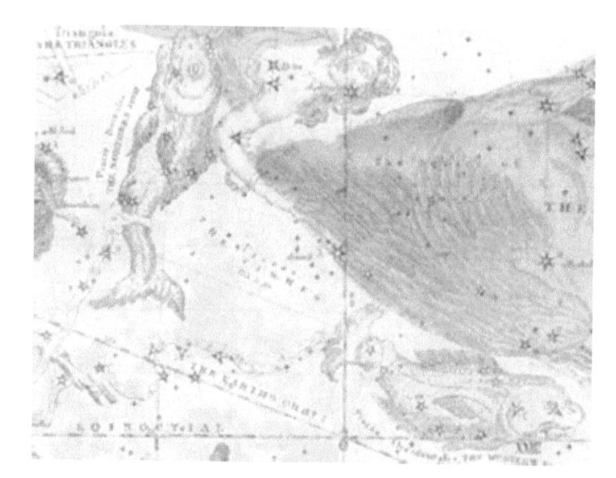

### Pisces: The Fishes

Pisces is between Pegasus and Andromeda and is a faint "V"-shaped group of stars that mark another of the ancient zodiacal constellations. The current location of the Vernal equinox is in this inconspicuous part of the sky.

## The Winter Constellations

We begin our "parade" of winter constellations from the European tradition with Orion and work our way around the Winter Hexagon (Fig. 2.20). The winter sky is home to the constellations of Orion, Taurus, Auriga, Gemini, and Canus Majoris, which together form the "Winter Hexagon." The Winter Hexagon starts at the foot of Orion (Rigel) and makes its way to Aldebaran in Taurus, Capella in Auriga, crosses to the bright pair of stars in Gemini (Castor + Pollux), then connects to Procyon and Sirius in the constellations Canus Minoris and Canus Majoris. These beautiful stars grace the long winter nights and can be easily seen from nearly all locations on the Earth.

**Fig. 2.20** The winter constellations, centered on the "Winter Hexagon," which connects Sirius, Rigel, Aldebaran, Capella, Castor + Pollux, and Procyon (figure by the author, using the Voyager IV computer program)

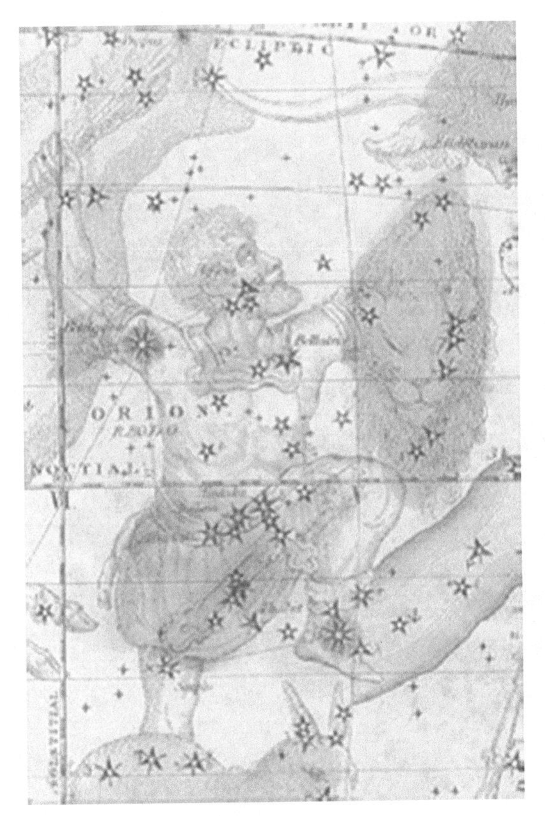

## Orion: The Hunter

Orion is one of the most striking constellations, and nearly every culture on the Earth recognized the three "belt" stars as significant. Orion was the son of Neptune and the nymph Euryale and was a gigantic and powerful hunter, who feared no animals until Gaia sent a scorpion to attack Orion. Ophiucus restored Orion to health and because of the attack, Orion appears on the opposite side of the sky as Scorpius. In Egypt, Orion was associated with Osiris, god of light, and lord of everything, while the Babylonians associated Orion with the hero Gilgamesh (Figs. 2.21 and 2.22).

**Principal Stars and Objects** Betelgeuse, Rigel (*upper left* and lower right stars), and the belt stars of Mintaka, Alnilam, and Alnitak. The Orion Nebula (M42) is located just below Orion's belt, in the middle of the three stars of his "sword."

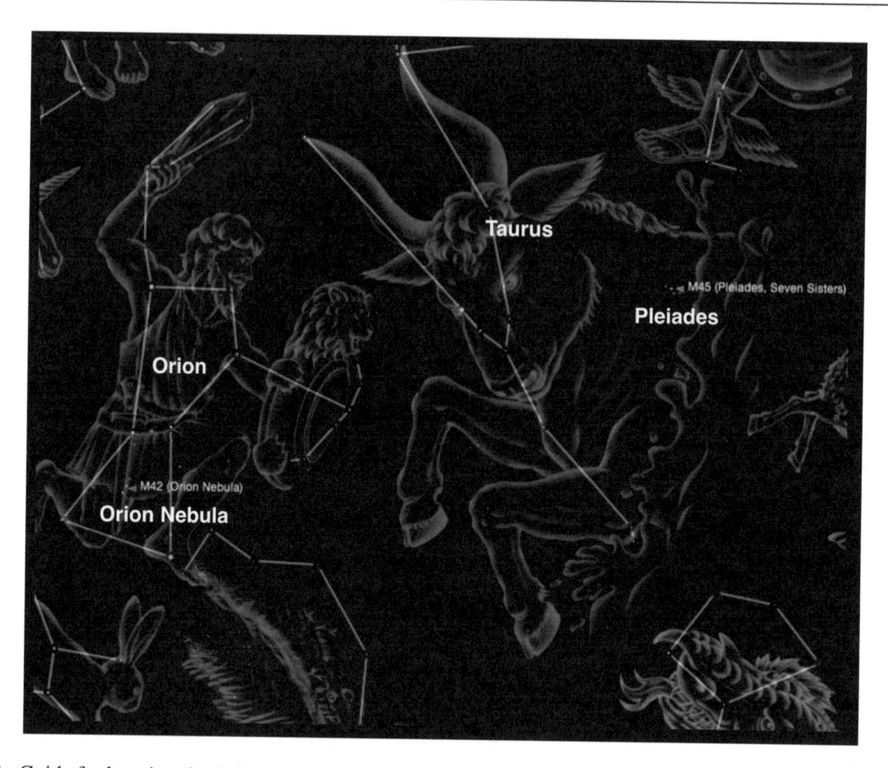


**Fig. 2.21** Guide for locating the Orion Nebula and Pleiades Cluster. The Orion Nebula (*left*) appears as a diffuse glow in the middle star within Orion's Sword, and the Pleiades cluster is located behind the V shaped Hyades cluster that makes up the head of Taurus

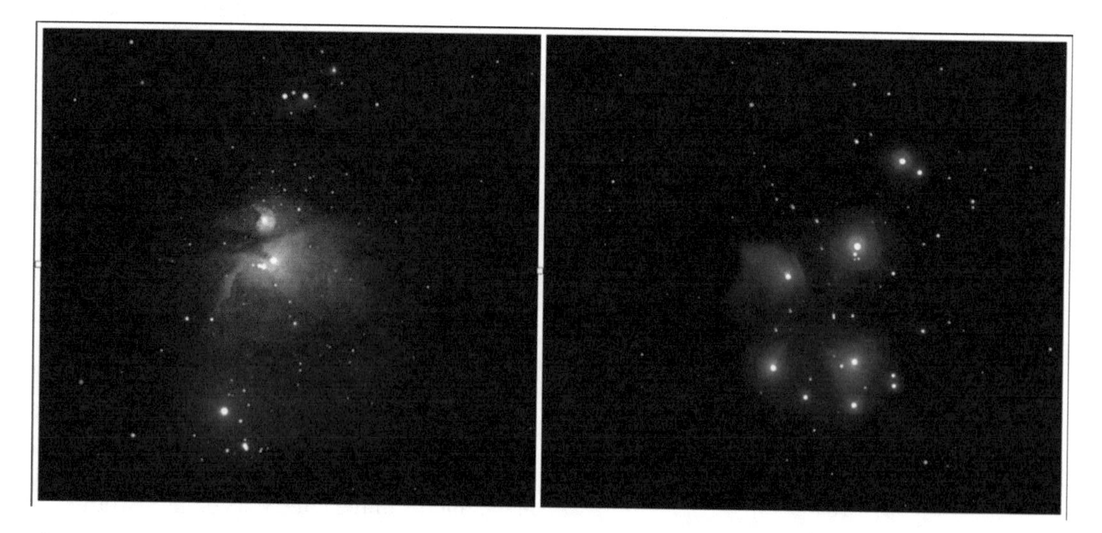

**Fig. 2.22** The Orion nebula through binoculars or a telescope (*left*) and the Pleiades cluster (*right*) both reveal dozens of stars, and clouds of glowing interstellar gas to the naked eye

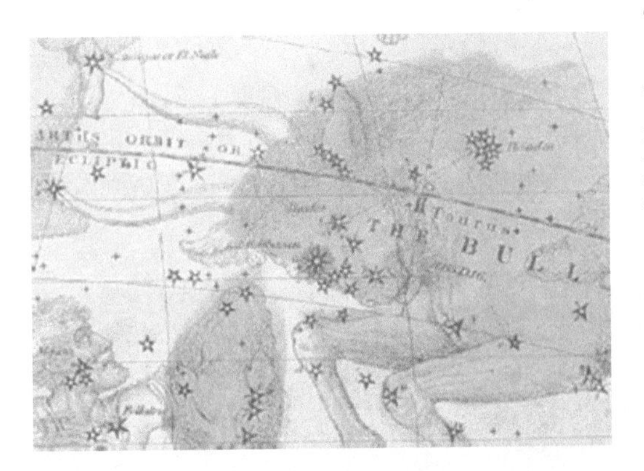


## Taurus: The Bull

Taurus is paired in the sky with Orion, where they face each other in conflict. Taurus is associated with Io, the daughter of the river god Inachos. Juno transformed Io into a cow to punish Jupiter for his affair with Io. The "head" of the bull in Taurus itself is a V-shaped star cluster known as the Hyades.

**Principal Stars and Objects** Aldebaran, the "eye of the bull," the Hyades and Pleiades star clusters, and the Crab Nebula (M1).

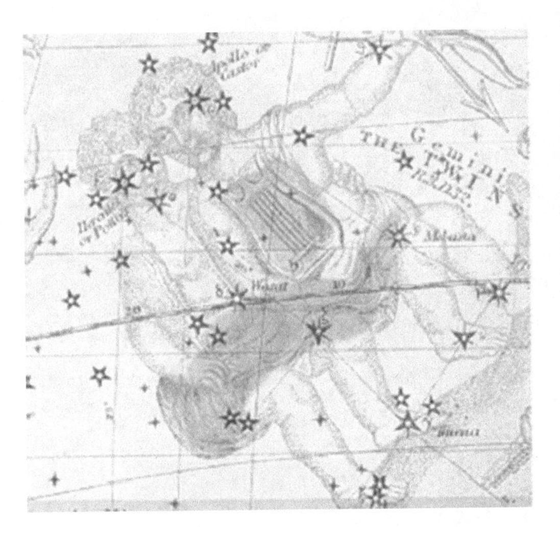


## Gemini: The Twins

Gemini appears just above Orion's left shoulder and includes the two bright stars Castor and Pollux. The twins Castor and Pollux were mythological heroes, who follow each other in the sky as they did in legend—Pollux was immortal, the son of Jupiter and Leda, while Castor was mortal. In legend, the two were placed into the sky together after Castor died, and Pollux asked Zeus to give Castor immortality. When Castor sets, Pollux follows his twin below the horizon.

**Principal Stars and Objects** Castor and Pollux, Geminid meteor shower.

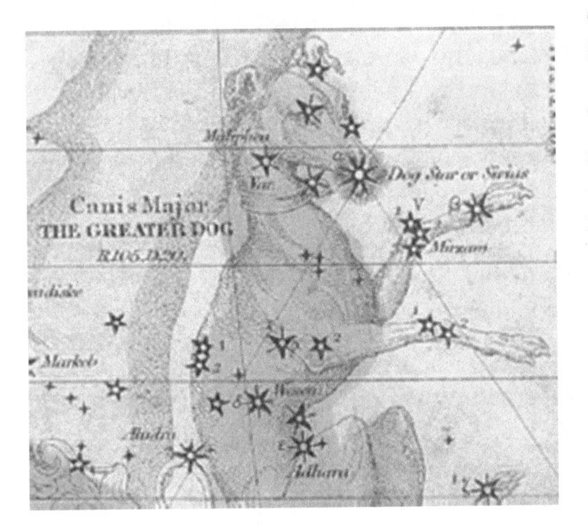


## Canis Minoris and Majoris: The Hunting Dogs

The two hunting dogs follow Orion through the sky and include the bright stars Procyon and Sirius. The "Dog Star" Sirius is the brightest star in the sky and includes the faint companion Sirius B, which is a white dwarf star. In Egyptian mythology, Sirius was Isis, the consort of Osiris (embodied by Orion), and followed him through the sky.

**Principal Stars and Objects** Sirius and Procyon. Sirius B, the extremely faint binary companion of Sirius, is the first white dwarf star ever discovered.

## The Spring and Summer Constellations

With the departure of winter comes a new cast of characters in the sky—led by Leo, and followed by the zodiacal constellations of Cancer, Virgo, Sagittarius, and Scorpius (Fig. 2.23). The summer sky brings the Milky Way into clear view, along with the constellations of Lyra, Aquila, and Cygnus that form the "Summer Triangle."

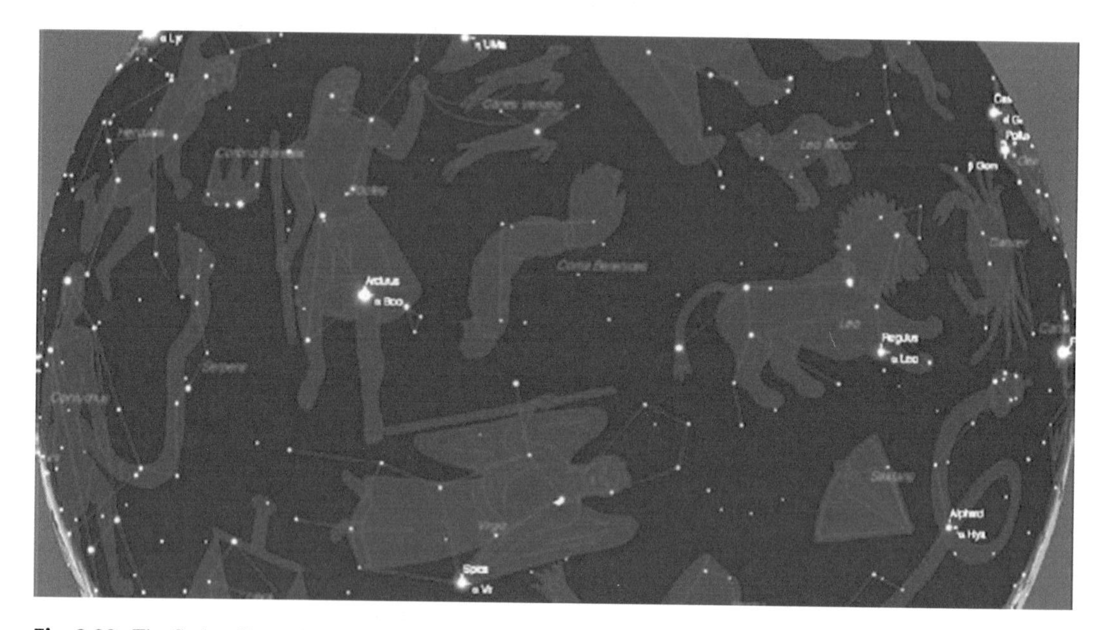

**Fig. 2.23** The Spring Sky, with Leo (*right*) and the constellations of Virgo, Bootes, and Hercules, as viewed from a northern location (figure by the author, using the Voyager IV computer program)

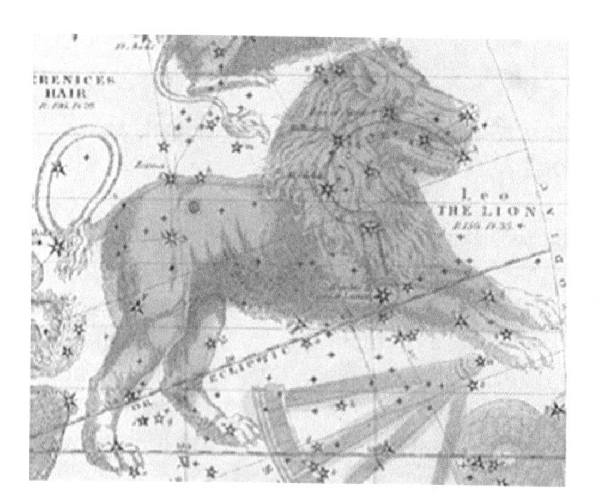


### Leo: The Lion

Leo is easily located in the spring sky and is shaped something like a backward question mark, which forms the mane of the lion. In myth, Leo was associated with the Nemean Lion, who menaced the ancient world until Hercules strangled it. The power of Leo was believed to bring the intense heat of the summer, as the Sun appeared in Leo near the summer solstice 4000 years ago.

**Principal Stars and Objects** Regulus and the Leonid meteor shower.

## Aquila: The Eagle

The Eagle has associations with royalty which survives to this day in modern symbols of state and power. Romans believed that careful study of the behavior of birds such as eagle enabled them to discern the will of the gods. Aquila flies across the Milky Way from Cygnus and features the bright star Altair, part of the Summer Triangle.

**Principal Stars and Objects** Altair, part of the Summer Triangle.

## Cygnus: The Swan

Cygnus is remembered for the tale of Jupiter assuming the form of a swan to seduce Leda, the mother of both Pollux and Helen. It is also known as the Northern Cross and appears vertical on the Western horizon near Christmas time. The bright star Deneb appears at the back of the swan (or the top of the cross) (Fig. 2.24), while a very interesting pair of red and blue stars known as Albireo appears at the front. Cygnus also appears near the middle of the Milky Way, in a region rich with star clusters and dust clouds, one of which is known as the "Cygnus Rift".

**Principal Stars and Objects** Deneb, Albireo, part of Summer Triangle.

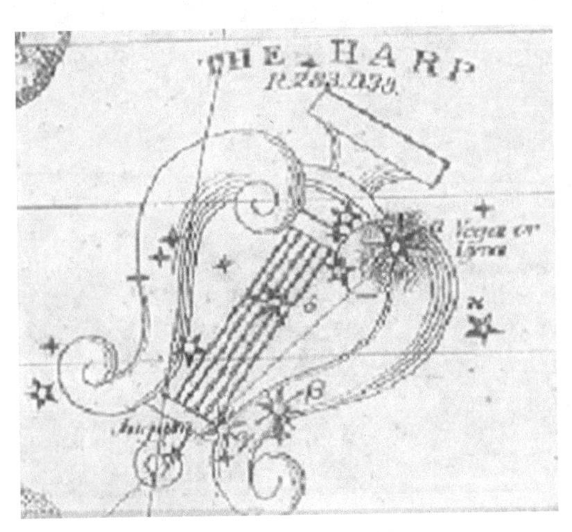


## Lyra: The Lyre

Lyra features the brilliant white star Vega and celebrates the invention of music, credited to Apollo and Mercury. Mercury stretched strings across the shell of a turtle one day and enjoyed the sound it created, and Apollo took this initial prototype to develop the first true musical instrument of myth—the Lyre (Fig. 2.25).

**Principal Stars and Objects** Vega, The Lyrid meteor shower, and the Summer Triangle.

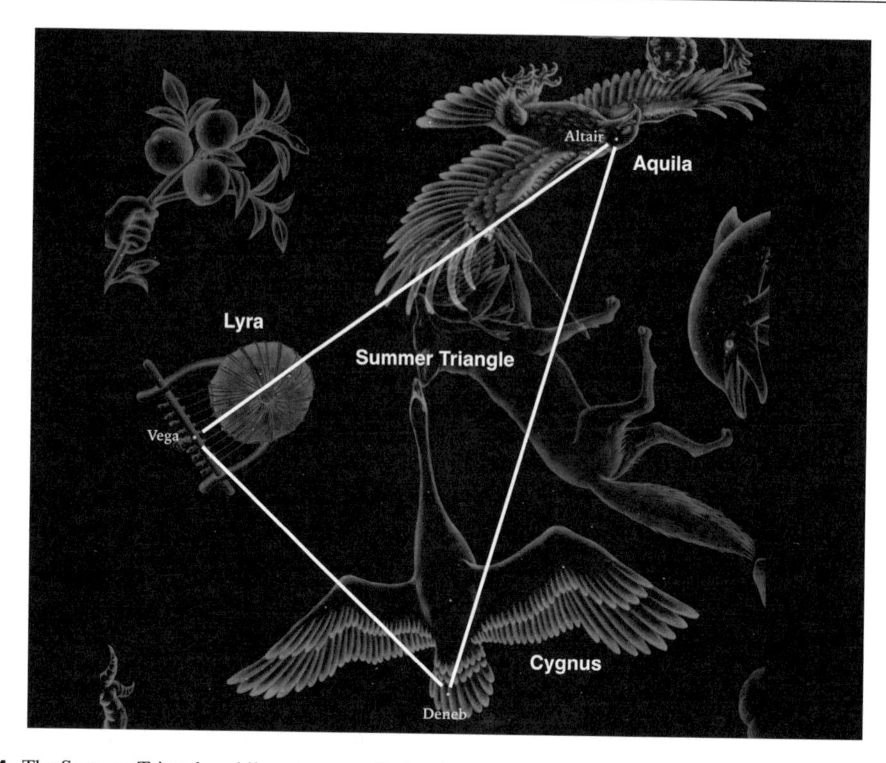


**Fig. 2.24** The Summer Triangle, while not a constellation, is a very convenient sky construction for finding three of the most interesting summer constellations—Lyra, Aquila, and Cygnus. At each vertex are spectacular bright stars—Vega, Altair, and Deneb, and with each constellation are numerous interesting nebulae and star clusters as this region is close to the Milky Way

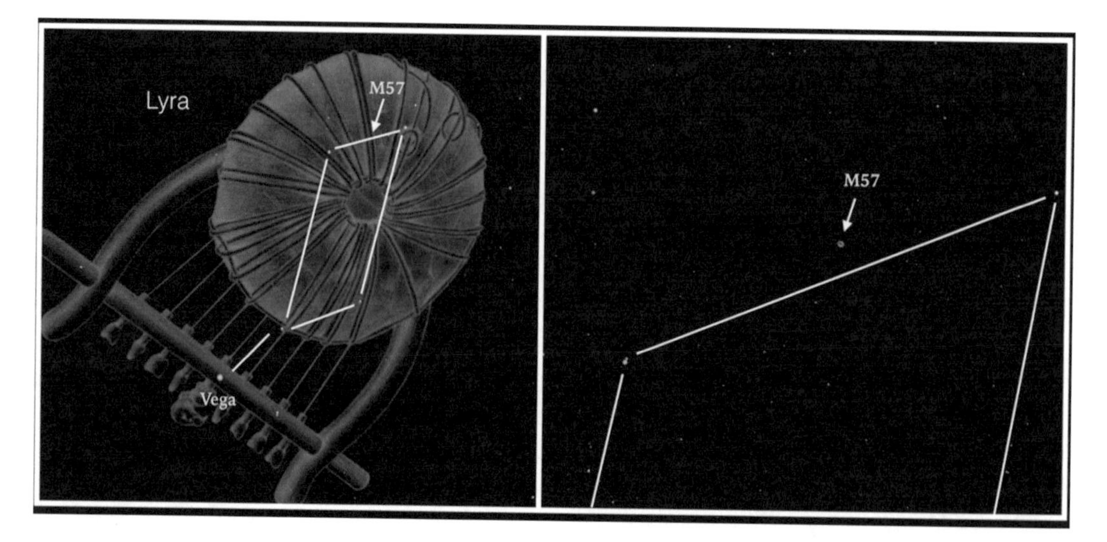


**Fig. 2.25** One of the most interesting objects in Lyra is the Ring Nebula, or M57. It can be found by looking in the middle of the two stars on the top of the parallelogram formed by the stars within Lyra. The two images show the view with the naked eye (*left*) and through a pair of binoculars (*right*)

## Ophiucus: The Healer

Sometimes considered the "13th" zodiacal constellation, Ophiucus is a healer, the name means "serpent bearer." The legendary healer Aesculapius is embodied in Ophiucus and bears the serpents as a symbol of his great ability to even restore life to dead creatures using an elixir of immortality, which was brought to him by a snake. For this reason our modern symbols for medical professionals sometimes includes a serpent wrapped around a staff.

**Principal Stars and Objects** Ophiucus lacks many bright stars but does appear close to the center of the galaxy and has many beautiful nebulae and star clusters.

## Hercules: The Hero

Hercules is the son of Jupiter and Alcmene, the wife of a famous military leader. His great strength is celebrated in the legend of his Twelve Labors, which took Hercules across the ancient world to battle monsters, clean stables, and do other impossible tasks for King Eurystheus, to win his freedom and achieve his well-deserved glory. Hercules is a difficult constellation to spot, but a trapezoid of stars can be spotted on clear summer nights, which is the center of Hercules' huge upside-down body in the sky.

**Principal Stars and Objects** One of the most beautiful star clusters, M13, is in Hercules and can be seen with a small telescope.

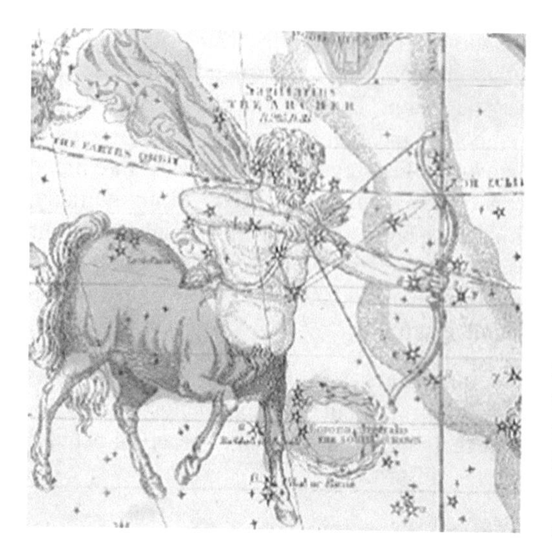


## Sagittarius: The Archer

The form of Sagittarius is half man, half horse—a form sometimes known as a Centaur. The location of Sagittarius coincides with the Sun's position during ancient hunting seasons and could be the cause of the name. Sagittarius is also the location of the center of our Milky Way galaxy, in which an enormous black hole churns behind thousands of light years of stars deep in the center of the dense star clusters of our galaxy's nucleus.

The Sagittarius region features many beautiful nebulae and bright star clusters. Sagittarius is probably most easily found by looking for the "teapot" shape, a box with triangles on the sides and top. The image is complete when you notice the "steam" that appears above the spout of the teapot from the many nebulae of our Milky Way (Fig. 2.26).

**Principal Stars and Objects** The Trifid Nebula (M20), The Lagoon Nebula (M8), and the Omega Nebula (M17), the galactic center.

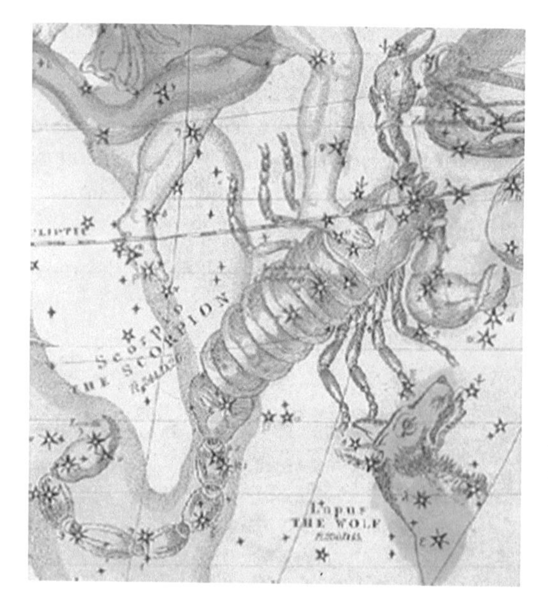


## Scorpius: The Scorpion

Scorpius is one of the few constellations that looks like its namesake—a long arc of stars forms the stinger and the blood red star Antares the heart of the monster scorpion in the sky. Legend tells of the scorpion being created by Gaia the earth goddess to humble mighty Orion, and after biting Orion, the Scorpion, and Orion eye each other warily from opposite sides of the celestial sphere. The fact that Scorpius appears near some of the densest dark clouds in the Milky Way can be explained by the common observation that Scorpions like dark places.

**Principal Stars and Objects** Antares and star clusters M6 and M7.

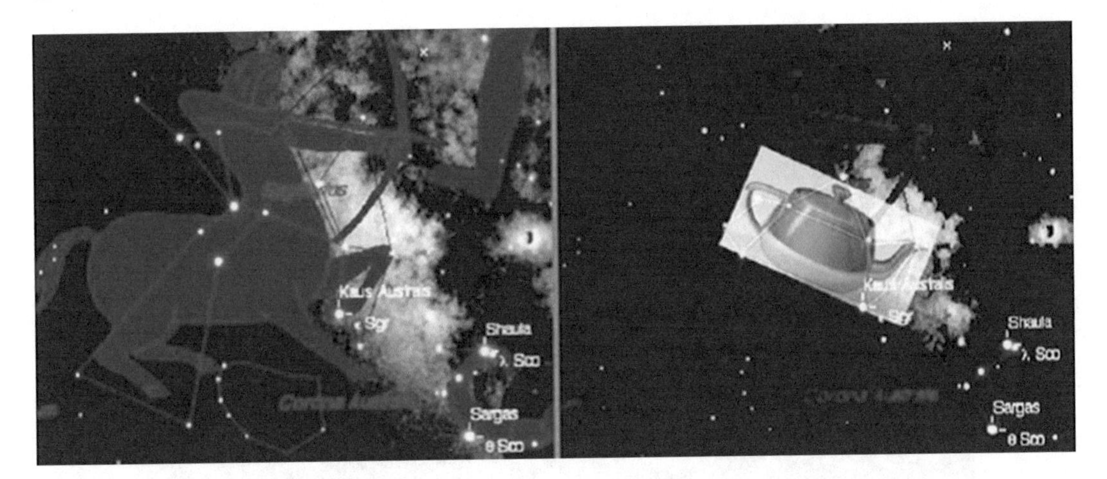


**Fig. 2.26** Sagittarius as a centaur (*left*) and as a teapot (*right*). The "steam" can be seen in the glowing nebulae of our Milky Way (figure by the author)

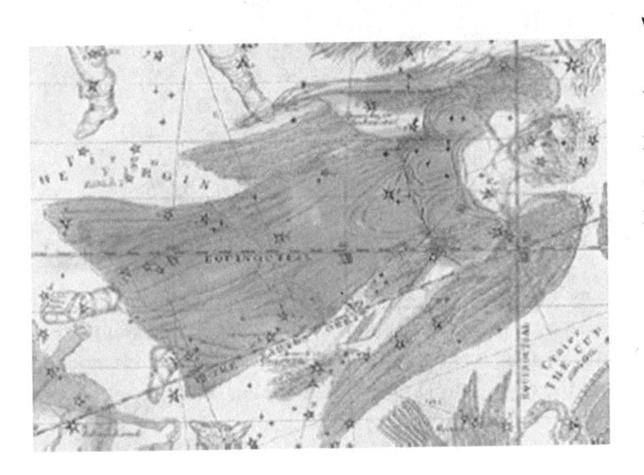

### Virgo: The Virgin

Virgo is one of the zodiacal constellations which is hard to spot, but is famous for its bright star Spica, and the large cluster of galaxies in the region named the Virgo cluster. In myth, Virgo is sometimes associated with Ceres, the goddess of agriculture, or her daughter, Proserpina, goddess of springtime.

**Principal Stars and Objects** Spica and the Virgo cluster, the nearest cluster of bright galaxies in the sky, and home to some of the most spectacular galaxies such as M102 (Sombrero galaxy).

## The Equatorial Sky from Around the World

### *The Hawaiian and Polynesian Sky*

#### Hawaiian and Polynesian Constellation Lore and Stars

The Hawaiian navigators were able to cross the entire Pacific Ocean, traveling between Tahiti and Hawaii in open boats for weeks at a time only by using exquisite star knowledge to create an original and highly effective system of celestial navigation. "Star compasses" were constructed from a hollowed gourd or a calabash, which was filled with water to help level the device and star positions on the horizon were carefully studied to give headings on the open sea. A network of 32 stars were observed, and the appearance, location in the sky, and rising and setting points of these stars were well known to the Hawaiian navigators through their training and oral tradition (Fig. 2.27). By using these stars, the Hawaiian boats could maintain a heading to within a few degrees. Further refinement of course and navigation was provided by careful observations of birds, water color, currents, and other cues from nature about the ocean landscape. Groups of nine basic compass headings were subdivided by additional stars to give 32 of the main guide stars (Johnson and Mahelona 1975, p. 65).

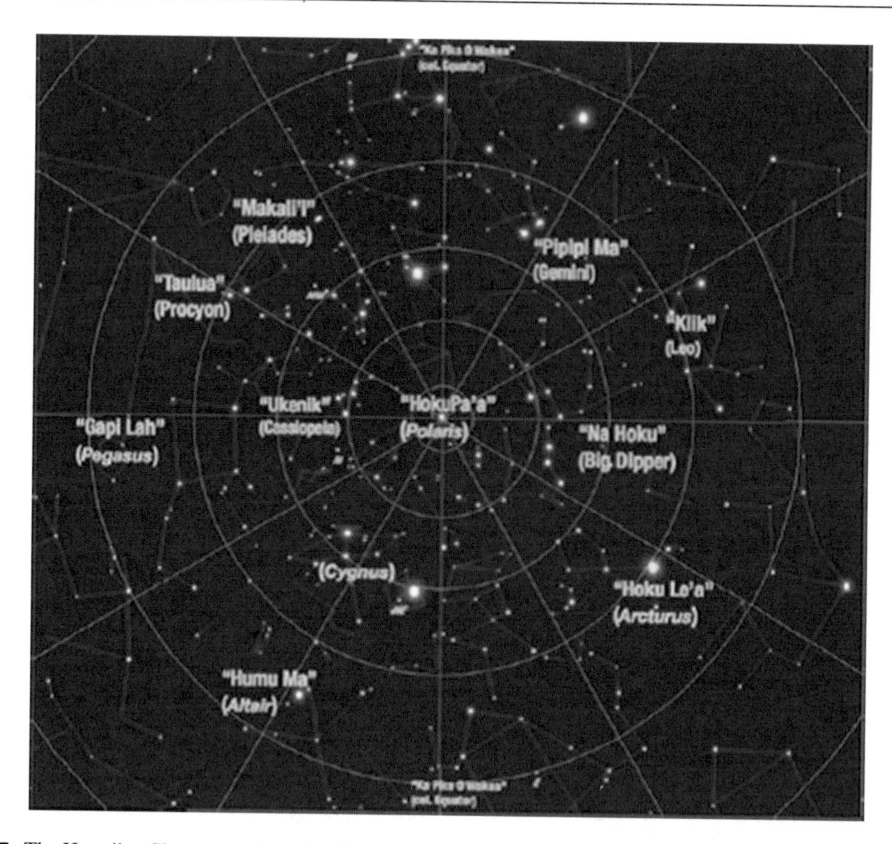


**Fig. 2.27** The Hawaiian Sky, centered on the circumpolar region. The ancient Hawaiians and Polynesians were master navigators and were able to travel thousands of miles in open boats using star lore and their keen observational skill to guide them. The observations included stellar sightings, looking for meteorological information and subtle changes in the ocean currents, colors, and animals to help them travel across the open seas (figure by the author, based on information from Johnson and Mahelona 1975)

A quote from an 1865 guide for constructing a gourd compass describes the basic system:

> Take the lower part of a gourd.. rounded as a wheel, on which several lines are to be marked. These lines are called '*Na alanui o na hoku hookele*' (the highways of the navigation stars), which stars area also called '*Na hoku-ai-aina*' (the stars which rule the land). Stars lying outside of these three lines are called '*Na hoku o ka lew*' (foreign, strange or outside stars). The first line is drawn from the '*Hoku paa*' (North Star), to the most southerly of '*Newe*' (Southern Cross). The portion to the right or east of this line is called '*Ke ala 'ula a Kane*' (The dawning, or the bright road of Kane), and that to the left or west is called '*Ke alanui maaweula a Kanaloa*' (the much traveled highway of Kanaloa). (Kamakau, in Hawaiian Annual 1891, quoted in Johnson and Mahelona 1975, p. 72)

The Hawaiian navigator would also include important markings on the edge of the gourd. These would include the northern and southern standstill points for the Sun and key navigation stars, as well as the positions of planets. The author describes the crossing of the equator and gives advice to the navigator:

> When you arrive at the '*Piko o Wakea*' (Equator), you will lose sight of the *Hoku-paa* (North Star); and then the '*Newe*' will be the southern guiding star, and the constellation of '*Humu*' (Altair) will stand as the guide above you. You will also study the regulations of the ocean, the movements of the tides, floods, ebbs, and eddies, the art of righting upset canoes.. All this knowledge contemplate frequently, and remember it by heart, so that it may be useful to you on the rough, the dark and unfriendly ocean. (Kamakau, in Hawaiian Annual 1891, quoted in Johnson and Mahelona 1975, p. 72)

The Hawaiian star compass was a powerful tool that provided both a navigating device and a calendar. The navigator could determine their latitude by viewing the North Star in its reflection on the water in the gourd, and could set course at night by using small holes on the edge and a net of strings above the compass to provide multiple sightlines to other stars on the horizon. Groups of stars were remembered for journeys that took place in each season, and their locations on the horizon for rising and setting were assigned to one of 32 "houses" or azimuthal directions. The 32 houses provided a set of 11° landmarks on the horizon, and groups of stars in each of these locations were used to provide fine course corrections for guiding the ship. Various versions of the star compass have been produced, showing the locations of key stars for navigation as known by different regions in Hawaii and Polynesia. The star compass is more of a mental map than a literal compass, as the Hawaiian and Polynesian navigators had an exquisite memory and eye for locating the positions of these stars.

In recent years, these star compasses have been used on voyages across the Pacific using both the vessels and navigational techniques honed over centuries, providing verification of these star-based techniques by modern Hawaiian and Polynesian navigators (see Bryan 2002; and PVS 2016). A British resident of the Gilbert Islands, Sir Arthur Grimble, reported that a local navigator named Biria, had memorized the seasonal rising positions and times of 178 stars, which would give a total of 356 locations on the horizon—amounting to a resolution of 1° in terms of navigational accuracy! (Huth 2013). These practices were not limited to Hawaii or even the South Pacific, but were also used by Navigators in the Bay of Bengal, and the Arabian sea—any location near the equator will work well since near the equator stars rise vertically from the horizon and can mark locations for an extended time. The star compass was also supplemented by a series of stars which appear directly overhead on particular islands—enabling a navigator not only to steer a course by the stars, but also to gauge when the ship is

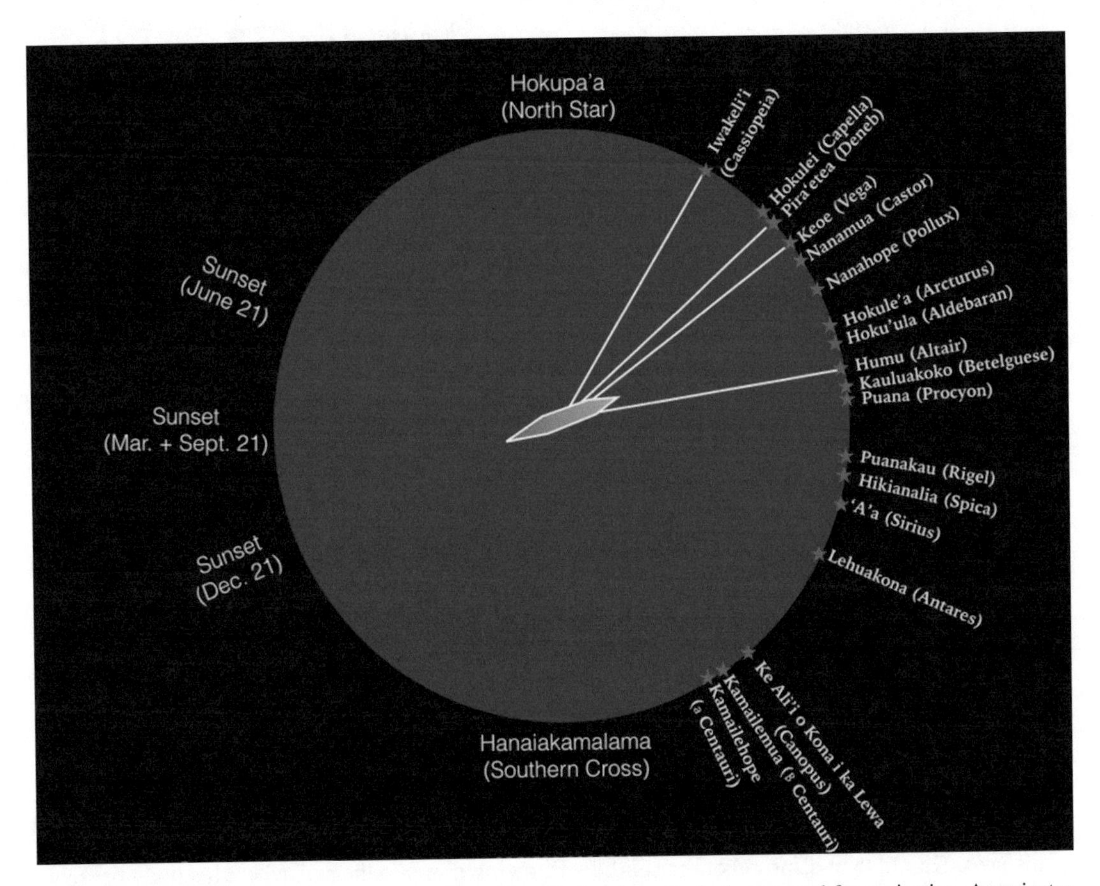


**Fig. 2.28** Representation of a Star Compass from Hawaii, showing principal stars used for navigation. A navigator could watch these stars rising on the horizon and guide the ship by maintaining a fixed orientation relative to a group of navigation stars (shown with *white lines*). The diagram is symmetric so the same locations for star setting will be visible on the opposite side, providing a complete set of reference points around the horizon. The angles on the horizon were also associated with one of 32 "houses" which were memorized along with the stars by navigators at an early age in their training. (Adapted from Bryan 2002, and PVS 2016)

close to arriving. These overhead stars were known as fanakenga in Tonga ("star that points down from the sky"), and Tahitian navigators referred to these stars as "pillars of the sky." (Huth 2013) (Fig. 2.28).

A list of constellations and major stars in Hawaiian is provided in Table 2.5. The stars of Hawaii were similar to those of Micronesia and Tahiti, and perhaps these people collaborated in designing their system of navigation. Hawaiian star names embody both mythological and navigational meanings. For example, Canopus, the second brightest star in the sky in Hawaiian, is known as the "chief of Kona in the heaven," and the name for the star Sirius translates to "star to guide canoe." One difference between the equatorial Hawaiian star lore and European constellations is the important role of

**Table 2.5** Hawaiian star (Hoku) names and corresponding translations (adapted from Johnson and Mahelona 1975)

| Hawaiian star (hoku) name | European equivalent | Translation |
|---|---|---|
| Humu | Altair | "To bind together" |
| Kao | Antares | "Dart" |
| Hokule'a | Arcturus | "Star of gladness" |
| Kaulua-koko | Betelgeuse | "Blood star" |
| Na Hiku | Big Dipper | "The seven" |
| Ke ali'lo Kona ika lewa | Canopus | "Chief of Kona in the heaven" |
| Hokuhei | Capella | |
| Iwakeli'i | Cassiopeia | "Little frigate bird" |
| Nanamua | Castor | "Look ahead" |
| Ke-ho'oea | Vega | |
| Hokupa'a | North Star (Polaris) | "Immovable star" |
| Na kao | Orion | "The darts" |
| Makali'i | Pleiades | "Tiny, fine, small-meshed" |
| Nanahope | Pollux | "Look back" |
| Puana-kau | Rigel | |
| Hokuho'okelewa'a | Sirius | "Star to guide canoe" |
| Hokuke'a | Southern Cross | "Star cross" |
| Keoe | Vega | "Sweet potato" |

Canopus and the Southern Cross, both easily visible from the location of Hawaii, but hidden under the southern horizon for North America and Europe.

The stars in Hawaiian culture are more than navigational aids. Many tales tell of the creation of the stars and their connection with mythological heroes. One creation tale tells of the time of deep darkness, in which the gods Kane, Ku, Lono, and Kanaloa first emerged. Kane, the god of creation, picked up a vast gourd floating in the sea and tossed it high into the air. Its top flew off and became the sky, two pieces broke off to become the Sun and Moon, and the seeds inside scattered about and became stars. After this creation of stars, the world was filled with life, people, and light (Thompson and Weisgard 1966, p. 11).

Another story tells the stars being formed when a Hawaiian mythical hero, Ka-ulu, encountered the chief of the sharks. In this tale, Ka-ulu's brother had been taken away, and Ka-ulu had to battle many different forces to bring his brother back. Ka-ulu defeated the spirits of Surf, Great Stone Man, Great Barking Dog, and then finally reached the Chief of the Sharks. When the Shark told Ka-ulu that his brother had been eaten, Ka-ulu took hold of the giant shark and rescued his brother by pulling him out of his jaws. Ka-ulu then disposed of the shark by throwing him into the sky, where his body broke up into millions of pieces which became the stars (1966, p. 44).

Another astronomical element of Hawaiian language is its strong emphasis on the Sun. Just as the Inuit are said to have many names for ice and snow, the Hawaiian language has dozens of words and phrases referring to the Sun and different positions of the Sun. The Sun in Hawaiian is known as "Ka La" but includes different words for sunrise (*pukana la*), early morning sun (*kahikole*), the before noon sun (*kahiku*), the noontime sun (*kau ika lolo,* lit. "sun rests on the brains"), sunset (*napo 'o ana o ka la*), a red sunset (*aka ula*). The word for the eastern dawn sky is a beautiful phrase in Hawaiian, "Ke alaula a Kana," or the "flaming road of Kane."

The Hawaiian creation chant, the *Kumulipo*, describes the initial creation of the Hawaiian islands, the Hawaiian people, and the stars. This chant consisted of over 2000 lines and was memorized and repeated verbatim at important festivals such as the Lono festival. In the very first ten lines of the chant, the power of the stars within the Pleiades (*Makali'i*) is mentioned, as part of the creative forces that bring life to the Earth:

> At the time when the light of the sun was subdued
> To cause light to break forth,
> At the time of the night of *Makalii* (winter)
> Then began the slime which established the earth,
> The source of deepest darkness.
> Of the depth of darkness, of the depth of darkness,
> Of the darkness of the sun, in the depth of night,
> It is night,
> So was night born.
>
> Kumulipo was born in the night, a male.
> Poele was born in the night, a female.
> A coral insect was born, from which was born perforated coral.
> The earth worm was born, which gathered earth into mounds,
> From it were born worms full of holes.
> The starfish was born, whose children were born starry.
> The phosphorous was born, whose children were born phosphorescent.
> The Ina was born Ina (sea egg).
> The Halula was born Halula (sea urchin).
> (Lines 1:18 of the Kumulipo, translated by Queen Liliuokalani 1897)

The rise of the Pleiades gives rise to the first life of the Hawaiian Islands in this creation chant, just as each season the rise of the Pleiades heralds the beginning of spring and summer, and the sprouting of seeds and the spawning of fish across Hawaii.

# Equatorial Sky from Around the World

## *The Navajo Sky*

### Navajo Constellation Lore and Stars

The Navajo people (or the *Dineh*) celebrate the stars in sand paintings, in healing ceremonies and in chants. For the Navajo the stars were vital in maintaining balance with their environment and in health. Constellations were viewed to set the agricultural calendar and therefore had many names related to agriculture. Some stars were viewed through crystals and were known as "igniting stars" for their power to cast rays of colored light within the crystal that could be used to divine the future. Stars were drawn within sand paintings to invoke the power of constellation to heal patients or to bring good fortune.

In the Navajo tradition, the pattern of stars in the sky that we call constellations was set in place by Black God himself, the creator god. The initial creation of the constellations by Black God began with the Pleiades, which was known as *Dilyehe* to the Navajo:

> Black God strode into the Hogan of creation with Dilyehe (Pleiades) on his ankle. He stamped his foot several times, and the star group jumped to a different location each time – his knee, his hip, his shoulder, and finally his left temple. "There it shall stay," he told the Creators. In this way he established his mastery over the stars and secured the right to arrange them in the sky.
>
> ("Starlore Among the Navajo" quoted in Miller 1997, p. 188)

Black God set to work creating the major Navajo constellations. The North Circumpolar region was seen as a couple, perhaps First Man and First Woman, revolving around a campfire. The male is known as "Nahookos Bika Ii," or the "Male One Who Revolves," and is associated with the stars of the Big Dipper. The female is known as "Nahookos Ba aadii," or "Female One Who Revolves," and is associated with the constellation Cassiopeia. The campfire or hub of the revolving stars is the North Star or Polaris. The Navajo circumpolar sky in many ways is a reflection of the strong value placed by the Navajo on the family and on single families recreating in their own home the perfect harmony of First Man and First Woman (Fig. 2.29).

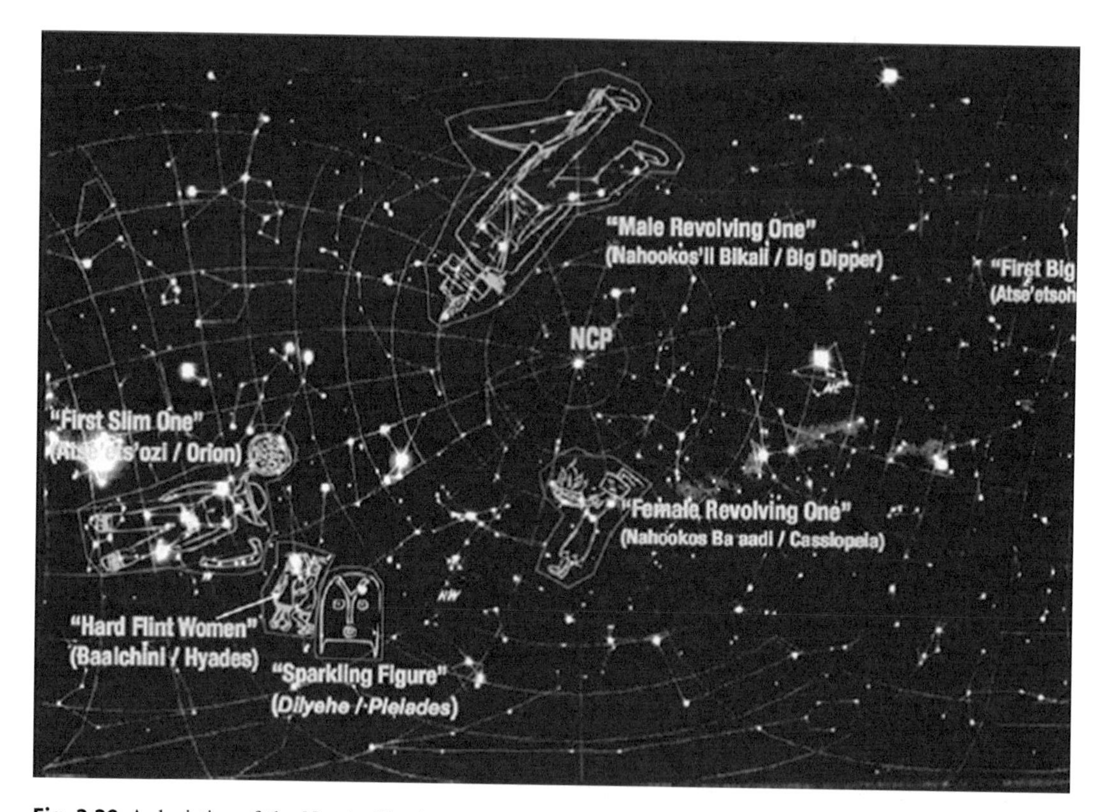


**Fig. 2.29** A depiction of the Navajo Sky, based on tradition in which the sky is seen as Father Sky. The body of Father Sky is decorated with the many Navajo constellations, including First Big One (Scorpio—*upper right*) and First Slim one (Orion—*lower left*). A key figure is *Black* God, which has on his head the Pleiades, known to the Navajo as Dilyehe or "Sparkling Figure" (based on information in Griffin-Pierce 1992) (figure by the author; artwork by Kim Aldinger)

Additional constellations created by Black God include ones with names such as "Man with Legs Ajar" and "First Big One" along with "Rabbit Tracks" and "Pinching Stars." Just as in European constellation lore, some of the constellations do not look much like their namesakes, but the stars serve as an aid to memory of tradition and culture.

The final phase of the Navajo constellation design involves Coyote, who is said to have interrupted Black God's work by spilling the remaining stars across the sky, creating the chaotic array of nameless stars visible on a clear dark night in Navajo country. Some accounts tell that Coyote held one star for himself, known as Coyote star, which appears in the south, known to the Navajo as *Ma'ii bizo*. This star is thought to be the source for confusion and chaos in the world and appears only for a few weeks each year. Naturally the exact identity of Coyote stars is in dispute, or confusion, but is often thought to be Antares or Canopus.

**Table 2.6** Navajo constellation names, translations, and corresponding European constellations (*sources*: Griffin-Pierce 1992; Miller 1997)

| Navajo name | European name | Translation | Function or significance |
|---|---|---|---|
| Nahookos Bika ii | Ursa Major/Big Dipper | "Male one who Revolves" | First Man |
| Nahookos Ba aadii | Cassiopeia | "Female one who revolves" | First Woman |
| Dilyehe | Pleiades | "Sparkling figure" | Black God's ornament, Hard Flint Boys, wife of First Slim One |
| Baalchini | Hyades | "Pinching stars" or "Hard flint women" | Daughters of Pleiades, mothers of "Hard Flint Boys" |
| Hastiin Sik'ai'i | Corvus | "Man with Legs Ajar" | Marks beginning of second month of Navajo year—"Parting of the Seasons"—helps people prepare for winter |
| "Atse'etsoh | Scorpius | "First Big One" | Old man with a cane and marks beginning of the month of Great Wind (December) |
| Yikaisdahi | Milky Way | "Awaits the dawn" | Symbolizes the corn meal sprinkled by First Woman as she prays in the morning |
| Gah heet'e'ii | Tail of Scorpius | "Rabbit tracks" | Marks month of Baby Eagle (February) which the plants and animals are awakened |
| Atse'ets'ozi | Orion | "First slim one" | This constellation is much like "First Big One" and appears to be a man with either a cane or digging stick |
| Ma'ii bizo | Canopus or Antares | "Coyote star" | Symbolizes confusion and disorder in world, placed there by Coyote |

We summarize some of the Navajo constellations in Table 2.6 with the Navajo star names, the translation of these names, and the corresponding European constellations. In the final column of Table 2.6 we give some of the symbolic associations these stars share with key elements of Navajo myth and culture.

## *The Chumash Sky*

The Chumash tribe inhabited Southern California in the region between coastal Malibu and Santa Barbara, as well as in the offshore Channel Islands. Archeological evidence places these people in their land for at least 12,000 years, making them some of the oldest Native American groups in North America. The Chumash attained a very high degree of linguistic diversity, with dozens of dialects spoken among more than a dozen different "city-states." These city-states were governed by leaders known as "poqwats," who served as leaders of several villages that were individually ruled by the local chief or "wot." The Chumash city-states were organized into a confederacy with active trade routes used with plank canoes or *tomols*, with shared cultural practices and beliefs. The anthropologist P. J. Harrington described the cosmology of the Chumash in the early twentieth century, through interviews with surviving elders. These interviews and additional insights into the astronomy and cosmology of the Chumash were written into two books, "Crystals in the Sky" by Hudson and Underhay (Hudson and Underhay 1978) and "December's Child" by Blackburn (Blackburn 1975), which provide a wealth of information beyond that described below (Figs. 2.30 and 2.31).

**Fig. 2.30** Chumash Rock art from the Condor Cave, depicting abstract figures drawn by ancient Shamans, in their quest to make contact with the sky people (image courtesy of Jennifer Perry)

**Fig. 2.31** Photograph of a Chumash painted cave in the Los Padres National Forest which may represent a map of the Chumash constellations (Hudson and Conti 1984. Photograph by Rick Bury; reprinted with permission)

## Chumash Constellation Lore and Stars

The Chumash constellation lore centers on the Milky Way, which was referred to by the Chumash as the "Path of the Pinon Bearers," who are often also associated with the departed dead who travel the road of the Milky Way on their journey to *Shimilaqsha*, the "Land of the Dead." Important asterisms for the Chumash include the Cygnus Rift, and the bright stars of Cygnus, known to the Chumash as "Scorpion Woman," and the bright stars of the Big Dipper, known as the "Boys who turn to Geese." (Fig. 2.32).

The Chumash were actually several different city-states, with different dialects and beliefs. Their constellation tales can take several forms and also can have several layers of meaning. One example can be found in the story of the "Boys Who Turn to Geese." The most common version of the story involves a group of boys who were rejected by parents and stepparents. They formed a band and live together, assisted by Raccoon in getting food. As they experimented in the outside world, they developed an interest in flight and attached goose feathers to themselves and rose to the sky during a ritual dance around the fire. One night they flew north and their mothers ran after them begging them to return. By then it was too

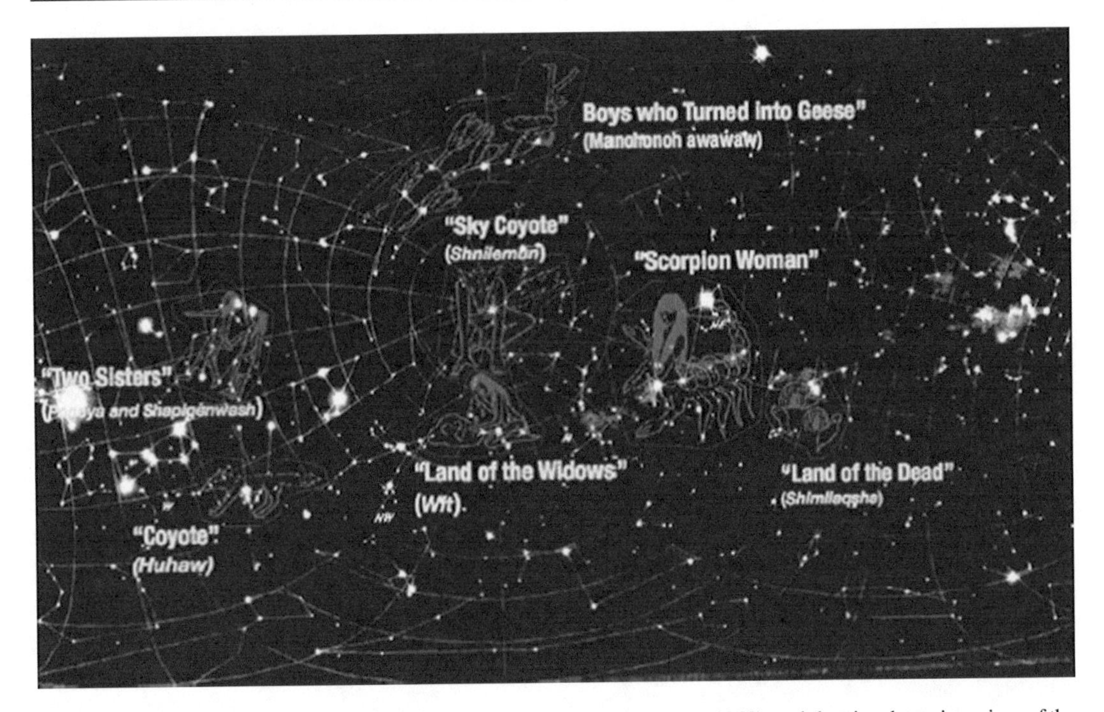


**Fig. 2.32** The Chumash constellations, as described in Hudson and Underhay (1978), and showing the main regions of the Milky Way which was believed to be the path of the departed people on their way to the Land of the Dead, or "Shimilaqsha," near Aquila (figure by the author; artwork by Kim Aldinger; an interactive version is included in the CDROM)

late, as they had been transformed into geese. Some have also suggested that the tale of the "Boys who Turn into Geese" celebrates the motions of the Chumash dance of Hutash, which involves circular dancing around concentric groups of people. Nearby tribes such as the Gabrieleno and Luiseno believed that the Geese symbolized the northerly direction, which made the connection to the Big Dipper appropriate. The Luiseno also performed a dance where people moving in a circle imitated the motions of the Big Dipper (Hudson and Underhay 1978, p. 105).

Two other Chumash stories relate journeys of the First People into the sky to become stars. The stars Castor and Pollux, known in the European tradition as Gemini, to the Chumash were *Ponoya* and *Shapiqenwash*, sisters who rose to the sky to visit their cousin, the Sun. The Pleiades to the Chumash are known as "the wives" as they were thought to be a group of women who rose to the sky after their husbands had hunted successfully but declined to share any of their meat with their wives. The women rose together to the sky in protest.

Perhaps the most interesting journey described in Chumash sky lore involves the journey of the souls of departed people as they rise to the sky. Departed souls were believed to rise into the sky at the edge of Chumash territory (perhaps near Pt. Conception) and there began a journey along the Milky Way. As they arrived to the sky they were greeted by the "Widows," a group of stars known to the Chumash as "Wit" (or Land of the Widows) in the European constellation of Cassiopeia. The widows were eternally youthful and bathed themselves in a shallow pool and did not eat, but instead inhaled their food. After leaving the widows, the souls crossed a deep ravine and encountered giant ravens who pecked out their eyes. The dead would replace their eyes with one of the many poppies that grew in this part of the sky. The ravine mentioned in the story might be associated with the stars within Cepheus and Lacerta. After the harrowing encounter with the raven, the dead were confronted with the formidable Scorpion Woman, known as "She Who Thunders" and represented by the stars of Cygnus. Scorpion Woman blocked their path and the souls had to get around her to cross a large body of water, which in the sky is the Cygnus Rift, a dark area of the Milky Way. The souls

**Table 2.7** Summary of the Chumash constellations and their Western equivalents (*source*: Hudson and Underhay 1978, pp. 99–110)

| Western name | Chumash name | Translation |
|---|---|---|
| Polaris | Shnilemun | "Star that Never Moves" or "Sky Coyote" |
| Aldebaran | Huhaw | "Coyote" |
| Little Dipper | Iluhui | "Woman leader" |
| Big Dipper | Manohonoh awawaw | "Boys who turned into geese" |
| Castor and Pollux | Ponoya and Shapiqenwash | "Two sisters who fled from the Earth and headed for sky to visit their cousin Sun" |
| Orion's Belt | Misq loka iyilike | "The three kings" or "The three steady persons" |
| Pleiades | Iyitaku | "The maidens" or "the wives" |
| Cassiopeia | Wit | "Land of the widows" |
| Cygnus + Vega (as stinger) | | "Scorpion Woman" |
| Aquila | Shimilaqsha | "Land of the dead" |

crossed this water over a very narrow bridge where huge monsters would taunt the souls and attempt to get them to fall off the narrow bridge. Only after crossing this frightful gauntlet of challenges did the souls reach their final destination, known as "Shimilaqsha" or "Land of the Dead." The path of these souls is visible on any clear night and the final destination is the brighter of the two parts of the Milky Way past the Cygnus rift, near the constellation of Aquila (Table 2.7).

## The Southern Sky

To the south are constellations first seen by Western eyes during the great age of exploration 500 years ago. The names of constellations in the south for Europeans therefore abound with names of explorers, ships, and nautical equipment. The Magellanic Clouds, the constellations of Carnia (stern), Sextans (sextant) are just a few examples. The lore of the southern sky for Europeans is therefore new and sparse, as it only has had a few centuries to develop (Figs. 2.33 and 2.34).

For additional tales of the southern sky we must instead look to those ancient cultures which are rooted in the Southern Hemisphere for thousands of years. The Aboriginal Australians, the Inca from Peru, Bolivia, and Chile, and the inhabitants of Africa, all were able to weave tales of the power for these southern stars, and we will explore their constellations and star tales in separate sections.

## The Southern Sky from Around the World

### *Star Maps for World Constellations: Southern Hemisphere*

In the sections below, we will provide maps of the Southern sky which feature constellation figures from the Aboriginal Australian, Incan or Andean Cultures, as well as from the Western cultures. Unlike the Northern Hemisphere, the Southern sky features no "pole star." However it does feature several fantastic constellations and bright stars, such as Canopus, the Southern Cross, and α and β Centauri. These last two can be used to locate the Southern Celestial Pole. By extending the main axis of the Southern Cross southward, and looking for its intersection with a line perpendicular to α and β Centauri, you can find the blank part of the sky where the South Celestial Pole is located. Figure 2.35 shows this imaginary triangle in the Southern sky, along with the other Southern constellations in the same projection used throughout our tour of constellations. Figure 2.35 also shows the Winter Hexagon, the ecliptic or zodiac, and the constellation Scorpius.

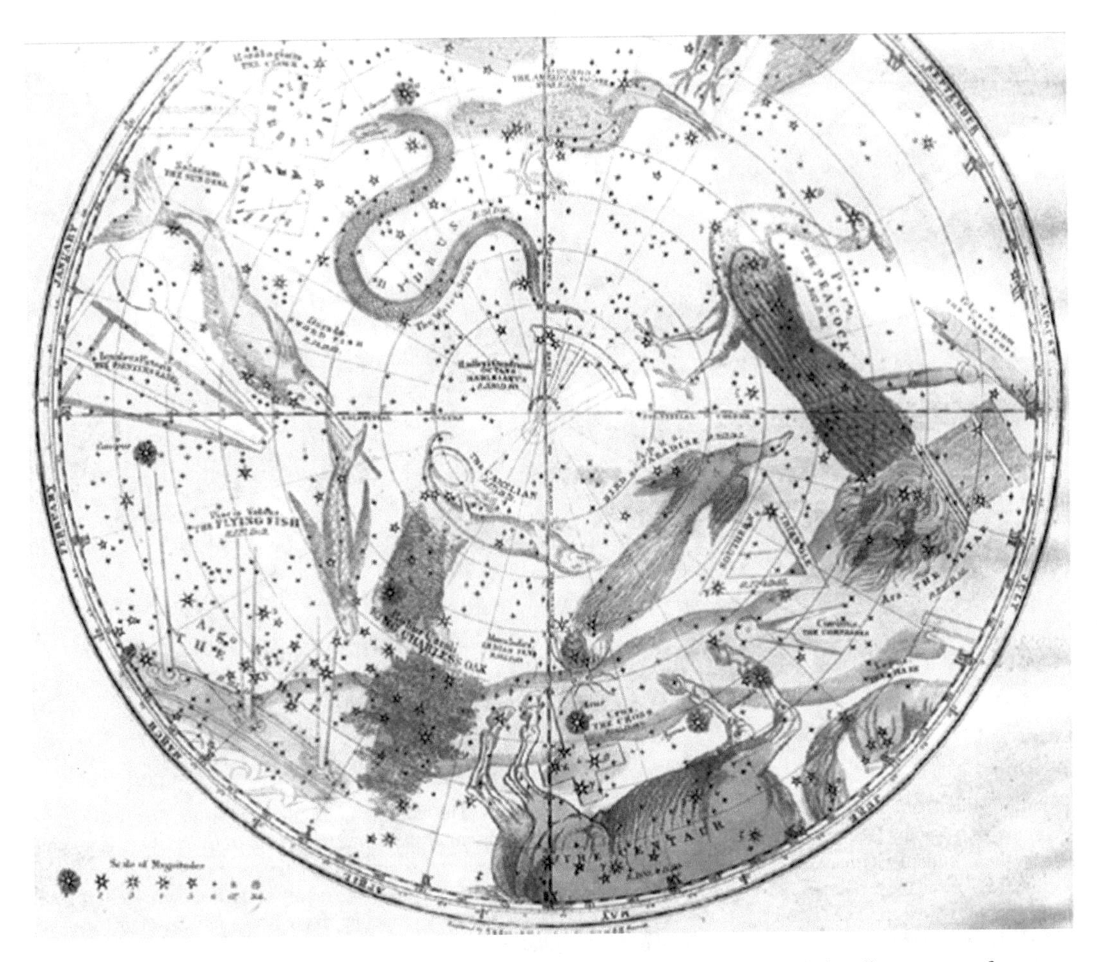


**Fig. 2.33** The southern circumpolar sky, showing western constellation figures that include a diverse array of creatures and nautical equipment from the age of exploration such as compasses, sextants, and parts of ships (from Burritt 1833)

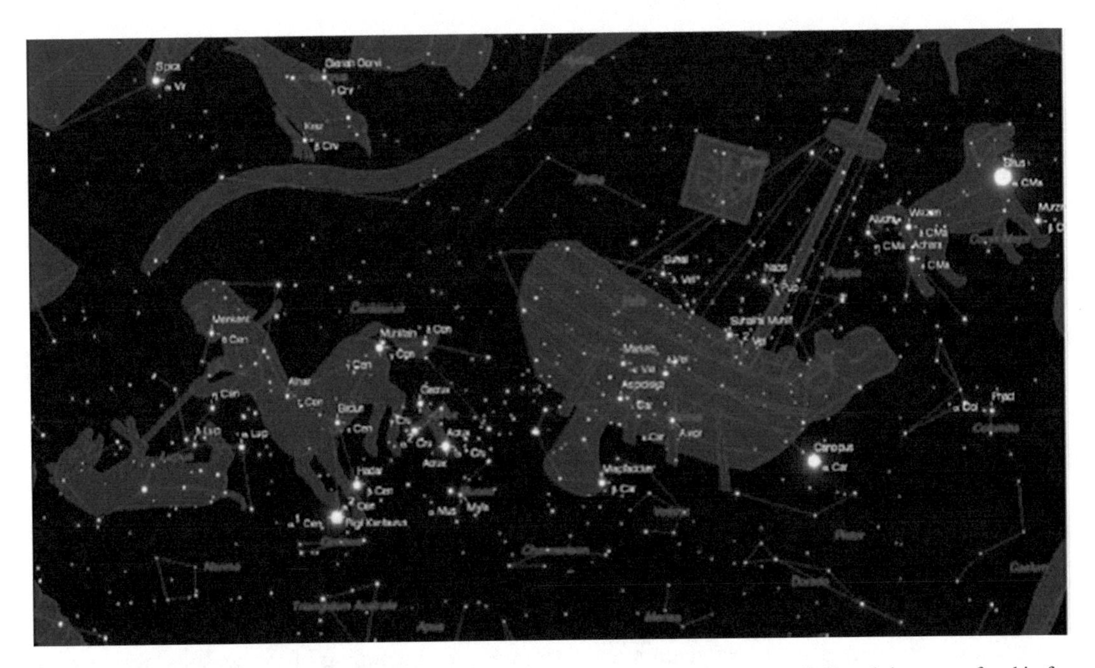


**Fig. 2.34** The southern celestial pole, with European constellation figures of a centaur (*left*) and the stern of a ship for carina (*right*). Other nautical constellations in the southern sky include Octans, the octant, a precursor to the sextant used by navigators, and Telescopium, the telescope (figure by the author, using the Voyager IV computer program)

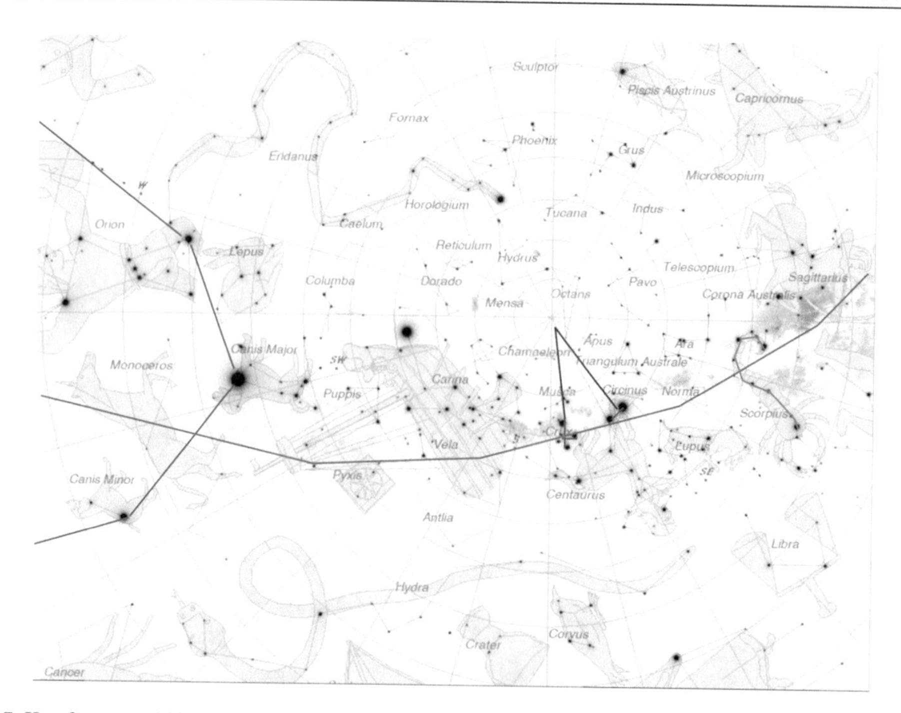


**Fig. 2.35** Key features within the Southern Hemisphere Star Maps. The familiar constellations of the Southern Cross, Scorpius, and α and β Centauri are shown in *Blue*, along with the bottom of the Winter Hexagon (*left*). As there is no pole star in the Southern Hemisphere, navigators used the Southern Cross and α and β Centauri to point toward the South Celestial Pole, and these pointers are shown with *red lines*. The Milky Way figures prominently in the Southern Hemisphere, and is shown with a *red line*. Two bright stars not visible in the North, but spectacular in the South are Canopus (*near the left* of *center* in the image) and Achernagr (just above the South Celestial pole in the map)

## *The Aboriginal Australian Sky*

### Aboriginal Australian Constellation Lore and Stars

The Aboriginal Australian culture is one of the most ancient on the Earth and has been estimated to originate before 20,000 BC. The wellspring of much Aboriginal myth and sky lore is "The Dreamtime," something of a parallel universe in which a set of spirits and creatures influence both the physical universe, the forms and behaviors of animals, and the culture of humans. The Dreamtime is a timeless space that predates our physical universe, continues to exist today, and will outlast any of our notions of time or space.

According to the Aboriginal Australians, our physical universe is one surrounded by a sky dome and like many ancient people, the Aboriginal Australians offered many explanations for both how the sky is propped up and how one travels beyond this world into the sky world. Some accounts tell of the sky being propped by solid wooden pillars being watched by an old man, while others talk of a canopy of stars being held up by star people. Some say the sky dome is held up by a giant tree or instead is made of a hard shell or crystal (Johnson 1998, p. 14).

The Meriam people of the Eastern Torres Strait (in northern Australia) describe constellations in terms of the giant *Tagai*, who rides in the sky in a canoe and who fills the sky with parts of his body, his canoe, and various crewmen who were thrown overboard for stealing. For these Aboriginal Australians stars were used for navigation and for setting the beginning of the seasons, which were timed along with shifts in the winds, the availability of various food sources such as turtles, yams, taro, birds, fish, and even the "flying fox," a large bat found in Australia (Fig. 2.36).

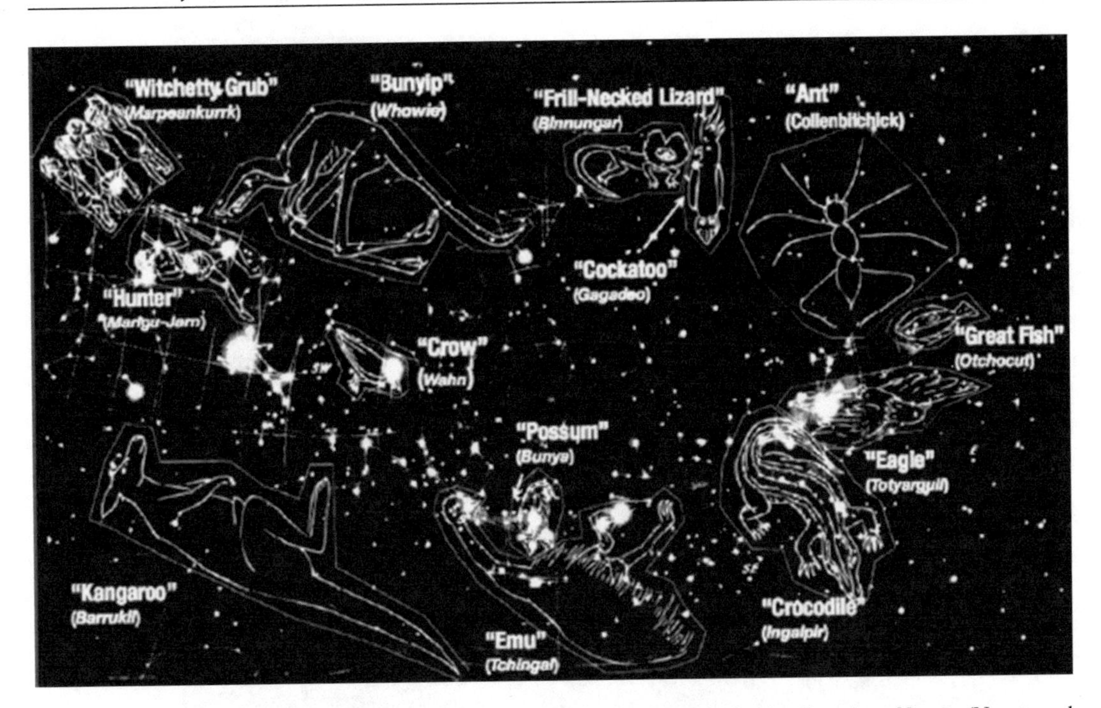


**Fig. 2.36** Aboriginal Australian constellations, as described by the Aboriginal elder Gaparingu Naputa (Naputa and Patston 1996), with the English and Aboriginal names indicated (figure by the author; artwork by Kim Aldinger)

Variations of the canoe constellation among sea going Aboriginal people exist, with some groups seeing the Sky Canoe in the Milky Way and others saw a canoe within the entire set of stars of Orion, Taurus, and environs. In this version the canoe is filled with fishermen, and their wives, who sit in a small group we call the Pleiades.

For those living in the Southern Hemisphere, the southern night sky offers some dramatic differences to the northern sky. First, the southern location means that the center of the galaxy (located near our constellation of Sagittarius) can cross close to the zenith of the sky, giving an unobstructed view of the spectacular dark clouds and dense star clusters on both sides of the center of the Milky Way. Another dramatic addition is the nearby galaxies we know as the Magellanic Clouds, a pair of large smudges of light between the Milky Way and the south celestial pole. The two clouds are known to astronomers as the Large Magellanic Cloud (LMC) and Small Magellanic Cloud (SMC), and they have about the same surface brightness of the Milky Way, even though they are 150,000 light years away. In the Southern Hemisphere many important stars such as the Southern Cross, α,β Centauri, and Canopus appear brightly and play an important role in Aboriginal Australian sky lore.

The Magellanic Clouds are often seen as groups of sky people by the Aboriginal Australians, who imagined the diffuse light to be that of many distant campfires (Fig. 2.37). Sometimes the Magellanic Clouds have been associated with pairs of individual heroes or animals, much as a pair of bears in European constellation lore represents Ursa Major and Ursa Minor. Since there are no bears in Australia, the Magellanic Clouds instead have been described to be a bandicoot (LMC) and kangaroo rat (SMC) by one group or as two snakes representing two separate heroes or as a man and a woman in the sky.

Dark clouds in the Southern Milky Way, such as the Coal Sack, play an important role in star lore as well. These dense clouds are much more distinct than the dark patches that are visible in the Northern Milky Way and make their way into Australian sky lore as a large rock fish, a plum tree, a mythical animal known as a *torong*, a waterhole used by some of the celestial creatures embodied by stars, as an emu, or as an evil spirit (Johnson 1998, p. 57).

**Fig. 2.37** The southern sky provides a spectacular array of sights not visible to northern stargazers, which includes a spectacular view of the Milky Way's center and its dark clouds, and the pair of Magellanic Cloud galaxies (the two glowing regions near the *top of the image*). This photo from Chile shows the Milky Way and Magellanic Clouds, with the CTIO 4-meter Blanco telescope in the foreground. Note the dense patches of dark clouds in the Milky Way, a part of the sky celebrated in the star lore of the Inca (*image credit*: Roger Smith/NOAO/AURA/NSF, reprinted with permission)

The Pleiades in Aboriginal Australian culture is often seen as a group of women escaping unwanted advances from a group of males (typically embodied in the belt of Orion). In one account, the Pleiades are sisters hunting in a canoe, known as *Djulpan*, being pursued by three brothers who were banished to the sky for eating forbidden flesh. Western Australian Aboriginal stories tell of the Pleiades coming about as a group of girls turn themselves into stars to escape the advances from a man or group of men. Other cultures affirm the imagery of Orion as a boat, and for the Milingimbi group of Aboriginal people Orion, the Hyades and Pleiades are known as the Tjlulpuna or the "canoe stars." (Bhathal 2010, p. 53). In the Torres Strait, the Pleiades are part of an asterism which forms a Shark's fin—and the changing location of the stars on the horizon between February and September is used to mark the times for planting crops, and also the changing of the seasonal winds (Gilchrist and Holland 2009). In the Western Desert, the Pleiades appearance in the dawn sky revealed the season for dingo breeding, and helped people time their hunts for dingo pups and was a cue for an annual feast. (Bhathal 2010, p. 45). Other Aboriginal people from the Melville and Bathurst Islands saw Orion as a set of Dingos pursuing a group of kangaroos, who are embodied in the Pleiades. (Bhathal 2010, p. 66).

Many star tales include the stars α and β Centauri, the two closest stars to the Earth, and the Southern Cross (Fig. 2.38). One tale tells of Tchingal the Emu and Bunya the Possum, who were turned into these stars after hunters chased Tchingal across the sky. Tchingal ran and frightened Bunya, who quickly ran into a tree and turned into a possum. The two are together in the sky, with Bunya hiding in the "treetop" of the Coal Sack and Tchingal the emu next to him as α and β Centauri. The Southern Cross stars are the spears thrown by the hunters at Bunya. The tale also explains why the possum is always scared and hiding in trees.

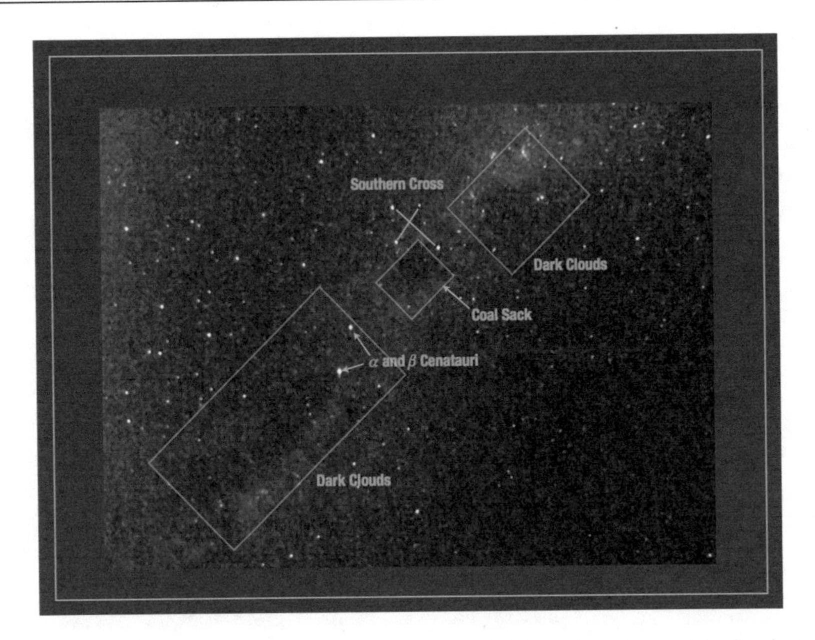


**Fig. 2.38** The stars α and β Centauri (*left and center*), the Coal Sack (*right of center*), and the Southern Cross (just *to the right* of the Coal Sack). These stars and dark clouds played an important role in Australian and Southern Hemisphere sky lore. In one tale, the Coal Sack is a treetop in which Bunya the possum is hiding (original image by B. Penprase, photo taken at Las Campanas Observatory, Chile)

From the Stradbroke Island people of Queensland comes the tale of the Southern Cross as Mirrabooka, a guardian ancestor spirit. Biami the good sky spirit wanted to find someone to watch over the tribes on the Earth. Biami found a good man named Mirrabooka and when Mirrabooka grew old, Biami gave him a spirit form and placed him into the sky as stars:

> Biami.. placed him in the sky among the stars, and promised him eternal life. Biami gave Mirrabooka lights for hands and feet and stretched him across the sky, So that he could watch forever over the tribes he loved. And the tribes could look up to him. from earth and see the stars which were Mirrabooka's eyes gazing down on them.
> (Oodgeroo and Bancroft 1993, p. 67)

Among the Yolngu community of Dhalinybuy in Arnhem Land comes the tale of the Milky Way as the Emu in the Sky. The Emu in the Sky stretches across the Milky Way, connecting the regions of Scorpius and Sagittarius (which form the body) to the Southern Cross. The dazzling expanse of dark clouds and shimmering Milky Way of the Southern sky form the image of the emu, with its body located in the center of the Milky Way (near Scorpius and Sagittarius), and its neck and beak stretching along the Milky Way past the Southern Cross (see Fig. 2.39). The Emu is not as much a constellation as the embodiment of an animal spirit in the diffuse glow of billions of stars from the Milky Way.

From the Aboriginal elder Gaparingu Naputa of central Queensland comes a set of 12 constellation tales from around Australia, organized by month. A brief summary of the names of these constellations, the European equivalent, and the meaning of the constellations to the Aboriginal people is presented in Fig. 2.36 and Table 2.8.

Some of the Aboriginal groups from this same region also describe the Milky Way as a river in the sky, with the glowing light coming from the fires of younger sky people camping on the riverside. The Magellanic Clouds come from an older couple forming a separate camp off to the side of the celestial river (Norris, p. 8). Other Aboriginal groups, such as those in the Central Australian regions, saw the sky less as distinct constellations and more as a pair of two seasonal sky maps—a winter sky that included Scorpio, Centaurus and the Southern Cross, and a summer sky that included Orion, the Pleiades and Taurus. These two groups were split by the Milky Way, and were thought of as two camps know as the

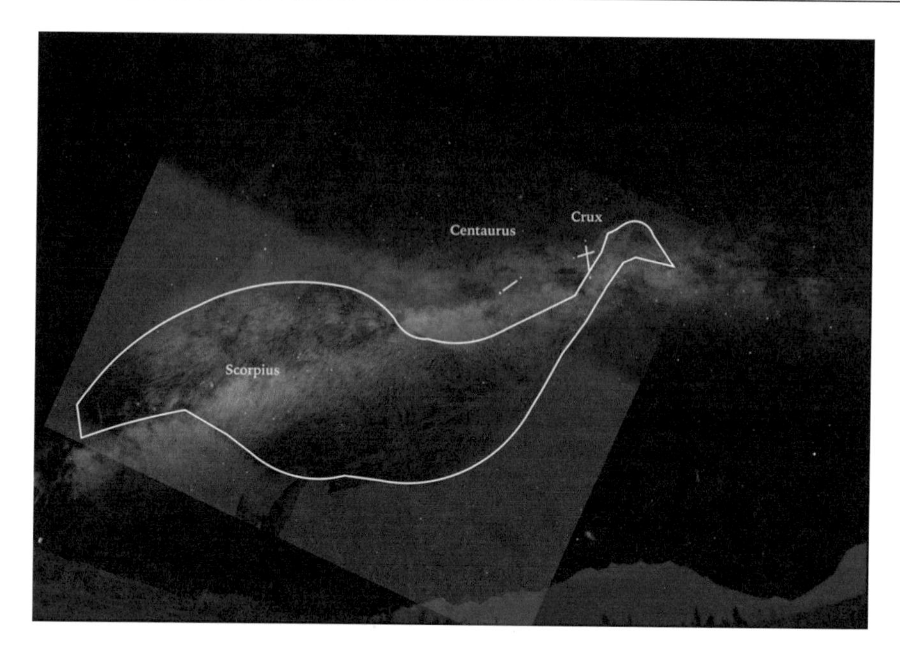


**Fig. 2.39** The Emu in the Sky, as seen by Aboriginal Australians in Arnhem Land. The emu connects most of the Milky Way in a single figure that uses the contours of the Milky Way's center near Scorpius to form the body, and the dark and light clouds forming the neck and beak of the emu (adapted from Norris and Norris 2009)

**Table 2.8** Aboriginal Australian months and constellations (*sources*: Naputa and Patston 1996; Johnson 1998)

| Month name | Aboriginal Australian name | English name | European equivalent | Meaning to Aboriginal people |
|---|---|---|---|---|
| January | Marigu-Jarn | The hunter | Orion | Jarn would hunt and play with his dancing sticks chasing his seven sisters through sky. Marks time in year when one can search for echidnas |
| February | Wurrawana Corinna | Tasmanian tiger | Region near Gemini | Tale of how tiger gets stripes; ghost tiger has stripes drawn by Sky Spirit Polana to honor his bravery in fight |
| March | Wahn | The crow | Canopus | The star Canopus is "War," the center of a crow named Wahn, who took fire and gave people the inspiration with help from the Sky Spirits for making fire |
| April | Barrukill | Kangaroo | Region near Corvus | Hunting kangaroo and dog useful for navigation in fall |
| May | Tchingal and Bunya | Emu and possum | $\alpha$ and $\beta$ Centauri and Southern Cross | Emu is hunted and runs across sky, scaring the possum Bunya into a sky tree (Coal Sack) |
| June | Marpean-kurrk | "Witchetty grub" | Libra + Virgo | Marks time of the year when one can gather grubs for food, during September |
| July | Ingalpir | The crocodile | Scorpius | Heliacal rising of this constellation marks trading season in summer (December) |
| August | Totyarguil | The eagle | Aquila | Eagle is a guardian spirit of the people, and one of first animals created by the sky spirit Biame |

(continued)

**Table 2.8** (continued)

| Month name | Aboriginal Australian name | English name | European equivalent | Meaning to Aboriginal people |
|---|---|---|---|---|
| September | Otchocut | The great fish | Delphinius | A brave fish who fought hard for his life against kingfisher and who was placed in the sky for his bravery |
| October | Collenbit-chick | The ant | Aquarius + Capricorn | A brave ant tried to rescue eagle from the bunyip, a mythical monster, and was rewarded for his bravery by the sky spirits |
| November | Binnungar and Gagadoo | Frill-necked lizard and cockatoo | Southern Fish and Cetus | Dreamtime creatures which aided in communication between tribes; the lizard sends message through the sand, while the cockatoo sends messages through the air |
| December | Whowie | The bunyip | Carina + Vela | A fearsome swamp monster from the Murray River, driven out from his caves by people, and his spirit still haunts us in the sky |

Aranda and Luritja camps (Bhathal 2010). Other large groupings of stars are seen in the Torres Strait, where the constellations of the Milky Way (Centaurus, Crux and Scorpius) make up a group known as the *Tagai,* which shows a man standing in a canoe with a fishing spear (Bhathal 2010, p. 57).

Another Southern Cross tale comes from the northern Flinders region. In this tale, two brothers were hunting emu, and after they caught the emu, they hung it to prepare it for cooking later. As flies descended on the emu in the late afternoon, the brothers decided to start a fire to drive the flies away. The fire grew stronger, and fed by a Northern Wind, blew burning sticks and embers across the land, causing a massive brushfire. The brothers sought higher ground, climbing ridges and mountaintops, but the fire still raged around them. Their only refuge was to fly into the sky, where they were turned into the pointer stars of the Southern Cross (Mountford, p. 34).

Across Southern Desert Australia the Southern Cross is seen by the Aboriginal people as tracks from the wedge-tailed eagle (from Southern Australia), with the Coal Sack as his nest, and the stars alpha and beta Centauri as his throwing stick. Coastal Aboriginal people associated these stars with fishing, and saw the Southern Cross as a stingray being pursued by a shark, embodied by alpha and beta Centauri (Bhathal 2010).

No matter if the Southern Cross is a treetop, and emu, a possum, or Mirrabooka—in all of these cases the shimmering stars of the southern sky inspired the Aboriginal Australians, and they continue to inspire us today.

## *The Incan Andean Sky*

### Inca/Quechua Constellation Lore and Stars

The Inca once ruled a region that included modern Ecuador, Peru, Bolivia, and Chile. The Inca had a system of roads which included more miles than the Roman Empire and were master stonemasons, creating some of the greatest structures of the ancient world. Because the Inca did not develop writing, much of their culture is lost to us. Some star lore has been recovered from the anthropologist Gary Urton, in the book "*At the Crossroads of Earth and Sky*." Urton's study (Urton 1981) reports the results from interviews with present-day Quechua people, the descendants of the Inca, and conveys their understanding of the night sky and constellations. The rich star lore of the Inca is built upon the unique constellations from the southern sky, and makes use of both the dark clouds of the Milky Way and the dazzling Southern stars (Fig. 2.40).

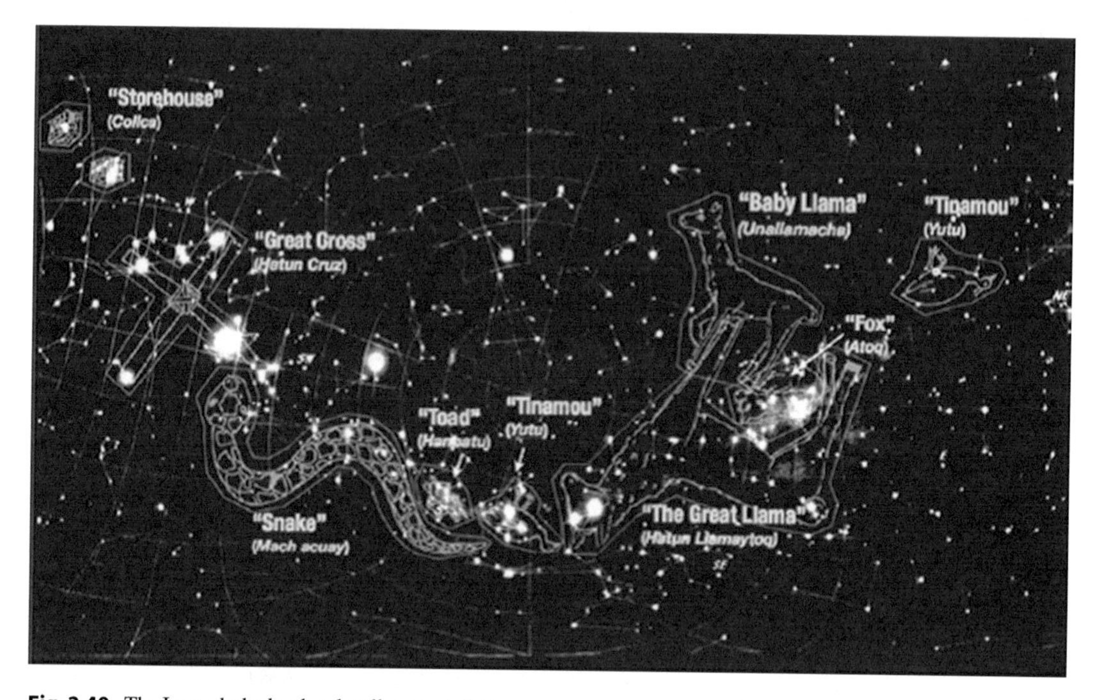


**Fig. 2.40** The Incan dark cloud and stellar constellations, superimposed on a modern map of the southern sky, centered on the South celestial pole (based on information in Urton 1981). Key constellations included the Llama (α and β Centauri), the Tinamou (Southern Cross), and a series of animals formed by the dark clouds of the Milky Way. The Inca were one of only a very few civilizations that recognized both stellar and dark cloud constellations and had a rich lore involving both (figure by the author; artwork by Kim Aldinger)

The astronomical practices of the Inca were described soon after the conquest by sources such as the chronicler John Santa Crus Pachacuti Yamqui. The Spanish chronicles described the importance of observations of Venus, the Pleiades, and the Sun to the Inca ruler, himself believed to be a descendant of the creator god Viracocha. Special chambers with windows aligned to horizon positions of the Pleiades and other celestial bodies were part of the palace and played a key role in seasonal ceremonies.

Incan practices included an elaborate series of ritual days organized along 12 sidereal months within a 328-day year. Unlike many lunar calendars, the Inca calendar was designed around the 27.3-day lunar sidereal month, instead of the 29.5-day lunar synodic month (Aveni 1997). Incan astronomers had to watch the skies carefully to determine the location of the Moon relative to the stars, in order to determine the beginning of the sidereal lunar month. The year included ceremonies for each day in which offerings were made at a series of shrines or *ceques* distributed around the major cities in a radiating pattern that mirrored the four quarters of the Inca year and were arranged in a clockwise direction.

Within the culture of both the present-day Quechua and ancient Inca empire existed star lore that celebrated the sky and its stars and dark clouds. These constellations include divisions of the sky into four quarters and groups of stars within each quadrant to help set the calendar for ritual celebration. Urton describes a system by which the Milky Way served as a celestial "hour hand" and marked the different parts of the year based on its orientation in the clear Andean skies. A set of Quechua constellations and stars regulated the calendar based on heliacal risings and reminded the people of mythological animals associated with each constellation.

Particularly interesting is the Quechua system of dark cloud constellations, which includes an entire menagerie of animals formed by the distinct dark clouds visible in the Southern Milky Way. Urton describes 16 separate dark cloud constellations that include a fox, condor, toad, two llamas, a fish, a toad, and many other creatures (Figs. 2.41 and 2.42).

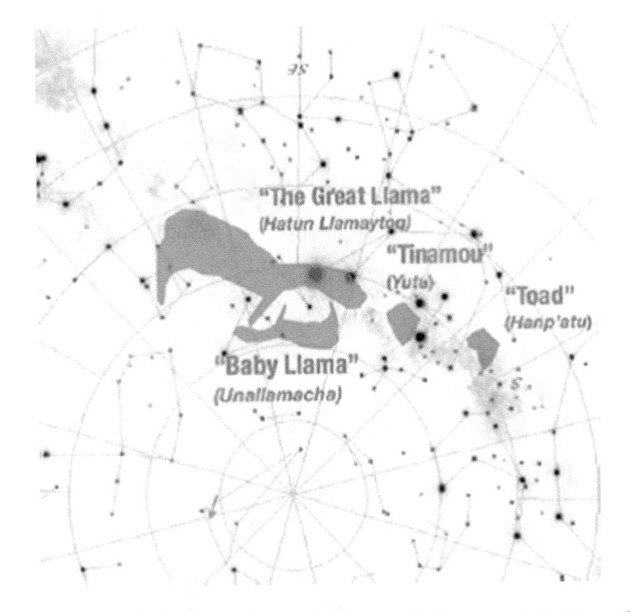


**Fig. 2.41** The southern circumpolar sky, showing the numerous dark cloud constellations of the Quechua, including the two llamas, and the region near the Southern Cross. The stars α and β Centauri are conveniently located to serve as the "eyes" for the large llama dark cloud constellation (figure by the author, with information from Urton 1981)

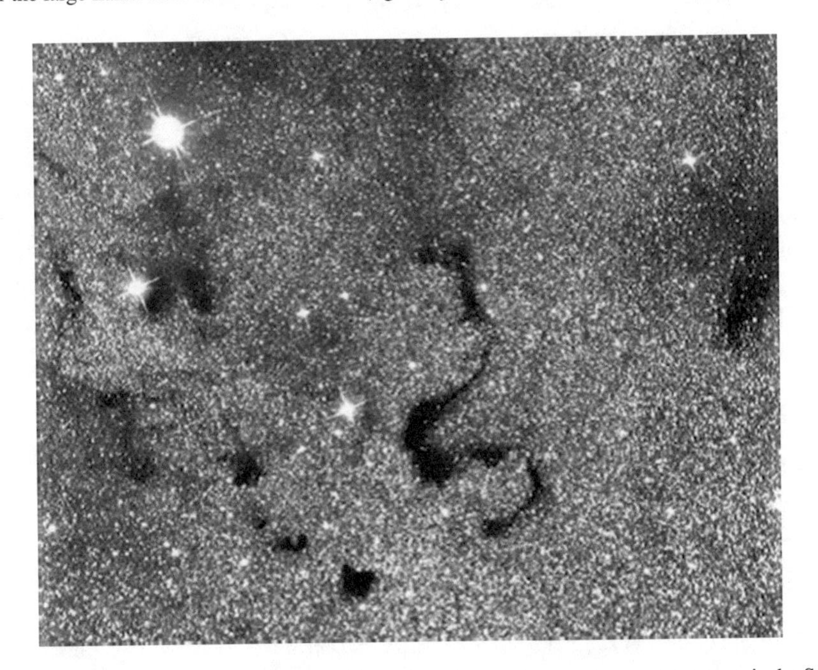

**Fig. 2.42** Another example of a "dark cloud constellation," as observed by northern astronomers is the Snake Nebula, a dark cloud region in the constellation Ophiucus. Such clouds are found in the central part of the Milky Way and are a key part of Incan sky lore (image from NOAO/AURA, reprinted with permission, from http://www.noao.edu/outreach/aop/observers/b72.html)

Probably one of the most original and distinctive of all constellations from the Southern Hemisphere is the Quechua dark cloud constellation of the llama (Fig. 2.41). The region of the Southern Cross, with its numerous dark clouds and α and β Centauri, forms a large mother llama and its baby on opposite sides of the celestial sphere from each other. The dark clouds form the outlines of both animals and the bright stars α and β Centauri form the eyes of the mother llama. A summary of Quechua constellations, with both stellar and dark cloud constellations, is presented in Table 2.9.

**Table 2.9** Summary of Incan constellations and translations (*source*: Urton 1981)

| Quechuan name | Translation | European name | Use or meaning |
|---|---|---|---|
| Haton Coyllur | "Large star" | Sirius | |
| Amaru contor | "Serpent changing to condor" | Scorpius | |
| Boca del Sapo | Sp. "mouth of the toad" | Hyades | |
| Collca | "Storehouse" | Pleiades | Useful for predicting weather and for timing agriculture |
| Contor | "Condor" | Head of Scorpio | |
| Llamacnawin | "Eyes of the llama" | α and β Centauri | The eyes of the dark cloud constellation of the llama |
| Sonaja | Sp. "tamborine" or "rattle" | Corona Borealis | |
| Atoq | "fox" | Dark spot between Scorpio and Sagittarius | |
| Contor | "Condor" | Dark spot near Scorpio | |
| Hanp'atu | "Toad" | Coalsack or region near Scorpio | |
| Hatun llamaytoq | "The great, large llama" | Dark streak between Coalsack and Scorpio | |
| Maya | "River" | The Milky Way | |
| Unallamacha | "Suckling baby llama" | Dark cloud near a Cen toward τ Sco, opposite large llama | |
| Yutu | "Tinamou" | Either Coal Sack or region near tail of Scorpio | |

We have completed our tour of the world's constellations and seen the imaginative and inspiring ways in which our ancestors have brought together the stars and nebulae of the night sky to tell stories, to help them survive, and to inspire them through the generations. Our next chapter will discuss the range of beliefs of ancient people in where it all came from—the Cosmology of the Universe and how the world of stars, the Earth and sky came to be.

## References


Allen, R.H. 1963. *Star names, their lore and meaning*. New York: Dover.

Aveni, A.F. 1997. *Stairways to the stars: skywatching in three great ancient cultures*. New York: Wiley.

Bayer J. 1603. Uranometria. http://www.lindahall.org/services/digital/ebooks/bayer/.

Bhathal, R. 2010. *Aboriginal astronomy*. Sydney: ARI.

Blackburn, T.C. 1975. *December's child: A book of Chumash oral narratives*. Berkeley, CA: University of California Press.

Blundeville, T., et al. 1594. M. Blundevile his exercises : containing sixe treatises … verie necessarie to be read and learned of all yoong gentlemen that haue not bene exercised in such disciplines and yet are desirous to haue knowledge as well in cosmographie, astronomie and geographie, as also in the arte of navigation, London.

Bryan, E.H. 2002. *Stars over Hawaii*. Hilo: Petroglyph Press.

Burritt, E.H. 1833. *Atlas, designed to illustrate the Geography of the heavens*. Hartford, CT: F. J. Huntington.

Flamsteed J. 1729. Atlas Coelestis. http://www.lindahall.org/services/digital/ebooks/flamsteed1729/.

Gilchrist, S., and A. Holland. 2009. *Shared sky*. Melbourne: National Gallery of Victoria.

Griffin-Pierce, T. 1992. *Earth is my mother, sky is my father: Space, time, and astronomy in Navajo sandpainting*. Albuquerque, NM: University of New Mexico Press.

Hevelius J. 1690. Firmamentum Sociescianum sive Uranographia. Gdansk.

Hudson, T., and E. Underhay. 1978. *Crystals in the sky: An intellectual odyssey involving Chumash astronomy, cosmology, and rock art*. Socorro, NM: Ballena Press.

Hudson, T., and K. Conti. 1984. The rock art of Indian creek, occasional paper #12, San Luis Obispo Country, Archaeological Society.

Huth, J.E. 2013. *The lost art of finding our way*. Boston: Belknap Press.

Johnson, D. 1998. *Night skies of Aboriginal Australia: A noctuary*. Sydney, NSW, Oceania: University of Sydney.

Johnson, R.K., and J.K. Mahelona. 1975. *Na inoa hoku: A catalogue of Hawaiian and Pacific star names*. Honolulu: Topgallant.

Krupp, E.C. 1983. *Echoes of the ancient skies: The astronomy of lost civilizations*. New York: Harper & Row.

Liliuokalani. 1897. *An account of the creation of the world according to Hawaiian tradition*. Boston, MA: Lee and Shepard.

Lionberger, J. 2007. *Renewal in the wilderness: a spiritual guide to connecting with god in the natural world*. Woodstock: SkyLight Paths.

MacDonald, J., et al. 1998. *The Arctic sky: Inuit astronomy, star lore, and legend*. Toronto, ON: Royal Ontario Museum/ Nunavut Research Institute.

Miller, D.S. 1997. *Stars of the first people: Native American star myths and constellations*. Boulder, CO: Pruett.

Mountford, C.P. 1969. *The dawn of time: aboriginal Australian myths*. Sydney: Rigby.

Naputa, G., and G. Patston. 1996. *Aboriginal sky figures: Your guide to finding the sky figures in the stars, based on Aboriginal dreamtime stories*. Sydney: Australian Broadcasting Corporation.

Needham, J., and C.A. Ronan. 1980. *The shorter science and civilization in China: An Abridgement of Joseph Needham's original text*. New York: Cambridge University Press.

Norris, R., and C. Norris. 2009. *Emu dreaming—an introduction to Australian aboriginal astronomy*. Sydney: Emu Dreaming.

Oodgeroo, N., and B. Bancroft. 1993. *Stradbroke dreamtime*. Pymble, NSW: Angus & Robertson.

PVS (Polynesian Voyaging Society). 2016. Web resource at http://pvs.kcc.hawaii.edu/ike/hookele/holding_a_course.html. Accessed May 2016.

Rajawat, D.S. 1990. *Astronomical observatory at Jaipur*. Jaipur: Delta Publications.

Rao, S.B. 2014. *Indian astronomy: Concepts and procedures*. Bangalore: M.P. Birla Institute.

Staal, J.D.W. 1988. *The new patterns in the sky: Myths and legends of the stars*. Blacksburg, VA: McDonald and Woodward.

Thompson, V.L., and L. Weisgard. 1966. *Hawaiian myths of earth, sea, and sky*. New York: Holiday House.

Urton, G. 1981. *At the crossroads of the earth and the sky: An Andean cosmology*. Austin: University of Texas Press.

# Chapter 3

# Creation Stories from Around the World

*To see the world in a grain of sand.*
(Blake)

The Absarkoe or Crow people tell the story of how the world began:

*Once there was only water, and Old Man Coyote. Coyote needed someone to talk with and soon found two ducks. He asked the ducks to check whether anything was under the water – one duck dove and disappeared for so long that Coyote worried he had died. But the duck emerged and informed Coyote that he had found the bottom. He went a second time to find a root, and the third time to bring up some earth. Coyote blew on the mud and it grew until it became the earth. Coyote planted the root on the earth, and plants and trees arose. Then Coyote began to make valleys, mountains, and lakes, and then took some of the clay to make people and more ducks. Within his earth, a smaller version of himself urged him to make all the other animals, as well as people with different languages who would fill the earth. In this way the bear, antelope, and other animals, as well as the dance, and war which came from people who could not understand each other.*
(Summarized from Leeming and Leeming 1995)

A second example is a Chinese folk tale of the universe emerging from a primordial creator Pan Ku (Fig. 3.1)




B.E. Penprase, *The Power of Stars*, DOI 10.1007/978-3-319-52597-6_3

*In the beginning was a huge egg. It contained chaos, yin-yang, and Pan Ku. Pan Ku was a giant who separated the opposite elements of the world, and grew each day for 18,000 years, until heaven and earth were separated by 30,000 miles. Pan Ku was covered with hair, and with his hammer he created mountains, canyons, and lakes. He also made the sun, moon and stars. When he died, 'his skull became the top of the sky, his breath the wind, his voice thunder, his legs and arms the four directions, his flesh the soil, his blood the rivers… the fleas in his hair became human beings.*

(From Leeming and Leeming 1995, p. 49)

**Fig. 3.1** Pan Ku creating the world, from the Tian-gon Yuan (1820), an illustration of the Chinese creation myth where Pan Ku hammers out the mountains and valleys of the world and then forms the bulk of its surface with his body (image from Library of Congress, "Heavens" online exhibit at http://www.loc.gov/exhibits/world/images/s33.1.jpg)

In both of these examples, from cultures across the Earth, we see a common theme of creation from a primordial creator, either Coyote or Pan Ku, who each make the world in a gradual process which accretes material until the vast expanse of our Earth is formed. Each creator brings to the task a set of traits which are valued by each people. For the Crow, Coyote brings a mysterious wisdom and omnipotence. The Chinese story explains both the creator and the system of *yin* and *yang*, which arise from Pan Ku, and also the dimensions of the cosmos as understood at the time. The stories include elements common to each civilization and place the foreground aspects of the landscape, water, etc., as the primary element from which all arose.

The tale of creation from a culture is itself a creative act. It draws upon the imagination, the landscape, and the values of its people. In ancient times, the cosmology and creation story was transmitted across the generations through oral tradition, or in elite priesthoods which kept sets of incantations, chants, and other knowledge within a small group who could draw on their power in times of crisis.

Today, most individuals believe in explanations of the origins of the universe that are a combination of a diverse mix of religious and scientific conceptions (Fig. 3.2). We often construct a personal belief system which integrates our experience or the local environment, and which may mix together our exposure to science, to the teachings of the Church, Temple, or Mosque of our families, and long-held myths from grandparents, and others. And even as scientists polish off their models of the initial

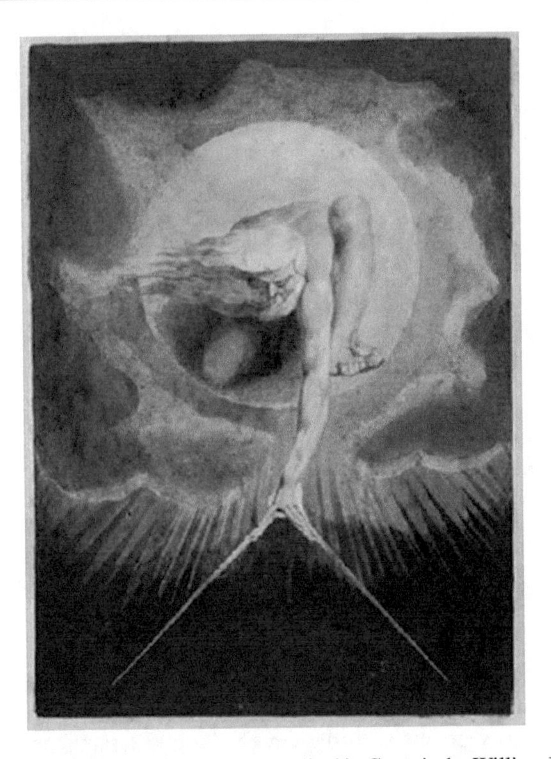

**Fig. 3.2** A depiction of the Judeo-Christian Creation, as described in Genesis, by William Blake's "Urizen as a Creator of the Material World" (1794) (image from Library of Congress, World exhibit, at http://www.loc.gov/exhibits/world/images/s21n.jpg)

$10^{-34}$ s of the Big Bang, with improved coupling constants for the strong and electro-weak forces, many people still hold the same beliefs that their ancestors held for thousands of years.

Each creation story and cosmology, no matter how diverse in its content, shares the same purpose: to provide a satisfying explanation to people about their origin and the origin of the universe, using characters, elements, and features of the environment familiar to a culture. These stories build with interconnections across time and space between people that transcend individual lives and that stretch imagination to encompass the vast realms of time to the beginning of everything. The human mind thirsts for knowledge of the First Person, the first elements, and for the answers of how all of these fit together, and each culture states this thirst with its own cocktail of religious, scientific, and moral teachings.

Leeming and Leeming in their "Dictionary of Creation Myths" (1995) divided creation stories into five main categories. These categories include creation stories where the universe originates: "(1) from chaos or nothingness, (2) from a cosmic egg or primal maternal mound, (3) from world parents, (4) from a process of earth diving, and (5) from several stages of emergence from other worlds."

They further identify "archetypical characters" in creation myths that include one of the following actors: (a) a creator, (b) a trickster, (c) a first man and woman, and (d) a flood hero.

From thousands of cultures separated completely by oceans and centuries, a relatively small number of patterns can describe the vast array of creation stories. This could be a result of the common features within the human subconscious, which Carl Gustav Jung, Swiss psychologist, would call the "archetype." Creation and cosmology can arise much as a dream does in an individual, but as a collective dream of a culture. In Jung's own words, "This whole dream-work is essentially subjective, and the dream is the theater where the dreamer is at once: scene, actor, prompter, stage manager, author, audience, and critic" (Jung 1974, p. 54).

To help gain perspective on both the diversity and commonality of creation stories, we present a very small set of examples of the creation stories of diverse cultures, organized by continent to illustrate some of the many ways that humans have explained their origins.

## European Creation Stories

Many traditional European creation stories exist, but we have chosen three representative examples—the Greek, Celtic, and Norse creation stories to represent some of these diverse European cultures.

### *Mythic Greek Creation and Cosmology*

The most familiar creation story from Europe is the Greek mythological creation, in which the universe arose from a series of ancient gods, who became the ancestors of the Greek and Roman Gods. These stories were incorporated into the Roman religion and give rise to many well-known tales, and even day names within our present-day society.

While the family of gods within Greek and Roman mythology is well known, the creation story that predates the Olympian Gods is less familiar and is presented by Hesiod in his Theogony from the period of 700 BCE. The story begins with the Greek sky god Uranus and the Earth Goddess Gaia. Each night Uranus "covers" Gaia, and soon she bore him children. From the two primordial gods arose the Titans, six sons and six daughters, along with an entire race of giants known as the Hecatonchires, and another race of one-eyed giants known as the Cyclopes, famous from Homer's Odyssey. When Uranus learned that one of his children would destroy him, he imprisoned his son Cronos along with his Titan siblings. Only through a violent attack was Cronos able to escape, and in the process he castrated Uranus (thereby creating the island of Crete). Cronos ruled for a time and tried to prevent his own violent overthrow by attempting to consume his son Zeus, but Zeus was able to avoid destruction by a warning from his mother Rhea and deposed his father Cronos to rule over the universe.

Many of our modern words for time (chronology, chronography) arise from the name of this early patriarch of creation. The cosmology of violent family strife of the Greeks and Romans perhaps mirrors some of the earthly events of the time. When Cronos is overthrown by Zeus, his son, the universe becomes a "family business" in which each part of the Earth and sky is ruled by one of Zeus' family—again, perhaps an appropriate arrangement for Greek and Roman culture in which family matters played a central role in politics, business, and religion.

Zeus ruled his unruly family pantheon along with his wife and sister Hera, deity of marriage and women, his brother Poseidon, who became god of the ocean, and Zeus' sisters Demeter, goddess of grains and crops. Zeus and Hera gave rise to their children Ares (god of war) and Hephasitos (god of crafts) who married Aphrodite (goddess of love, allegedly created during the violent act of Uranus' castration).

Zeus' extramarital activities, which play such a large role in star tales, also bring about many key gods and goddesses; his daughter by way of a Titaness was Athena (goddess of wisdom), and several other out of "wedlock" offspring of Zeus include Apollo (god of music, medicine), Hermes (messenger of the gods), and Dionysus or Bacchus (god of wine). Within the sphere of these family members, an ancient Greek or Roman could find a god to appeal for help in nearly every aspect of their life. To help summarize the lineages of the Greek gods, a family tree of the Greek Gods is provided in Fig. 3.3, which includes the basic outline of the main figures within the Greek creation myth.

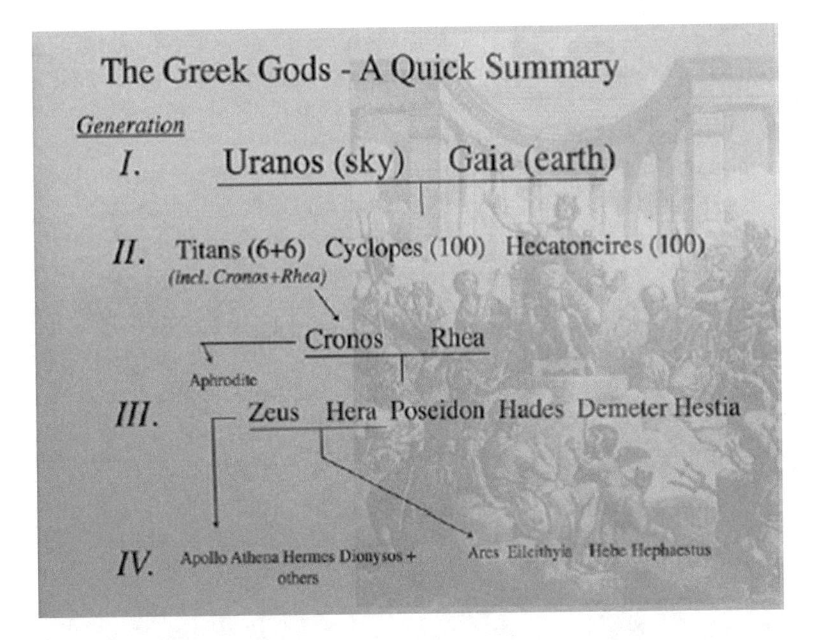


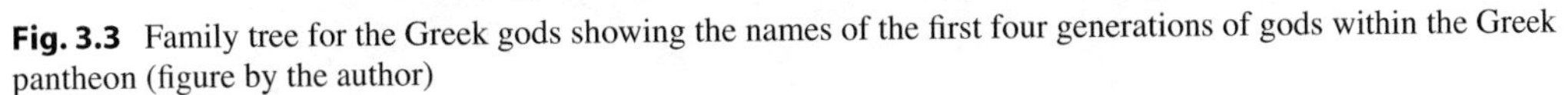
**Fig. 3.3** Family tree for the Greek gods showing the names of the first four generations of gods within the Greek pantheon (figure by the author)

## *Norse Creation and Cosmology*

> He sang who knew
> tales of the early time of man,
> how the Almighty made the earth,
> fairest fields enfolded by water,
> set, triumphant, sun and moon
> for a light to lighten the land-dwellers,
> and braided bright the breast of earth
> with limbs and leaves, made life for all
> of mortal beings that breathe and move."
>
> *Beowulf, Creation Song*
> (Gummere 1909)

Just as our modern calendar is a blending of Roman, Greek, and Norse elements, many of our popular conceptions of ancient cosmology include elements from the Roman, Greek, and Norse cultures. The Norse cosmogony appears on the surface to be similar to the Greek cosmology and creation. A genealogy of gods rule the universe, with their power apportioned to each of several elements of power within the Norse world. Approximate analogs between Norse and Greek/Roman gods exist and play a role in our day names. Odin (namesake of our day "Wednesday") bears some resemblance to Zeus in his role as head god and progenitor of many of the pantheon of the Norse. His wife Frigg (namesake of our day "Friday" and the goddess who knows the fates of all men) also plays many of the same roles as Hera. Odin also has a reputation for treachery, with numerous tales of affairs with giants, combat with mythical creatures, and betrayals and thefts from giants and heroes. The god Thor (namesake of our day "Thursday") descends from Odin, and Jord, an earth goddess.

Thor bears some resemblance to the Greek God of war, Ares, as he is known to battle the enemies of the gods that can include giants, monsters, and mysterious forces. Thor is armed with an unusual arsenal that includes his famous hammer Mjölnir, iron gloves, and a fantastic belt of strength, which together give him superpowers, even for a God.

The lineage of Norse gods began when the frost giant Ymir formed from primordial vapors, whose essence was imparted to the stones of the Earth through a primordial cow which gave rise to the patriarch of the Norse gods, named Buri; Buri's lineage included Bor, and (by way of a giantess wife), the trio of gods Odin, Vili, and Ve. Odin's consort Frigg gave birth to their son Baldr, and Odin also gave rise to Thor by union with an Earth goddess named Jord, and to Baldr's blind half-brother named Hor. A summary of the lineage of the Norse gods is shown in Fig. 3.4.

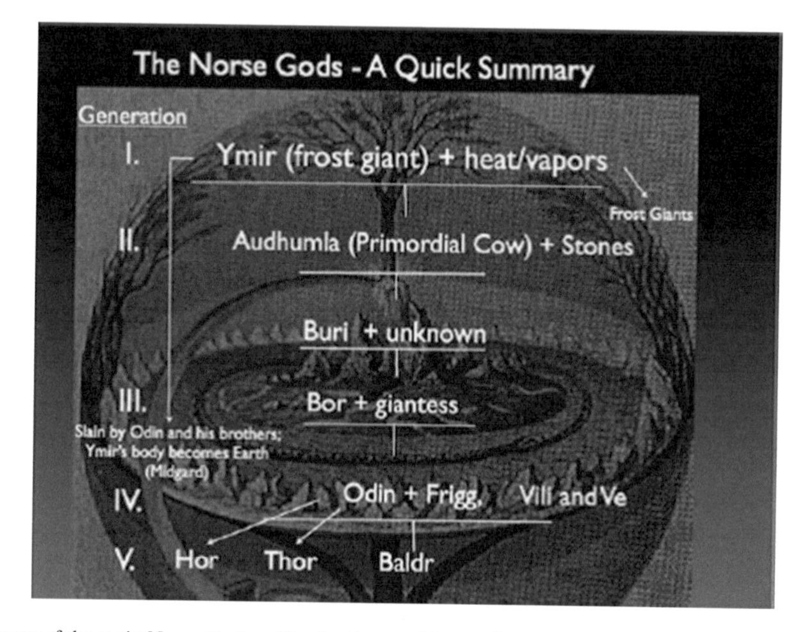


**Fig. 3.4** Lineage of the main Norse Gods, with a background image from a painting by Oluf Olufsen Bagge, from an 1847 English translation of the Prose Edda (figure by the author)

A deeper examination of the Norse universe and its pantheon reveals complexities unique to the Norse culture. The gods Odin and Thor, in particular, have a more complex relationship than Zeus and Ares (their Greek equivalents). Odin's complex nature can be surmised from his many names are listed in the poem Gylfaginning, as related by Snorri Sturluson's work Prose Edda from 1220:

> 'I call myself Grim, Thunn, Unn, and Ganglari, Helblindi, Har, Herian'…, continuing with a list that includes over 20 names that include meanings as diverse as 'the masked one', 'the high one', 'weak-eyed', 'father of victory', 'the one with a magic staff', and the god of 'terror', 'wind' and of 'men.'
>
> (Sturluson 1966)

Odin is also depicted as blind or one eyed, in reference to his loss of an eye in return for a drink from *Mimisbrunn*, a well that provides wisdom to those who drink from it. Drinking also plays a role in many of Odin's exploits, as he is famed for stealing the "mead of poetry" from the giant Suttung, and in the process also succeeds in seducing his daughter, the giantess Gunnlod, after entering Suttung's mountain abode Hnitbjorg in the form of a snake (Lindow 2002, p. 156).

Thor is also much more than a "God of War." Thor was described in a temple at Uppsala (according to Adam of Bremen, a medieval scholar) as "the most powerful... he commands the air, he governs thunder and lightning, winds, and rainstorms, fair weather and crops" (Page 1990, p. 46). In his role as a foe of the "enemies of the gods" Thor appears as a champion of all the gods and restores cosmic order. The examples include his slaying of the evil giant Hrungnir and destroying the "World Serpent" named Jormungand who stretched around the world, swimming deep in the mighty ocean that surrounded Yggdrasil, the "World Tree."

Thor emerges not as a subordinate and loyal son to Odin but as a bitter rival in many stories. For example, the hero Starkad was a favorite of Odin but was hated by Thor. Odin loved Starkad so much he gave him three lifetimes, but Thor undermined all of these lives by making Starkad commit evil deeds in each. Thor added to the insults to both Starkad and Odin by making sure Starkad suffered wounds in the battles and by preventing Starkad from remembering the lines of beautiful poetry created with Odin's gift of composition (Lindow 2002, p. 281). Thor and Odin also shared responsibility for those who die, with Odin entertaining the best fighting men killed in battle in his palace Valholl (after choosing the best of the dead with assistance from his demi-goddesses known as *valkyries*), and Thor taking responsibility for the "race of slaves" (Page 1990, p. 61).

The emphasis on violent conflict, deceit, stolen magical powers, and the ambiguous nature of both Thor and Odin's role in commanding the universe are key features of the Norse pantheon which mirror the conditions and mindset of the medieval Norse society. The creation of the present universe, according to the *Voluspa* or the Wise Woman's Prophesy, is also forged in conflict. A hint of an earlier, more peaceful, universe is described in which gods met in a place known as *Idavoll* and began a civilized life including a mysterious board game known as the "Game of the Gods" which determined the fate of the world and built houses, metal tools, and other implements. This peace was broken when three giant maids arrived from *Jotunheim*, who brought with them "corruption" which led to strife (Mackenzie 1912).

In what seems a common theme within the Norse, Chinese, and Greek cosmogony, Odin slew his ancestor Ymir and carried the body of Ymir to a great void known as Ginnungagap, where his body became the earth (or Midgard, a middle world of the Norse cosmos) and his head the heavens. As described in the Eddic poem *Grimminsmal*:

*From Ymir's flesh was the world fashioned,*
*And from his blood the sea.*
*Crags from his bones, trees from his hair,*
*And the vault of heaven from his skull.*

*And from his brows the genial gods*
*Made Midgard for mankind.*
*And from his brains were all those harsh storm clouds created.*
(Page 1990, p. 58)

## Celtic Stories

Written accounts of Celtic creation stories are less extensive than those for the Norse, as they arise from an oral tradition extant between 600 BC and AD 400, which is still celebrated in present-day Ireland and Wales. The Celts during this period included in their territory most of continental Europe, including Spain, Northern France, and a swathe of land extending all the way to Hungary. The medieval Church regarded much of the early Celtic lore as "pagan," and so our written record of this rich Celtic tradition is incomplete.

Fortunately, some early Irish writings exist, including the "Mythological Cycle" and the "Ulster Cycle" in which monks of twelfth century AD compiled some of the folk history of Ireland. The Welsh literature includes sources such as the "Four Branches of the Mabinogi," and the Tale of Culhwch and Owen, also belonging to the twelfth century AD.

The early Celts trace their origin to a series of invaders, which began with a kingdom under the mythical king Partholon, who came to Ireland after a flood. Later came the "People of the Goddess Danu" who brought to Ireland a set of four magical tools that endowed the land with a sacred power. These included the Stone of Fal, which could detect righteousness in rulers, the Spear of Lugh, which could not be defeated, the Cauldron of Daghda, which could satisfy all who fed from it, and the Sword of Nuadu, which could not be escaped. The first mythical people, known as the "Tuatha De Danann," worshipped a god known as Daghda, a male deity equipped with a magic club which could kill on one end and restore life with the other (Aldhouse-Green 1993). An additional trio of Gods embodied the arts of the time; Goibhniu was the god of the smithy, Dian Cecht was a god of healing, and Manannan was a sea-god protecting all the sailors and providing them with endless provisions. The final invaders were the Gaels who defeated the godly Tuatha De Dannan and drove them into a world beneath the Earth. The Gaels also were said to have made a pact with the goddesses of the land, Banbha, Fodla, and Eriu, the last of which became the namesake of Ireland, and who promised the Gaels that the land would be theirs for eternity.

The rich Celtic tradition offers a diverse retinue of heroes, kings, and warriors utilizing magical weapons, spells, and animals to achieve victory. Yet little record exists of Celtic answers to the question of where the Sun, the Earth, and sky came from. Gods from Ireland (Lugh) and Wales (Lleu Llaw Gyffes or "Bright one of the skillful hand") appear to be associated with the sky and the Sun, and some evidence of a sun cult exists in the form of statues carved with Sun symbols (Fig. 3.5). The Celtic people believed that special powers were associated with nearly every aspect of the natural

**Fig. 3.5** The lush Irish countryside near Knowth (above), an important megalithic site, shows the dense forests so central to the Celtic creation lore. Few writings from the ancient Celtic stories survive, but many of the surviving art forms such as these spiral patterns carved in rock at Newgrange attest to the complexity and beauty of this ancient culture (images courtesy of Jennifer Perry, reprinted with permission)

landscape. Mountains, trees, and springs were associated with spirits, kingship, and fertility. The sacred landscape of the Celts still exists, yet we are left to wonder at the full extent of the lore and myth of some of the long-lost people who lived there.

## Egyptian Creation Stories

The ancient Egyptian culture is the best studied of the ancient African cultures and the only one with extensive written literature. The ancient Egyptian creation literature derives from an earlier oral tradition which still exists throughout Africa. Separate creation stories were developed in each of the earliest religious centers of the emerging Egyptian state, and two of the stories that survive arose from the ancient cities of Heliopolis and Memphis.

The Egyptian Pharaohs of Heliopolis (now Cairo) traced their descent from the primordial god Atum, the "Lord of Heliopolis." From the primordial waters known as Nu arose the god Atum, containing within all the potential for creation of everything, and all the universe and its Gods were said to arise from Atum. Atum created from his own seed a pair of gods, known as Shu and Tefnut (associated with air and moisture) who gave birth to their children Geb and Nut who became the Earth and Sky. Geb and Nut in turn gave birth to Osirus and Isis. Osirus was the god associated with Kingship and sometimes was referred to as the "Lord of Everything," while his sister and consort Isis, was the "Goddess of Magic" and a powerful cult Goddess during the Roman period.

Osirus and Isis gave birth to the hawk-headed god Horus, who was said to be the direct ancestor of the first kings of Egypt. To complete the strong dualistic nature of the Egyptian Cosmology, Osirus and Isis were paired with Seth and Nephtys, who provide a counterpoint of disorder and trickery to the authority of Osirus. A summary of the origin of these Gods is shown in Fig. 3.6.

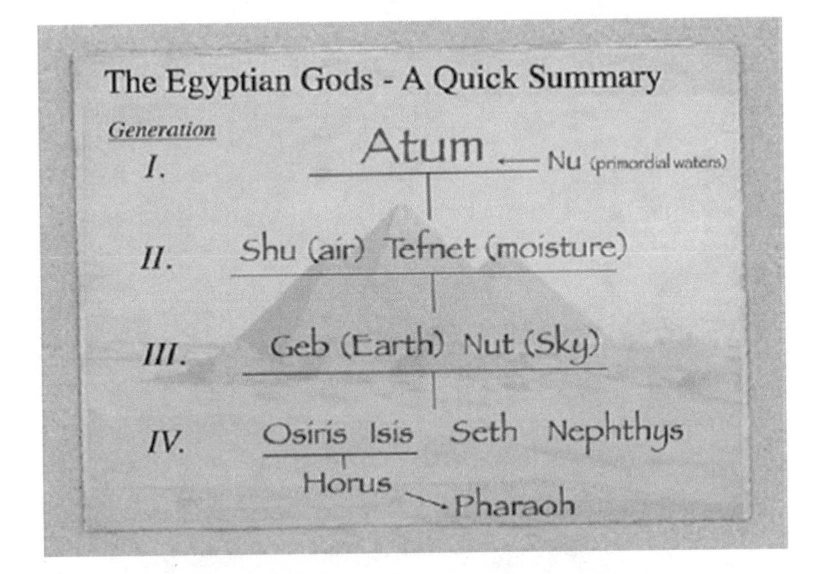


**Fig. 3.6** Summary of the lineage of the principle Egyptian Gods (figure by the author)

One of the most well-known Egyptian myths describes a battle between Osiris and Seth, in which Osiris is tricked into being chopped up into 28 pieces by Seth and sent down the Nile River before being restored to life by Isis. This fanciful story may also be connected with a description of the lunar phases (as the Moon is also "chopped" into 28 pieces by the lunar month). The emphasis on death and resurrection came naturally to the ancient Egyptians and gave rise to the practice of mummification and the *Book of the Dead* used to aid the departed in their journey into the afterworld (Krupp 1983).

The later kingdoms of Egypt, based in Thebes, offered an alternative creation myth in which Atum directly forms the universe out of his own body. The elemental forces known as *Ogdoad* were both made of Atum and formed Atum, according to the Theban priests. Once he arose Atum gave the primordial mound of the universe known as Ptah form and motion and also created the Sun, Re. It is said that all gods were manifest from Atum and were part of his body. The Sun god Re was his face, Ptah was his body, and further gods formed directly from Atum. In the kingdom of Memphis, Ptah (rather than Atum) is thought to have given birth to the gods with his tongue and heart, much as a spell or word which animates the universe (Blacker et al. 1975). Another account of creation has the gods arising from the tears of a creator god. In the papyrus entitled "The Book of Knowing how Re Came into Being, and of Overthrowing *Apepi*," we learn that humans arose from the tears of a creator god, known as *Neb er Djer*, or "Universal Lord" who is quoted in the papyrus as saying:

> Now after the creation of Shu and Tefnet I gathered together my limbs. I shed tears upon them. Mankind arose from the tears which came forth from my eye... The gods I created from my sweat, but mankind is from the tears of my eye.
>
> (Blacker et al. 1975, p. 31)

One very early pre-dynastic creation myth comes from the ancient city of Memphis. The Phoenix, or Bennu bird, is said to carry creation within its breath of life that emerges from its throat and gives rise to the gods from a primeval darkness. Herodotus in 500 BC describes a Phoenix-inspired funeral ritual, where a recently deceased bird is placed in a cocoon of incense and myrrh, and then is reborn in the Temple of the Sun. Further stories from Egypt attribute creation to a series of Gods and Goddess who arose from primordial waters in the form of frogs and snakes, and a variant of the Heliopolis creation of Geb and Nut posits that Nut the sky goddess took the form of a celestial cow, standing above the Earth with the stars and other celestial objects on her belly. Figure 3.7 shows a representation of the creation story of Hermopolis, with the Sun in a boat, traveling alongside the edge of the world. The sky above is represented as a ceiling supported on the four corners of the Earth, from which heavenly bodies are suspended by "cords" (Warren 1909).

**Fig. 3.7** A depiction of the Egyptian creation from Hermopolis, in which the world sits beneath a ceiling supported from the four corners of the world, where the heavenly bodies hang "from cords" from the sky (image from Warren 1909, p. 63)

Sun myths from Egypt are rich and diverse in metaphors. Various myths describe the travel of the Sun through the sky as that of a celestial calf being born each morning, or of a "glittering hawk" soaring to heaven, or of a newborn son of Nut who is born each morning and grows to maturity in the day and dies at night. Other Egyptian myths describe the Sun's motions arising from the actions of a giant scarab beetle, who would roll the Sun across the sky much as these insects roll small balls of dung across the Egyptian desert.

## Hindu Creation Stories

The Indian Hindu tradition has not one but an entire panoply of creation stories. The creation of the universe has many authors in the Hindu literature, just as the Rig Veda—the primary text for many of the stories. The diversity of creation stories within the Rig Veda spans the entire range of Creation stories found across human cultures. The Rig Veda includes *ex nihilo* creation from nothing—by way of a primordial creator Aditi. The female creator Aditi gives birth to the make creator Daksha, who in turn creates Aditi, in an interesting cycle of creation. As recounted in the Rig Veda:

> In the earliest age of the gods, existence was born from non-existence. After this the quarters of the sky, and the earth, were born from her who crouched with legs spread, and from the earth the quarters of the sky were born. From Aditi, Daksha was born, and from Daksha Aditi was born. After her were born the blessed gods, the kinsmen of immortality…
>
> Rig Veda - quoted in Doniger (2014)

Gods also play vital roles in creating the Universe. The ancient Vedic gods Indra (chief Vedic god, and god of the rain and storms), Varuna (god of the sky and celestial waters), and Agni (god of fire) themselves were born from an earlier creature known as Purusha. Purusha, a male primordial giant, releases Indra and Agni from his mouth, and Purusha's dismemberment creates many of the other gods, the earth, and the four groups of humans within the castes or *varnas*. This dismemberment of Purusha describes his mystical form, and how the earth came from his body—much as in the Chinese Pan Ku story:

> The Primeval Male has a thousand heads, a thousand eyes, a thousand feet. With three quarters the Male rose upwards, and one quarter of him still remains here. From this (quarter on earth) he spread out in all directions, into that which eats and that which does not eat. From the Male, the female was born, and from the female came the Male.
>
> When they divided the Primeval Male, his mouth became the Brahmin; his arms were made into the Raja; his thighs the people [Vaishya]; and from his feet the servants [Shudras] were born. The moon was born from his mind; from his eye the sun was born. From his navel the middle realm of space arose; from his head the sky evolved. From his two feet came the earth, and the quarters of the sky from his ear.
>
> Rig Veda - quoted in Doniger (2014)

Once Indra and the Earth were created, Indra brought order to the universe, created rivers and streams on Earth with his sacred axe, and ruled the skies from his throne in the clouds with his wife Indrani. Even today when we hear thunder, according to this myth, we are hearing the sounds of Indra battling demons in this sky world. Varuna, god of the sky and celestial waters, extends Indra's creation by imbuing in all creatures his life spirit into the universe:

> Varuna has extended the air above the trees; he has put strength in horses, milk in cows, will power in hearts, fire in waters, the sun in the heaven, and soma upon the mountain.
>
> Varuna poured out the leather-bag, opening downward, upon the heaven and the earth and the mid-region. Thereby does the lord of the whole creation moisten thoroughly the expanse of breath, as rain does the corn.
>
> Varuna - Rg Veda V. 85

The Vedic gods Indra and Varuna were supplanted by the gods Vishnu and Brahma, who play the key role in creation within Hinduism. Brahma is described in one passage in the Rig Veda as the creator who emerges from a cosmic egg:

> It appears that originally, i.e. before creation, there was nothing but darkness everywhere. Then water came into existence. From that sprang a fiery golden egg consisting of the two parts of the shell, viz. heaven and earth. Out of this arose Brahman, Creator of the universe, with the luminaries (the Sun and the Moon) as his eyes.
>
> RV, 10.190, 1-3, translated by Subbarayappa and Sarma

Vishnu, in his ten different incarnations, plays a key role in creating the universe or saving the world multiple times. The best known episode is Vishnu's creation of the universe while floating in the cosmic ocean. While dreaming, a lotus grew from his navel, and Brahma emerged from this lotus. Vishnu showed Brahma the three worlds to be created—all of which are created during the Vishnu dream. As described in the Rig Veda:

> When the three worlds were in darkness, Vishnu slept in the middle of the cosmic ocean. A lotus grew out of his navel. Brahma came to him and said, 'Tell me, who are you?' Vishnu replied, 'I am Vishnu, Creator of the Universe. All the worlds, and you yourself, are inside me.
>
> Rig Veda - quoted in Doniger (2014)

Vishnu also saves the earth several times, in multiple incarnations. In the first of these episodes, Vishnu in the form of a fish or *Matsya* rescues Manu the first man from the cosmic flood, and carried him to safety, to assure the survival of the Vedas. As the tortoise *Kurma*, Vishnu is able to assist the gods in recovering the *amrita* or the elixir of immortality which was lost in the Ocean of Milk which arose after the great flood. Vishnu and his tortoise shell provide the base upon which Mount Mandara pivoted as it churned the Ocean—with the giant serpent Vasuki as a rope. Indra pushed down upon Mount Mandara, and an assembled host of asuras, or demigods, pulled on both ends of the great serpent, and for a thousand years churned the ocean. This team effort finally producing the elixir of immortality and several other by products, such as *Dhanwantari*, a physician to the gods, *Lakshmi*, the goddess of fortune and beauty, *Chandra*, the moon, and several other key mythological figures (Roveda 2005) (Fig. 3.8).

**Fig. 3.8** The great sculptured frieze from Angkor Wat in Cambodia. The frieze depicts Vishnu in his incarnation as the tortoise *Kurma* assisting in the churning of the great Ocean of Milk. On both sides, legions of *asuras* pull on both sides of the giant serpent *Vasuki*, which is wrapped around Mount Mandara, with the god Indra pushing down from the top. After 1000 years the churning produced the *amrita*, or the elixir of immortality (figure from the author)

In his third incarnation as the Boar or *Varāha*, Viṣṇu recovered the earth after it had been thrown into the ocean by the demon Hiraṇyākṣa. In some artistic representations, such as the frieze at Mammalapuram in Tamil Nadu (see Fig. 3.9), the world is personified as the earth goddess Bhudevi, and Vishnu is shown lifting her carefully after rescuing her from sinking into the depths of the cosmic ocean.

**Fig. 3.9** Vishnu represented in stone at a temple in Mammalampuram, Tamil Nadu, India. Vishnu is shown in his incarnation as the boar *Varāha*, and is tenderly lifting the earth (personified as the earth goddess Bhudevi) from the depths of the cosmic ocean (figure by the author)

Vishnu rescues space itself in his fifth incarnation as a dwarf *Vāmana*. After the demon king Bali threatened the gods by trying to take over all of space, Vishnu appeared to him as a dwarf and offered to give him all of space, and allow the earth and sky to exist within just three steps of his small form. The demon Bali agreed, and Vishnu then began to take his three strides, growing to colossal size with his first two strides, and stretching across the sky. With the third step he covered the universe, and thereby contended Bali to the region outside (Sharma 2001).

## Mesopotamian Creation Stories

Like the Norse Greek and even Egyptian creation stories, the Mesopotamian universe began in a watery chaos, and the ruling god and order only came through familial combat. The Mesopotamian civilization includes the invention of cuneiform writing as one of its great legacies, which allows us to read the creation stories Gilgamesh and the Enuma Elish 5000 years after their writing. Within these accounts is a rich mythology that rivals any ancient culture for family strife, dangerous monsters and demons, and exciting battles in forming and managing the many-layered universe of the Mesopotamian cosmology.

In the earliest times, all that existed were primordial waters known as *Mammu* (similar to the Greek *chaos*) and a sea monster known as *Tiamat*. In the Middle Babylonian creation tale, the *Enuma Elish*, these waters were gods named *Apsu* (a freshwater god and male; sometimes called *Mummu*) and *Tiamat* (the Babylonian word for sea or salt water). These two watery gods form the first couple who together give birth to the remaining gods. The first lines of the Enuma Elish read:

> When heaven above had not yet been named
> Earth below had not yet been called by name,
> Apsu, the first, their progenitor,
> Creative Tiamat who bore them all,
> They mixed their waters together,
> When no meadow was matted, no canebreak was found.
> (Horowitz 1998, p. 109)

*Apsu* and *Tiamat* gave birth to pairs of gods such as *Lahmu* and *Lahama* (silt gods), and to *Anshar* and *Kisar* (the horizon), and to the sky god *Anu* and the Earth god *Ea*—thought to be the lord of the waters and keeper of all secret magical knowledge. *Ea* murders *Apsu* and with his wife *Damkina* gives birth to *Marduk*. *Tiamat* then attempts to kill her all of her offspring, and only *Marduk* dares to fight her and after a fierce battle, cuts her into half (Figs. 3.10 and 3.11). As described in the Babylonian "Epic of Creation":

> Bel (i.e. Marduk) rested, surveying the corpse, to divide the lump by a clever scheme.
> He split her into two like a dried fish, one half of her he set up and stretched out as the heavens…
> He crossed over the heavens, surveyed the celestial parts, And adjusted them to match the Apsu, Ea's abode.
> (Blacker et al. 1975, p. 55)

**Fig. 3.10** Marduk, the chief God of the Babylonians, who famously separated the Sky and the Earth from the primordial monster Tiamat, as shown in an image from an ancient cylinder seal (*image source*: http://commons.wikimedia.org/wiki/File:Marduk_and_pet.jpg)

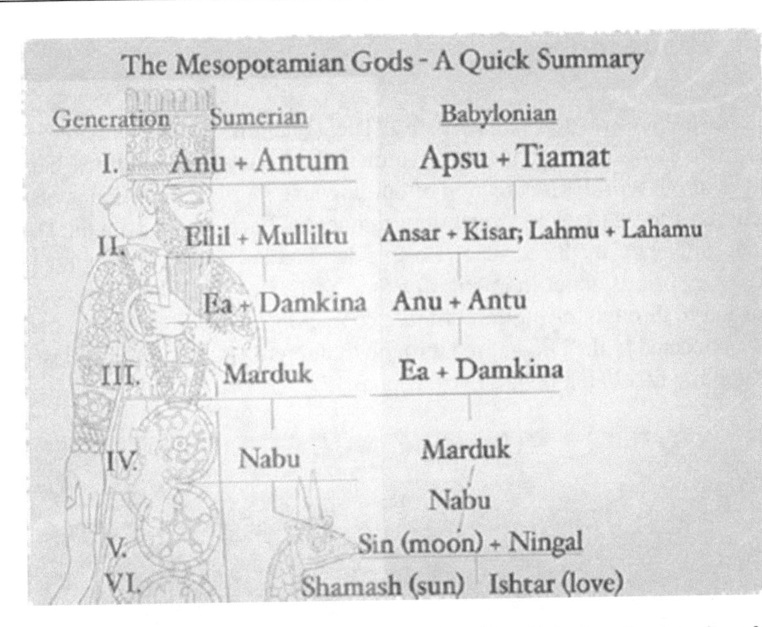


**Fig. 3.11** The Mesopotamian Gods are summarized above in the table which gives the genealogy from both the Sumerian account of creation (left) and the Babylonian creation described in Enuma Elish and others (sources: Blacker et al. 1975; McCall 1990)

With this violent raising of the sky, Marduk set in place the sky god *Anu,* the underworld god, *Ea,* and the Earth god *Enlil.* Marduk also set the Sun and the Moon and planets on their courses and brought order to the watery chaos of the early universe. Each celestial object was set on a different level of the sky and traveled across the waters of the sky world in celestial boats, as the embodiment of their ruling god. The Sun was the god Shamash, who was associated with the judgment of heaven and the Earth. The Moon was the god Sin, father of the goddess Ishtar.

Marduk's task of organizing the sky and setting the stars and moon in motion is described in the Enuma Elish:

> He (Marduk) revealed the year demarcating its segments; he set up three constellations for each of the twelve months… He made the luminary [the moon] shine forth, entrusting to [him] the night. He revealed him as a phenomenon of the night to make known the days. Monthly without ceasing, he made him form shapes with [his] crown: At the beginning of the month, rising over the land, you will shine with horns [i.e. be crescent shaped] to signify six days, and on the seventh day, a half crown.
>
> (Hetherington 1993, p. 49)

Marduk put the stars and planets into their places and their motions set the rhythm for the emergence of the Babylonian state, with its heavy emphasis on observational astronomy, planetary cycles, and predictive astrology.

The Mesopotamian creation story also contains significant astronomical details. The story describes particular constellations for marking the months and bands of the sky dedicated to each of the three gods. Anu, the sky god, rules over the equatorial sky (between +17 and −17 declination); Enlil, the god of winds and the Earth was given rule over the Northern sky (above +17 degrees declination); and Ea, the god of the underworld was given the Southern sky (South of −17° declination). Marduk kept for himself a special middle layer of the sky (probably where the planets were located) and held the "tablet of destinies" which revealed the fates of all men. From this Babylonian creation story came the astrological belief that the motions of the planets through the Zodiac could be used to read the fates of humans due to the influence of Marduk, and his associated sky gods, as they write the "Tablet of Destinies."

## Australian Creation Stories

The Aboriginal Australian creation comes from "The Dreamtime" and creation is known as the Dreaming. In the Dreaming, spirits not only exist outside of our space and time, but also determine much of the nature of our world. The creation of our universe in this conception is ongoing but determined by the actions of these timeless Dreamtime characters. Each action from the Dreamtime leaves an imprint on the universe, in the form of notable landmarks, and the movement of the earliest Dreamtime characters brings about the notion of the "Songline" in which groups of landmarks are connected in our world through the motion of these characters through the Aboriginal land. Each living being is also connected to the Dreamtime through their eternal nature, which exists independently before and after the life of a being on the Earth (Fig. 3.12).

**Fig. 3.12** Aboriginal Australian markings from the Carnarvon National Park, Queensland, Australia. These markings are related to the ancient practice of connecting with the Dreamtime, where spirits of departed ancestors can be reached (*figure source*: http://commons.wikimedia.org/wiki/File:Aboriginal_art_Carnarvon_Gorge.jpg)

One very compelling creation myth from Aboriginal Australia is told in the book "Stradbroke Dreamtime" a collection of tales recounted by Oodgeroo Bancroft from memories of her childhood growing up in an Aboriginal community (Oodgeroo and Bancroft 1993). In the creation tale, Rainbow Serpent awakened from a deep sleep, and emerged from the Earth's crust, pushing away any stones in the way. As she explored the slumbering earth, her body carved the canyons and rolling hills of the landscape. Rainbow Serpent called to the animals sleeping into the crust and brought them forth into this world. She first called to the frogs at the place of her emergence:

> The frogs were very slow to come from below the earth's crust, for their bellies were heavy with water which they had stored in their sleep. The Rainbow Serpent tickled their stomachs, and when the frogs laughed, the water ran all over the earth to fill the tracks of Rainbow Serpent's wanderings – and that is how the lakes and rivers formed.
>
> (Oodgeroo and Bancroft 1993, p. 61)

Soon afterward the other animals, including birds, kangaroos, wallaby, and emus, emerged and lived together. The Rainbow Serpent decided to reward the creatures that obeyed her laws with human form, and now the tribes of the Earth bear the totems of their original creatures. Those that broke her laws were turned into stone, in the form of mountains, to watch over the tribes of the Earth. Rainbow Serpent also set in place a law that prohibits each human tribe from eating the animals of their own totem, so that all would have food.

The Creation of the universe has been described in diverse ways by the different regional Aboriginal societies in Australia. In one account from Nullarbor people, the first day comes when the Great Spirit Baiame awakens the goddess Yhi. When she awakened her light bathed the earth, and as she walked across the land the light brought life to plants and animals. Yhi then created the Sun to mark the beginning of the day, and gave the Moon to her for her husband. As this celestial couple moves across the sky, they give birth to the myriads of stars we see.

The First Dawn of the world is described by some of the Aboriginal people as the work of the Magpies, which are a common bird in Australia. At the beginning, the sky was on top of the earth, and shut off all the light. Everyone had to crawl around in the darkness looking for food to eat. The Magpies, being smart birds, realized that if they worked together, they could lift the sky up. They picked the sky up and raised it with sticks, raising it higher until people could walk upright, and then as they stretched the sky it ripped open, revealing the First Dawn. The Magpies were overwhelmed with the beauty of this sunrise, and greeted it with their distinctive loud call, which they do to this day at dawn (Mountford 1969).

The light of the sky has also been seen by the Aboriginal people as arising from the discovery of fire by the creator spirits Jurumu and Mudati (associated with the eagle and forked-tail kite, respectively). As they were rubbing two sticks they burst into flames. One of the creator spirits, Purrukuparli, realized that this light could be used for light and to cook food, and used it to create two torches which he gave to his sister and his friend, who became the sun and moon (Bhathal 2010).

The moon, with its changing face, has been envisioned by one Aboriginal Australian group to be the Moon-man Japara, with a scarred face from his battle with the Creator spirit Purrukuparli. His insatiable appetite for mangrove crabs is so great that he eats them until he sickens, causing him to wither and die each month. Others describe the moon-man Japara as carrying a torch across the sky as he follows the sun-woman Wuriunpranilli, with her brighter torch. As told by the Tiwi Islanders, Taparra/Japara was punished for misdeeds with the wife of the Creator spirit Purrukuparli, and transformed into the moon as punishment. The Creator spirits son died from neglect during the affair, and from this misdeed came death in the Tiwi Islands, with the death of the moon each month as a reminder (Gilchrist and Holland 2009). The moon is also seen as a symbol of life, as its waxing has been associated with pregnancy by the Torres Strait Aboriginal people. Their names for the moon phases include the *dang mulpal*, or "unmarried" New Moon, the *tipi lag* or "married" First Quarter Moon. The Full and Third Quarter moon have the names *kaiza ipilaig* or "big one married" and *kaza laig* to "person with child or pregnant" (Bhathal 2010).

## Creation Stories from the Americas

The diverse creation stories of North and South America are impossible to summarize in a few paragraphs or a few pages. They reflect the great diversity of the people that inhabited the American continents. Many of the Native American conceptions of the universe involve multiple worlds, in which our world is one of a succession of universes, which either were created and destroyed by Gods (such as for the Aztec and Mayan conceptions) or in which the people migrate from one universe to the next, and emerge into the present world (such as for the Navajo and Pueblo creations of the universe). Other tribes describe a creator god (such as *Maheo* of the Hopi, the Great Spirit of the plains Sioux, and Viracocha of the Inca). In many cases, groups of animal spirits work together to create the present universe, and in some cases these "First People" live among us as present-day animals (such as in the Chumash, Abenaki, Luiseno creation tales).

One of the most recurrent motifs includes a disaster such as a flood extinguishing a previous race, and/or a succession of races of people and worlds existing before the present one.

For example, the Navajo creation story includes this account of the migration (and evolution) of the people from the earlier worlds to the present world:

> The First World, an island floating in the endless oceans, domed by the hogan-like sky, and secured by the four directions, was populated by the insect people. The insect people were quarrelsome, and the gods made a great flood which covered the first world, and they flew up to the hard sky; through a hole in the east they entered the second (blue) world. The Swallow people populated the Second World. After a while the Insects were expelled, and following a Locust South, emerged in the Third (yellow) world. In the third world, the Insect people discovered the Grasshopper people, who lived in holes in the ground. They were expelled and flew up to the sky, and were led to an opening in the west. The Fourth world was black and white. From these came First Man and First woman. They emerged into the Fifth world with the help of the gods and mysterious ceremonies, and then made the world of Dinehtah (Navajo land).
>
> (Navajo Creation Tale, from Leeming and Leeming 1995)

Native American cultures also typically recognize a "fourness" to the universe, with four directions, four colors, and four primary elements, often arising in the first moments of time. One example is the creation story of the Skidi Pawnee of the great American Plains:

> In the beginning there was space itself, Tirawahat. It was he who organized the gods in creation. He placed Sun in the east and Moon in the west. Evening Star was to be the mother and was placed in the West... Then Tirawahat made four other stars the supporters of the four corners of the world.
>
> He gave the wind, the thunder, the lightning, and the clouds to Evening Star, and they sang, and danced as Tirawahat created the earth. He dropped a pebble into the clouds and there was water. The earth supporters struck the water with their clubs and earth was formed. To populate the Earth, Evening star took Morning Star as her husband and they produced Mother of Humanity. Sun and Moon produced Father of Humanity. Evening star then made the sacred bundle, and the elements – clouds, wind, thunder and lightning, and taught the new people the sacred songs and dances they still do, calling it Thunder Ceremony.
>
> (Pawnee Creation Tale, from Leeming and Leeming 1995)

Our sample of creation tales from the Americas includes those of the Maya, Aztec, Hopi, Chumash, and Inuit, with an additional table summarizing several more creation stories from North American Native Americans.

## *Mayan Creation*

The Quiche Maya in their book *Popul Vuh* presents a mythic landscape of gods, demons, and animal spirits. The creation of the universe is described in the beginning of the Popul Vuh:

> There is not yet one person, one animal, bird, fish, crab, tree, rock, hollow, canyon, meadow, and forest. Only the sky alone is there; the face of the earth is not clear. Only the sea alone is pooled under all the sky; there is nothing whatever gathered together. It is at rest; not a single thing stirs...

The silence is broken by the Plumed Serpent, who arises along with the "Bearers" and the "Begetters" who are also in the water. The Plumed serpent, sometimes referred to as *Gucumatz*, in Mayan has the name *Q'ukumatz* or "Quetzal serpent," and is the primary creation god who is associated with the feathers of the Quetzal bird, a type of parrot (Fig. 3.13).

A discussion ensued in the darkness among "Plumed Serpent," "Heart of Sky," and *Hurrican*, the early sky and weather gods. When their thoughts joined and their words joined the Earth arose:

> And the earth arose because of them, it was simply their word that brought it forth. For the forming of the earth they said, "Earth." It arose suddenly, just like a cloud, like a mist, now forming, unfolding. Then the mountains were separated from the water, all at once the great mountains came forth.
>
> (Tedlock 1996, p. 65)

The Earth was soon populated with the Lords of the Underworld, and the first ancestors known as the "Hero Twins," Hunahpu and Xbalanque. Hunahpu and Xbalanque descended to the Maya underworld Xibalba, to avenge their father's death at the hands of the "Lords of Xibalba" who rule over the layers of underworld. After fantastic encounters with a diverse range of gods and demons, the twins defeat the Lords in the Mayan ballgame, resurrect their father, who arises to the Earth and is

**Fig. 3.13** The Mayan World Tree *(center)* and key figures in their creation story are depicted in this ink drawing of carvings at Palenque (from Antonio del Río (fl. 1786–1789). Illustrated by Ricardo Almendáriz. "Colección de Estampas Copiadas de las Figuras … de Chiapas, una de las del Reyno de Guatemala en la América Septentrional" [Palenque, Mexico: 1787]; image from Library of Congress online exhibit of the Kislak collection)

associated with the Maize god in some Mayan art. The twins themselves also arise and become celestial objects—the Sun and the Moon on the first day of the universe:

> And then the two boys ascended this way, here into the middle of the light, and they ascended straight into the sky, and the sun belongs to one and the moon to the other. When it became light within the sky, on the face of the earth, they were there in the sky.
>
> (Tedlock 1996, p. 141)

Additional characters in the Popul Vuh are also commemorated in the sky; Seven Macaw, associated with the Big Dipper, is said to have triggered a deluge that destroyed one of the four earlier races of men. The stars of the Big Dipper are appropriately at their lowest point in the sky during the hurricane season. The four hundred boys, who died fighting Seven Macaw's son Zipacna, were placed in the sky in the form of the Pleiades. The planet Venus is associated with the Hero Twin Hunahpu and is given this name when it appears in the morning sky.

The mythology of the Popul Vuh also contains within it a tale of the destruction of previous races of humans. In the first creation, various animals such as deer, jaguars, and serpents are created, who lack language, and instead of praising the gods and praying, only squawked and howled. The creators punished these early people and sent them into the wild to be hunted. A second race was created from clay but these beings fell apart when wet, and a third wooden race of people were made from wood but lacked souls and understanding. These people were destroyed from a flood and a vicious army of demons, and a few of the survivors became monkeys who remind us of the attempt at creation. The final and present race of people arises from the creation of people from yellow and white maize from the mountains of Paxil and Cayala. These people can give prayers of thanks and retain the knowledge needed to maintain the count of days needed for religious festivals and for managing the Mayan Calendar.

Cycles of destruction and creation are central to the Mayan sensibility, and are also built into the Mayan calendar, which posits a complete destruction and re-creation of the universe in regular intervals, of which the latest cycle began in 3013 BC, and is destined for destruction in AD 2012, which is discussed further in Chap. 4.

## Aztec Creation

The Aztec mythology arose from the Nahuatl speaking tribes that invaded present-day Mexico City in about AD 1300. The Aztec shared the Mayan belief in a Plumed Serpent creator, and also believed that a series of universes were tied together within a cycle of creation and destruction embodied in the form of the sacred calendar. The Aztec creation has some key differences from the Mayan creation, such as a more family based pantheon of creation gods (complete with the family rivalry so common in European mythology) and a shorter timeframe of cosmic cycles.

The original Aztec God was known as *Ometeotl* (God of Duality), who became the couple *Tonacatecuhtli* and *Tonacacihuatl*, and gave birth to four brothers, including *Quetzalcoatl* and *Tezcatlipoca*, who become "Plumed Serpent" and "Smoking Mirror." The fourth brother, *Huitzilopochtli*, is the patron god of the Aztecs. Quetzalcoatl (Plumed Serpent) is the primary creator god, while Tezcatlipoca (Smoking Mirror) is a god of conflict. Other key gods for the Aztec include *Tlaloc*, a god of rain and lightning, and his wife *Chalchiuhtlicue*, an Earth Goddess associated with rivers and water.

From the conflict between Quetzalcoatl and Tezcatlipoca comes a succession of Worlds or "Suns" in which the animals and people of the present world were created. Tezcatlipoca (Smoking Mirror) presided over the first world or "First Sun," in which a race of giants lived. Quetzalcoatl created a race of jaguars that devoured the giants, leaving their bones in the form of fossils, which are prevalent in the area around Mexico City. The Jaguar race dominated the "Second Sun" until Tezcatlipoca got his revenge by casting these jaguar creatures away with fierce winds, through an alliance with the weather god Tlaloc. The remnants of the Second Sun were thought to become monkeys, and Aztecs associate the Jaguar with the Big Dipper, perhaps to commemorate the Second Sun. The rain god Tlaloc ruled over this third world, the "Sun of Rain." A swift revenge followed when this third world was destroyed by a rain of fire (probably a reference to a volcanic eruption). The former inhabitants of the Third "Sun of Rain" are embodied today in the form of the turkey. Quetzalcoatl then formed an alliance with Tlaloc's wife, Chalchiuhtlicue (Jade Skirt), who ruled over the fourth world, the "Sun of Water," until a massive flood transformed the inhabitants into fish and made way for the "Fifth Sun" of the present universe.

To the Aztec, our present world of the "Fifth Sun" was created in a dramatic act of self-sacrifice, when the god Nanahuatzin jumped into a ritual sacrificial pyre. With this sacrifice the world was lit, and the god Nanahuatzin was resurrected as Tonatiuh, the Sun god of our universe (Taube 1993) (Fig. 3.14). The Fifth Sun is populated with people (the Aztecs) who were made from the bones of earlier races mixed with the blood of the Gods. These people were placed on the Earth with the seeds of white, black, yellow, and red maize and other edible plants derived from the four directional weather gods that form the basis of the Aztec diet.

## Hopi and Pueblo Creation

Like many Native American cultures, the Hopi and Pueblo civilizations offer exceptional levels of intricate detail in their creation myth, and like most of the Native American tales, they reflect the common emphasis on harmony with the land and sustainability. The main characters in the Hopi creation story are Tawa, or Taiowa, the Sun God, and Spider Woman, who are a form of Earth Goddess. As in both the Aztec and Egyptian creation stories, these early gods split themselves into multiple gods. Tawa divides into Tawa and Sotuknang, a new god and nephew of the creator, with the following charge:

> I have created you, the first power and instrument as a person, to carry out my plan for life in endless space. I am your Uncle. You are my nephew. Go now and lay out these universes in proper order so they may work harmoniously with one another according to my plan.
>
> (Waters and Fredericks 1969, p. 3)

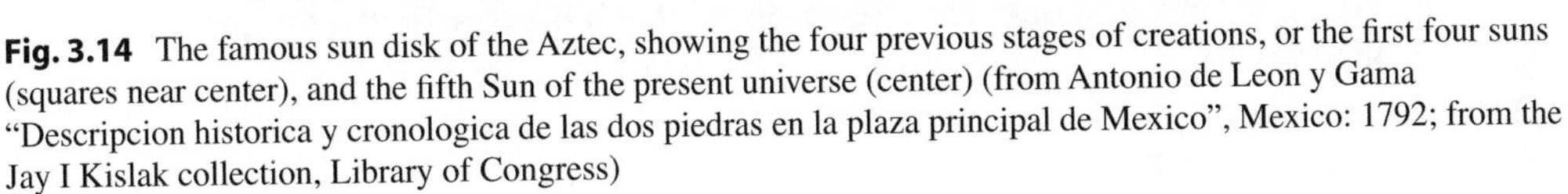

**Fig. 3.14** The famous sun disk of the Aztec, showing the four previous stages of creations, or the first four suns (squares near center), and the fifth Sun of the present universe (center) (from Antonio de Leon y Gama "Descripcion historica y cronologica de las dos piedras en la plaza principal de Mexico", Mexico: 1792; from the Jay I Kislak collection, Library of Congress)

Sotuknang begins by bringing together the elements of the universe into "nine universal kingdoms" in which two were for the creator gods and seven for the life to come. Sotuknang gathers endless space, which becomes water and divides it among the worlds and then brings the air to each world. Then Sotuknang gathered from the first world, and made Kokyangwuti, or Spider Woman. Spider Woman is given her charge to make life, and she begins by making a pair of gods from the Creation Song, a "cape of wisdom" and the Earth mixed with her saliva. These two creator gods are then charged with making the mountains and binding the Earth with the Sounds of the Creator. When they finished, they resided on the North and South Poles of the Earth "to keep the world properly rotating."

The animals and people of the world were created in a similar fashion. Together with Tawa, and Sotuknang, Spider Woman gathered the Earth in four colors: yellow, red, white, and black, and mixed them with saliva and covered them with the cape of wisdom and then sang to them to bring them life. As recounted in the Hopi Creation Myth:

> Spider Woman and Tawa had the Sacred Thought, which was of placing the world between the Up and the Down within the Four Corners in the void that was the Eternal Waters. The Thought became the first song: 'Father of all, Life and Light I am', sang Tawa. 'Mother of all, receiver of Light and weaver of Life I am' sang Spider Woman. 'My Thought is of creatures that fly in the Up and run in the Down and swim in the Eternal Waters', sang the God. 'May the thought live', intoned Spider Woman and she formed it of clay. Together the first gods placed a sacred blanket over the new beings and chanted the song of life. The beings stirred into Life.
>
> (Leeming and Leeming 1995, p. 124)

From this initial creation of life, Tawa and Spider Woman also carefully fashioned a man and woman, and breathed life into these First People. These First People and the animals multiplied, and were then led by Spider Woman through the four worlds, into the present-day world though a spider hole or *sipapu*. The *sipapu*, or Place of Emergence, is thought to be near the Grand Canyon. This emergence into the present world is symbolically re-created regularly in Hopi and Pueblo settlements, as people would rise from within their kivas on a ladder into the brilliant sun of the outside world (Williamson 1987, p. 64).

Spider Woman nurtured the people and taught them the arts of civilization, while Tawa made a daily journey over them, giving warmth and life to the people. The pattern of life and tradition, which included agriculture, kivas, and kinship, and family relations were set into place directly by Spider Woman, who taught them to the First People. The Great Serpent, who had the immense power to create the rain needed to preserve the crops, was also placed at the call of the people by Spider Woman, who then descended to her current home in the underworld. Within the Hopi creation tales are moral lessons on the need to live within the confines of the traditional culture, to respect elders, and to preserve the balance between people and animals and between agriculture and the environment.

## Chumash Creation

Like the Hopi, the Chumash people of Southern California believed a female creator goddess named Hutash took a direct role in the formation and nurturing of people. The story begins with a great flood that killed a number of the early spirits, the Nunashish, and transformed them into the mix of present-day animals that populate Southern California. The Sun God, Morning Star, the Moon, and Slo'w the Great Eagle all met to design new creatures to inhabit the new Earth after the deluge. The Luiseno (a neighboring tribe to the South) tell the tale that Coyote and Lizard were listening in on the proceedings, and Lizard beat Coyote in casting his handprint on the sacred White Rock which is why human hands are shaped something like Lizards. The Luiseno call the stars near the North Pole as the "Lizard's Hand" for this reason. After the gods designed the First People, they were placed on *Limuw* (now known as Santa Cruz Island, offshore of Ventura) by Hutash, the Earth Goddess, and creator.

As the people grew in numbers, their noise began to disturb Hutash and her husband Sky Snake (a god associated with the Milky Way and the first fire), and so Hutash made *Wishtoyo*, the Rainbow Bridge, to lead the people from the tallest mountain on Santa Cruz Island to the mountains of the mainland. As the people crossed, some looked down despite Hutash's warning, and fell. These First People who fell became Swordfish and Dolphin, who live in the ocean and protect the people when they are at sea by guiding them to fish and safety (Legends 2003).

## Inuit Creation

The Inuit creation stories are diverse, but often include the Raven as a central character who instigates the act of creation by bringing together other animals. In one story, raven convinced the animals to dive into the primeval waters to bring back dirt which he then gathered to form the Earth. Raven is also said to have acted something like a gardener, growing the First People and animals from vines and magic incantations.

One Inuit creation story involves an improbable conversion of two men into the first man and first woman who are also brother and sister. After the brother makes advances on the sister, they chase each other through the sky as the Sun (the angry sister carrying a bright torch) and the Moon (the brother running ahead with a fainter torch) (Leeming and Leeming 1995, p. 87). A slightly different version of this same story describes the creation of the Sun and the Moon as the result of sadness and guilt from this incestuous act:

> One night the brother sneaked over to the sister's bed and forced her to submit to him. The sister, wanting to find out who was violating her, marked the man's face with ashes. The sister awoke early… and saw the ashes on her brother's cheek. …
>
> She went home, picked up an Eskimo stone lamp, and started to walk in a circle counterclockwise. Slowly she rose up into the air. The people shouted at her to come back but she would not listen. The brother was riven with sorrow. He picked up his tool bag and started to walk in a circle counterclockwise. Soon he too rose into the air. As the sister continued rising into the sky, her lamp became warmer and warmer, giving out heat and light to the country. As the brother rose he became bright, but he never became hot. From that time on the moon, the brother, pursued the sun, his sister. But he never caught up with her, he was always behind. The heat from the sun comes from the Eskimo stone lamp, and the dark spots on the moon look like a man carrying a tool bag.
>
> (MacDonald et al. 1998, p. 277)

The explorer Knud Rasmussen recounted one Inuit creation story in which the universe began in darkness. The first parts of the Earth dropped down from the heavens. Soon after the soil, hills, and rocks fell together, the First People arose like bushes from them. The people soon decided they wanted dogs, and walked out with dog's harnesses and shouted out to the Earth. Soon dogs rose from tiny mounds. There was no Sun, but people could burn water, and did so inside their houses.

In a later part of the story, because there was not death, men grew very old, and overfilled the Earth until a huge flood came and drowned many of them, perhaps a warming against over-population:

> Two old women considered the predicament of people and said to each other: 'Let us be without the daylight if at the same time we can be without death.' As they said these words, the first light came along with death. As the first people died, the women learned how to bury them in a stone sleeping place. Along with death came the Sun the Moon and the stars, and the story goes that when people die, they 'go up to Heaven and grow luminous (perhaps becoming stars).'
>
> (MacDonald et al. 1998, p. 260)

## Further Native American Creation Tales

Many more Native American creation tales can be told, and rather than offering a long series of these tales, we summarize the essential elements of a selection of creation stories for tribes from across North America in Table 3.1. For each of these cultures we offer a very concise breakdown of the creation tale and list the culture, its approximate location in North America, and the primordial elements and actors of the creation. For each of the stories, the creation can also be described in the system posed by Leeming and Leeming, in their *Dictionary of Creation Myths*—this includes models such as "World Parent," "Emergence," "Earth Diving," and "Creation from Nothing," which cover a majority of the Native American stories.

**Table 3.1** Summary of a sampling of additional Native American creation stories, showing the story types and the agents essential for the creation of the world (Leeming and Leeming 1995)

| Tribe | Agent | Type | Representative quote |
|---|---|---|---|
| Apache (New Mexico) | Four people, Black wind, Black thunder | World parents/ creator | Let's make bones for the earth This way they made rocky mountains and rocks sticking out of the earth. These are the earth's bones… Then they made the sun so it travelled close over the earth from east to west |
| Jicarilla (New Mexico) | Buffalo, water, owl, mountain lion | Emergence | It took the help of the buffalo to get the people up the hole. They contributed their long straight horns to be used as a ladder, and it is because of the weight of the climbers that the buffalo's horns to this day are curved |
| Cherokee (SE USA; Florida) | Great spirit + water; water beetle, buzzard, Star Woman, turtle | Earth diving | Everything was covered by water… Desperate for more space, the animals sent Water Beetle out to explore. He dove to the bottom of the waters and came back with a bit of mud. The mud made earth-island, which the Great Spirit fastened to the rock with four rawhide cords stretching from the four sacred mountains of the four sacred directions |
| Eskimo/Inuit (Alaska/Yukon) | Raven; First Man with "pods" and "vines" | World parents/ creator | In the beginning first man was asleep in a pod. Eventually he broke out of the pod and fell to the ground fully grown. Suddenly Raven came to him and asked the man where he had come from. Many more men grew in the pods, and Raven made lots of animals and plants for them to eat and plant. Man was lonely; Raven went off and formed a beautiful figure out of clay and blew life into it with his wings; This was Woman |
| Navajo (Arizona) | Black God; insect people, swallow people, grasshopper people, First Man, and First Woman | Emergence | In the third world, the Insect people discovered the Grasshopper people, who lived in holes in the ground. They were expelled and flew up to the sky, and were led to an opening in the west. The Fourth world was black and white. From these came First Man and First Woman. They emerged into the Fifth world with the help of the gods and mysterious ceremonies, and then made the world of Dinehtah |
| Pawnee (great plains) | Tirawahat + space | Creation from nothing | In the beginning there was space itself, Tirawahat. It was he who organized the gods in creation. He placed Sun in the east and Moon in the west. Evening Star was to be the mother and was placed in the West… Then Tirawahat made four other stars the supporters of the four corners of the world. He gave the wind, the thunder, the lightning, and the clouds to Evening Star, and they sang, and danced as Tirawahat created the earth |

## References

Aldhouse-Green, M.J. 1993. *Celtic myths*. Austin: University of Texas Press.
Blacker, C., et al. 1975. *Ancient cosmologies*. London: Allen and Unwin.
Doniger, W. 2014. *On Hinduism*. Oxford: Oxford University Press.
Gilchrist, S., and A. Holland. 2009. *Shared sky*. National Gallery of Victoria: Melbourne.
Gummere, F.B. 1909. *The oldest English epic, Beowulf, Finnsburg, Waldere, Deor, Widsith, and the German Hildebrand*. New York: Macmillan.
Hetherington, N.S. 1993. *Encyclopedia of cosmology: Historical, philosophical, and scientific foundations of modern cosmology*. New York: Garland.
Horowitz, W. 1998. *Mesopotamian cosmic geography*. Winona Lake, IN: Eisenbrauns.
Jung, C.G. 1974. *Dreams*. Princeton, NJ: Princeton University Press.
Krupp, E.C. 1983. *Echoes of the ancient skies: The astronomy of lost civilizations*. New York: Harper & Row.
Leeming, D.A., and M.A. Leeming. 1995. *A dictionary of creation myths*. New York: Oxford University Press.
Legends I. 2003. Chumash Indians creation myth. From http://www.indianlegend.com/chumash/chumash_001.htm.
Lindow, J. 2002. *Norse mythology: A guide to the gods, heroes, rituals, and beliefs*. New York: Oxford University Press.
MacDonald, J., et al. 1998. *The Arctic sky: Inuit astronomy, star lore, and legend*. Toronto, ON: Royal Ontario Museum/ Nunavut Research Institute.
Mackenzie, D.A. 1912. *Teutonic myth and legend an introduction to the Eddas & Sagas, Beowulf, the Nibelungenlied, etc*. London: Gresham.
McCall, H. 1990. *Mesopotamian myths*. London: Published for the Trustees of the British Museum by British Museum Publications.
New World Encyclopedia. 2008. "Varuna". http://www.newworldencyclopedia.org/p/index.php?title=Varuna&oldid=795292.
Oodgeroo, N., and B. Bancroft. 1993. *Stradbroke dreamtime*. Pymble, NSW: Angus & Robertson.
Page, R.I. 1990. *Norse myths*. Austin, TX: University of Texas Press.
Roveda, Vittorio. 2005. *Images of the gods: Khmer mythology in Cambodia, Thailand and Laos*. Bangkok: River Books.
Sharma, A. 2001. *Classical Hindu Thought: An introduction*. Oxford: Oxford University Press.
Sturluson, S., S. Snorri, et al. (1966). The prose Edda: tales from Norse mythology. Berkeley, University of California Press.
Taube, K.A. 1993. *Aztec and Maya myths*. Austin: University of Texas Press.
Tedlock, D. 1996. *Popol vuh: The Mayan book of the dawn of life*. New York: Simon & Schuster.
Tian-gon Yuan, "Pangu Kaitian Pidi" from Tui Bei Quan Tu, 1820, copied by Wu-Yi Chao Xie, Circa 1900. Manuscript. Chinese Rare Book Collection, Asian Division, Library of Congress.
Warren, W.F. 1909. *The earliest cosmologies: The universe as pictured in thought by ancient Hebrews, Babylonians, Egyptians, Greeks, Iranians, and Indo-Aryans: A guidebook for beginners in the study of ancient literatures and religions*. New York/Cincinnati: Eaton & Mains; Jennings & Graham.
Waters, F., and O.W.B. Fredericks. 1969. *Book of the Hopi*. New York: Ballantine Books.
Williamson, R.A. 1987. *Living the sky: The cosmos of the American Indian*. Norman: University of Oklahoma Press.

# Chapter 4

# World Systems: Models of the Universe Throughout Time

*Material objects are of two kinds, atoms and compounds of atoms. The atoms themselves cannot be swamped by any force, for they are preserved indefinitely by their absolute solidity…*

Excerpt from the Lucretius, "The Nature of the Universe," relating theories of the earlier pre-Socratic thinker Democritus (ca 50 BC)

*Eyes and ears are bad witnesses for men if they have souls that cannot understand their language.*

Heraclitus, commenting on the "language of the cosmos," 500 BC

Cosmology is the study of the structure of the universe in its largest scales—it is an exploration of the structure of the Earth, planets, and stars and our place within this system. Through the centuries, in all continents, humans have developed world systems to describe the universe. Their description draws from the world known to each civilization—and often their worlds are made of water, or ice, or fire, or dirt. The stars, planets, and sky play a special role in nearly all of the world systems, since the immense power of these celestial objects is apparent to nearly all cultures, regardless of their time and place on the Earth. Some cultures describe the universe with layers, or nested spheres, or with spirits living in mountains, depending on the terrain and the aesthetic sensibility of the culture (Fig. 4.1). Each cosmology attempts to define the world in terms of what is precious and what is unique to each culture. In constructing a cosmology, a civilization gives both a model of the universe and a mirror to better understand what it means to be human.


B.E. Penprase, *The Power of Stars*, DOI 10.1007/978-3-319-52597-6_4

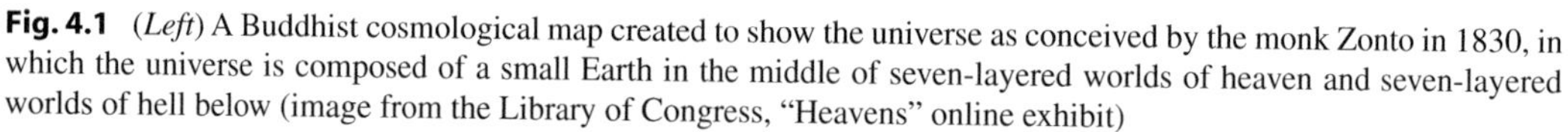

**Fig. 4.1** (*Left*) A Buddhist cosmological map created to show the universe as conceived by the monk Zonto in 1830, in which the universe is composed of a small Earth in the middle of seven-layered worlds of heaven and seven-layered worlds of hell below (image from the Library of Congress, "Heavens" online exhibit)

In this chapter we will consider models of the universe from many cultures, including those from the Egyptian, Babylonian, Native American, Chinese, and Incan civilizations.

The structure of ancient cosmological systems is often described in magnificent detail. In some cases, the descriptions are quantitative and include some of the best estimates from the science of the time of the scale of the Earth, the height of the sky, and the separations of the planets. In the case of ancient Chinese and Greek civilizations, these estimates were close to our modern values, as they were measured using precise observations of shadow lengths and geometric phenomena presented in lunar and solar eclipses.

In other cases, ancient cosmologies may appear to us as fantasies, such as the physical geography of 13 levels of underworlds (as in the case of Mayan and Babylonian cosmologies) or the existence of nested universes in which ours is the latest one in which people have emerged (such as in the Aztec, Mayan, and many Native American cosmologies). While some of these cosmologies may be difficult to understand literally, they all offer valuable insights into their cultures and are a credit to the depth and range of the human mind.

## Cosmology of the Egyptians

The Egyptian cosmology included both a physical description of the universe beyond the Earth and a metaphysical landscape of gods and forces to be navigated by the deceased in the afterlife. The journey to the universe beyond the Earth is undertaken at death, where the departed endures judgment, and then wanders to the outer edges of the Egyptian universe among the stars for eternity. To aid the departed traveler, many of the Egyptian tombs and coffin lids included detailed star maps and other celestial information.

The first part of the journey of the afterlife involves surviving judgment by Osirus. From the *Book of the Dead*, written in 1285 BC, we learn of the judgment after death in which the deceased has their heart weighed against the feather of *Ma'at* (the goddess of truth), in the presence of Osiris and a panel of divine judges from each province. Anubis, the dog headed god of the underworld, weighs the heart, and the results are dutifully recorded by Thoth, the god of writing. A light heart entitles the deceased soul to become part of Osiris' domain and a heavy heart (indicating the presence of bad deeds in life) condemns the deceased to be devoured by the hybrid crocodile and lion monster Amemit, waiting patiently at the right (Fig. 4.2).

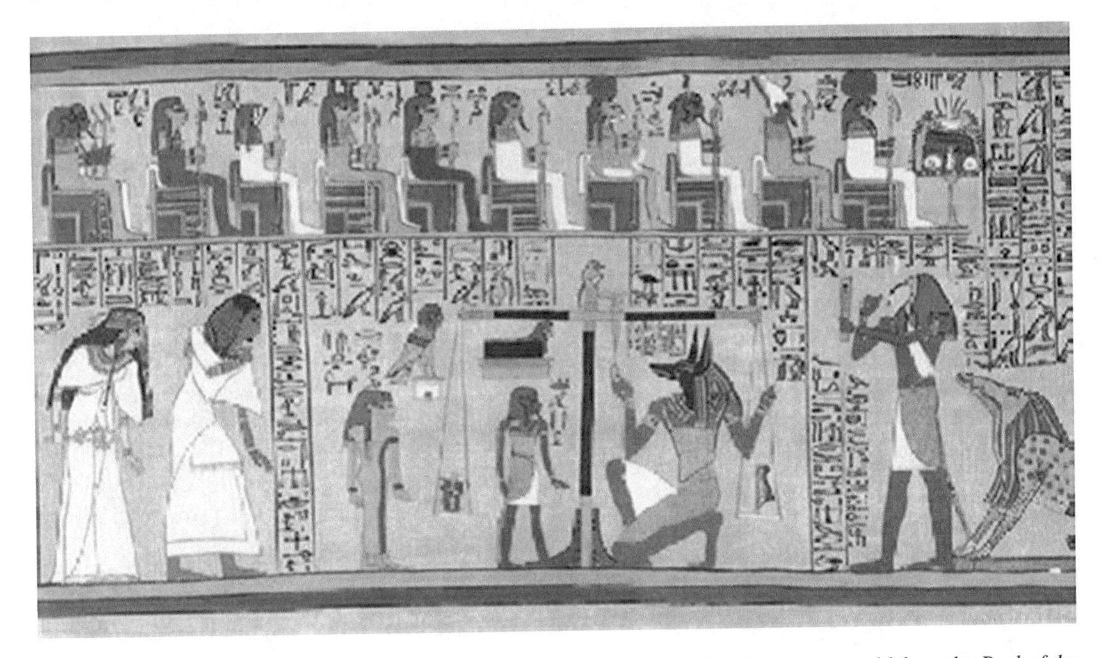

**Fig. 4.2** Panel of the judgment in the afterlife, as depicted in the Papyrus of Ani, known more widely as the *Book of the Dead*, which dates from the Nineteenth Dynasty of Egypt (1240 BCE). Osirus, followed by his consort Isis, is shown in the *lower left*, while a panel of judges view the proceedings below, where Anubis weighs the heart of the deceased against the Feather of Truth known as Ma'at (image from http://upload.wikimedia.org/wikipedia/commons/a/a6/Bookdead.jpg)

The Egyptian universe beyond the Earth includes the underworld, inhabited by Anubis, various monsters and demons, and the overworld of stars. Different versions of Egyptian myth disagree on the details, but wide agreement exists that the universe is animated by gods at all levels, which include the earth god Geb, supporting the sky god Shu, who in turn holds up the sky goddess Nut.

Additional details of Egyptian cosmology in some accounts include watery layers for an underworld, an upper watery ocean for the sky world, and a disk-shaped world surrounded by a "Great Circular Ocean" or the "Great Circuit." The watery layers are remnants of the primordial waters of creation, as provided by the first god Nu, who formed the universe (with Atum also created from the primordial water). At the edge of the disk earth were thought to be four peaks or posts on each of the cardinal directions that were in contact with the air. Some accounts even describe a celestial cow (perhaps an oblique reference to the goddess Nut) with four legs resting on each of these peaks at the corners of the Earth (Blacker et al. 1975) (Fig. 4.3).

The Nile River bisects the Earth in the Egyptian cosmos, and the sky is likewise bisected by a corresponding sky river (the Milky Way) that reflects the glow of the earth's Nile River. The Egyptian cosmos is one of the layers, with different waters bearing the disk-shaped earth, and providing passage to the planets, as they travel through the sky in boats each night. Figure 4.4 shows visualizations of the Egyptian universe, which includes both a set of layers for the Earth, and the underworld, and a watery sky world, as well as a representation of the Earth and the goddess Nut, with the Nile and sky rivers shown.

**Fig. 4.3** Schematic of Egyptian cosmology, in which the world is surrounded by a circular river and mountain range and was bisected by the Nile River below, and a celestial river above, with a watery iron-colored sky supported on the four corners of the Earth. The Sun was thought by early Egyptians to be carried on a boat around the circular river at night, to resume its westward journey in the morning (original figure by the author)

**Fig. 4.4** The Egyptian Cosmos—Ancient Egyptians could simultaneously entertain often mutually inconsistent pictures of the universe. The universe of the Egyptians includes layers of waters for underworlds, the Earth, and the sky, in which planets cross the sky in boats each night. The goddess Nut can be seen at right stretching over the disk-shaped earth, with orthogonal rivers in earth and sky visible (original drawing by Amarillys Rodriguez)

## Babylonian Cosmology

Like the Egyptians, the ancient Mesopotamians believed that the Earth was surrounded by a great ocean, which extended to a watery sky world, and underworld. The Mesopotamian cosmology matches the geography of the Tigris–Euphrates valley well, as the region is underlain with several freshwater springs, which form the basis of the watery underworld known as the "Apsu." The main triad of "earth gods," Anu, Enlil, and Ea, were associated with the elements of sky, earth, and water, respectively, and were considered to literally exist as three separate layers in the universe.

Some sources subdivide both upper and lower levels into additional layers. The sky world, ruled by Anu, was thought to be divided into separate layers for the various sky gods, or Igigi, who in some accounts number as many as 300. The middle layer was ruled and occupied by Marduk, below which was a third and lowest layer of sky for the stars and planets.

Marduk's throne was located in the central layer or "middle heaven" in this cosmology, to rule over the planets and to survey his domain of heaven and earth. Marduk's layer was described as a jeweled lapis lazuli palace with a lamp of "elmesu stone" and below Markuk's layer was a jasper layer for the stars:

> The upper heavens are of luludanitu-stone, of Anu.
> He settled the 300 Igigi therin.
> The middle heavens are of saggilmut-stone, of the Igigi.
> Bel sat therin on the lofty dais in the chamber of lapis lazuli, he lit a lamp of elmeu stone.
> The lower heavens are of jasper, of the stars.
> He drew the constellations of the gods thereon
>
> (Blacker et al. 1975, p. 58)

Marduk was in place to rule over the universe and people below could attempt to read his will in the motion of the planets across the "lower heavens of jasper" which housed the power of the stars for the Babylonians (Fig. 4.5).

**Fig. 4.5** Schematic of the Babylonian cosmology, featuring a layered, watery universe, which includes the Apsu (a fresh water sea surrounding and supporting the Earth), layers for each of the celestial objects, and layers for Marduk, Anu, and the Sky Gods, known as the Igigi (original figure by the author)

Below the heavens, on Earth, Marduk placed mankind, and below the Earth Ea ruled over the Apsu (fresh water) and the many under worlds. Locked below the Apsu were said to be 600 Anunnaki or underworld gods. The multilayer universe of Mesopotamia would therefore include at least three sky layers, a layer for earth and at least two underworlds (the Apsu and the home of the Anunnaki), bringing the total to at least six layers.

## Greek Cosmology

The Greek cosmological models evolved through a period from 500 BCE to the medieval period from the early musings of the "pre-Socratic" philosophers, such as Thales of Miletus, Pythagoras of Samos, and Empedocles of Acragas, to the classical philosophers Plato and Aristotle. The concept of a primordial, non-divine element underlying all things was a common theme of the Greek thought. For different thinkers this element could be water, fire, or a mixture of water, fire, air, and earth. Aristotle's metaphysics added to the mix of primordial elements the new ingredient called *aether*—the celestial medium of the planets and stars with their unchanging circular motions. Another common theme of Greek cosmological theory is an emphasis on geometric figures—either the "Platonic Solids" forming nested surfaces of the planets or of cylinders, bowls, and spheres of the pre-Socratic philosophers. The Greek shift from gods to inanimate elements and the shift from a living cosmos to a geometric cosmos set the stage for the explosion of the modern scientific worldview within Europe and beyond (Fig. 4.6).

### *Thales and the Milesian School*

One of the first and most famous of the "pre-Socratic" philosophers is Thales of Miletus, who was active in the period from 624 to 547 BC, founding the "Milesian school" which attempted to explain the physical nature of the universe. Thales is credited with saying that the most difficult task was to "know thyself" while the easiest was to "give advice." He is also famous for his practical skill when he was able to make a fortune from buying a large number of olive presses before a large harvest, proving that sometimes philosophers could become wealthy as well as wise.

Thales was able to explain the universe as a disk floating on a sea of water and suggested that water could be the primordial substance that formed the diverse range of materials from the Earth (Fig. 4.7). As described by the Greek historian Theophrastus in his work *Physical Opinions*:

> …warmth lives in moisture. Moreover the seeds as well as the nourishment of things are moist; and it is natural that what gives birth to a thing should also be its means of nourishment. Since water is the first principle of the moistness of anything, they declare it to be the first-principle of everything; hence they argue that the earth must rest upon water.
>
> (Wheelwright 1966, p. 50)

Thales' theory was also able to provide a non-supernatural explanation for earthquakes, explaining them as waves on a subterranean ocean. He reasoned that the Earth is spherical, and the Moon shines from the light reflected from the sun. Thales also posited that solar and lunar eclipses were formed by the Moon's shadow and the Earth's shadow, respectively. This knowledge was put to use when Thales was given credit for predicting an eclipse during a battle between the Medes and the Lydians on May 28, 585 BC, a stunning display of intellectual prowess that threw the battlefield into chaos and confusion.

Anaximander, a student of Thales, went further to hypothesize a different primordial substance he named *aperion*, which translates as "boundless" or "the eternal." This substance was thought to produce all the elements we see today such as fire, water, and earth. Anaximander explained thunder and lightning as a result of clouds being split in two. He imagined that enormous rings of fire, obscured by a foreground mist, circled the universe, and the planets and stars appeared through gaps of this mist (Fig. 4.8).

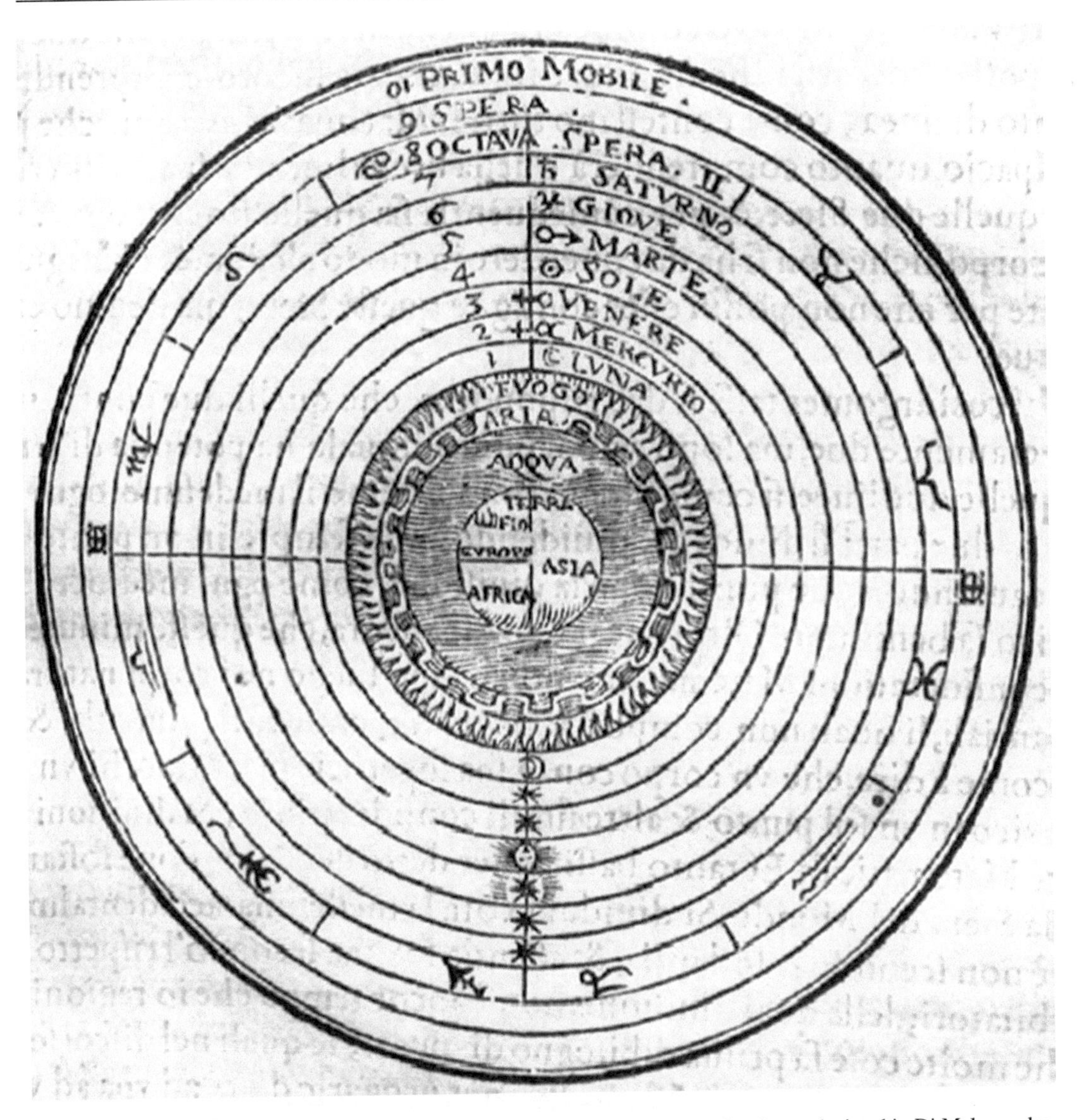


**Fig. 4.6** The European medieval model of the universe, as inherited from the Greeks, as depicted in Di Malessandro, "Del Mondo" (1579). The figure includes the medieval "T" map of the world (*center*), a depiction of the four primary elements from Greek thought (*Terra, Aqqua, Aria, Fuogo* or earth, water, air, and fire), which are depicted in the "sub-lunary sphere" below the remaining spheres for the planets, stars, and heaven or "Primo Mobile" (image courtesy of Claremont Colleges Special Collections)

Anaximander believed the earth "hangs free" as it is equidistant from everything, perhaps a precursor of the "celestial sphere" picture of the later Greeks. According to the classical work *Stromata*, attributed to Plutarch:

> Anaximander, an associate of Thales, said that the earth is cylindrical in shape, its depth being one third its breadth. And he said at the beginning of the world there separated itself out from the eternal a something capable of producing heat and cold. It took the form of a flame, surrounding the air that surrounds the earth, like the bark of a tree. This sphere became broken into parts, each of which was a different circle; which is how the sun, moon and stars were generated.
>
> (Wheelwright 1966, p. 57)

Another of the Milesians, Anaximines, propounded a pair of fundamental elements which included air and water and speculated on the transformations of elemental phases, noticing that air is able to

**Fig. 4.7** Prototypical cosmology of early Greece, showing the Earth surrounded by regions of air, water, and ether, with stars on a crystalline sphere. A circular ocean surrounds the disk-shaped earth. Beneath all is the mysterious underworld named Tartaros (original figure by the author)

**Fig. 4.8** A representation of Anaximander's universe, showing the Earth surrounded by a circular ocean, and space beyond earth filled by hoops of fire, which broke into parts to form the Sun, the Moon, and stars (original image by the author)

transform into rain and ice, depending on the amount of heat available. As described by Anaximines, "Air, when it condenses becomes successively wind, cloud, water, mud, earth and finally stone; in rarifying it becomes aether… then pure fire" (Wheelwright 1966, p. 60).

Anaximines also extended Thales' initial model of the Earth to include disks of the Sun and Moon suspended within a hemisphere above the cylindrical earth (Fig. 4.9). The stars formed a canopy over the earth like "a cap moving over one's head." The figures and separations of the heavenly bodies were prescribed by numerical ratios, with the Earth as a cylinder with diameter of three times its height, and the stars, Moon, and Sun in outer regions 3, 6, and 9 earth diameters away, respectively.

According to the ancient commentator Hippolytus, describing Anaximines' cosmology:

> The stars are made of exhalations which arose from the earth and became attenuated into fire as they ascended into the sky… the stars do not move under the earth at night, as others had supposed but travel along its outer edge, as a cap is turned on the head.
>
> (Wheelwright 1966, p. 63)

**Fig. 4.9** The universe as imagined by Anaximines is slightly modified from the universe of Anaximander, with a "cap" above the Earth which blocks our view of the fiery outer world except through small holes, which are the stars (original figure by the author)

## *Heraclitus*

A rival school of philosophy in Ephesus produced the philosopher Heraclitus, who developed a theory of the universe that posited fire as the prime element of the universe. According to Heraclitus, fire could be transformed into other elements through a "downward road" and elements could be converted back to fire via an "upward road." Heraclitus believed that change is the second fundamental element of the universe. According to Heraclitus, "panta rhei" or "everything changes."

Heraclitus presented an imaginative picture of the stars as "bowls turned with their hollow sides towards us, in which bright exhalations are collected and form flames." In Heraclitus' cosmology, the entire range of celestial phenomena is explained by distant celestial bowls turning to different angles to expose different quantities and types of fire. An eclipse is a view of the bowl from the bottom, while lunar phases result from the bowl slowly rotating, revealing different parts of the lunar "fire." According to the ancient commentator Diongenes:

> Eclipses of the sun and moon occur when the bowls are turned upwards. The monthly phases of the moon take place when as its bowl is gradually overturned. Day and night, months and seasons of the year are due to different exhalations.
>
> (Wheelwright 1966, p. 83)

## *Xenophanes*

The philosopher Xenophanes gave an explanation of all things deriving from assemblies of water and the Earth, in a vast universe similar to our modern conception in its allowance of multiple suns and planets. According to the commentator Hippolytus, discussing Xenophanes' philosophy:

> The sun is formed anew each day out of small particles of fire joined together. He (Xenophanes) declares the earth to be without limit downward, as against the view that it is entirely surrounded by air or aether. An unlimited number of suns and moons exist and all things come out of the earth.

Xenophanes was also intrigued by the discovery of seashells far inland, and fossils of fish in the mountains of Syracuse, and explained both with a form of plate tectonics, in which the Earth would rise and fall and the land would be turned into mud. According to Hippolytus:

> He (Xenophanes) says that at length the earth will sink into the sea and become mud again, at which time mankind will be destroyed and a new race will begin to be. A similar transformation he regards as taking place in all worlds.
>
> (Wheelwright 1966, p. 39)

## Empedocles

Empedocles (480–430 BCE), from Akragas, Sicily, provided one of the most elaborate of the pre-Socratic cosmologies. Empedocles believed that all compounds and substances were composed of four primary elements—air, fire, earth, and water—and that mixtures of these elements gave substances their apparently diverse and separate properties. Empedocles reasoned by analogy that all things can come to be in combination of these elements and they form the "root" of all material objects.

The diverse materials of the universe can be made from the four elements much as a painter can create a palette of colors from the three primary colors. For Empedocles, "earth, water, air and fire exist, and have always existed." Change happens when these elements are mixed or separated by the agent of the two primary forces of the universe—love and strife. From Empedocles we have the first physical unified theory, complete with a set of primary elements, and fundamental forces! The ancient commentator Theophrastus describes Empedocles' theory:

> Empedocles of Akragas considers the elements of things to be four: namely fire, air, water and earth. They produce change, he says, by combining and separating. What sets them in motion is Love and Strife, alternately the one bringing them into union and the other dissolving them.
>
> (Wheelwright 1966, p. 150)

Empedocles constructed a detailed geocentric model of the universe, with a spherical earth in the center, surrounded by the stars that fill a large sphere surrounding the Earth with their mixtures of fire and air. Within the sphere, the sun and planets are solid objects that reflect the fire from the outer universe. Empedocles believed that the Moon was composed of "solid air" much like hail and reflected the Sun's light. From ancient Greek writers we get a hint of Empedocles' ideas about the structure of the universe:

> The sky, he says, is formed by air being congealed by fire into crystalline form, and it embraces whatever is of the nature of fire or of air on either of the earth's hemispheres…
>
> Empedocles regards the moon as a mass of air congealed in the same way as hail, and contained within a sphere of fire.
>
> (Aetius and Plutarch, quoted in Wheelwright 1966)

## Pythagoras

The pure mathematical underpinning of the cosmos, so prominent in modern physics, was first imagined and taught by Pythagoras of Samos between 585 and 565 BCE. Pythagoras had a vision of a cosmos filled with music and harmony from the numerical properties of objects. This included the planets, that moved in such a manner that their motions produced a harmonious sound, a notion that greatly inspired Johannes Kepler over a thousand years later as he searched for the "Music of the Spheres." The integers had magic properties for the Pythagoreans and integers and their ratios provided the fundamental reality of the universe instead of more prosaic and material substances.

Even numbers were considered to be feminine, while odd numbers were considered to be masculine. Numbers were associated with perfect shapes (such as the sphere, the tetrahedron, and the cube), and also with particular attributes (marriage, the soul, justice). A perfect square (such as the number nine) or a sum or consecutive numbers (1 + 2 + 3 = 6) were considered particularly powerful, and given special importance to the Pythagoreans (Wilson 1995). A summary of the Pythagorean numbers and some of their attributes is shown in Fig. 4.10.

Pythagorean Numbers and their Meaning

| Number | Shape/form | Attribute |
|---|---|---|
| 1 | Point | Mind / Reason |
| 2 | Line | Femininity |
| 3 | Plane | Masculinity |
| 4 | Tetrahedron | Justice |
| 5 | Sphere | Marriage |
| 6 | Cube | The Soul |
| 7 | | Opportunity / Health |
| 8 | | Love and Understanding |
| 9 | Water | Justice |
| 10 | Cosmos (Tatractys) | Perfection |

**Fig. 4.10** A summary of some of the meanings of whole numbers in Pythagorean philosophy. For the Pythagoreans, whole numbers were the fundamental source of structure in the universe, and the individual whole numbers and their ratios determine the motions and separations of the planets, musical scales, and fundamental realities of human life (Wheelwright 1966; Lloyd 1971; Wilson 1995)

The most perfect of all the numbers for Pythagoras was the number 10, and therefore Pythagoras theorized that there had to be 10 heavenly bodies, even if it required inventing a new planet, known as the "counter-Earth" to give a 10th object. These 10 celestial objects include the spherical Earth, Moon, and Sun, the five known planets of the ancients, along with the "counter-Earth" and the "stars" which were counted as one object.

Pythagoras was able to surmise the possibility of a non-geocentric solar system 1100 years before Copernicus. Pythagoras believed that the Earth orbited about a "central fire," separate from the Sun, and that the counter-earth sometimes blocked this central fire to form eclipses (North 2008, p. 70) (Fig. 4.11). Pythagoras thought deeply about the dimensions of space and speculated upon the possibility of a vacuum and an infinite universe. Philolaus, a Pythagorean, stated "Nature in the universe was fitted together from the Non-limited and the Limiting, both the universe as a whole and everything in it" (Sambursky 1963, p. 29). Pythagoras also spoke broadly about the origin and fate of men's souls, which were thought to be immortal and reincarnated in a series of lives that could include both men and animals. These souls were also associated with the stars by the Pythagoreans. According to Aristotle, the prominent Pythagorean Alcmaeon of Croton made the connection between the soul and the stars explicit:

> First there is Alcmaeon of Croton who held that the soul must be a god because of the resemblance its always being in motion gives it to the heavenly bodies.
>
> (Aristotle and Lawson-Tancred 1986, p. 30)

**Fig. 4.11** The Pythagorean universe, which posited a system whereby the Earth together with the other planets was arrayed around a mythological "central fire" (image from Library of Congress from Thomas Wright, "A Synopsis of the Universe, or, the Visible World Epitomiz'd " (1742))

## *The Cosmologies of Plato, Aristotle, and Ptolemy*

Reality is not in the visible world but in the world of ideas
Plato, Timeus

The later Greek cosmologies from the work of Plato, Aristotle, and Ptolemy synthesized the work of pre-Socratic philosophers with newly acquired measurements and geometric calculations of the sizes and distances of the planets (Fig. 4.12). The result is an evolutionary refinement of cosmology beyond the earlier pre-Socratic musings, which was based on a small set of fundamental assumptions, while incorporating new constraints from observation. The Greek cosmologists employed many of the best features of the scientific method, despite the fact that by a modern perspective their geocentric cosmos was "wrong." The subsequent dominance of the cosmology from Aristotle and Ptolemy is a testament to both the quality of their initial system and the corrosive effects which come from a lack of free exchange of ideas when the line between science and religion is lost.

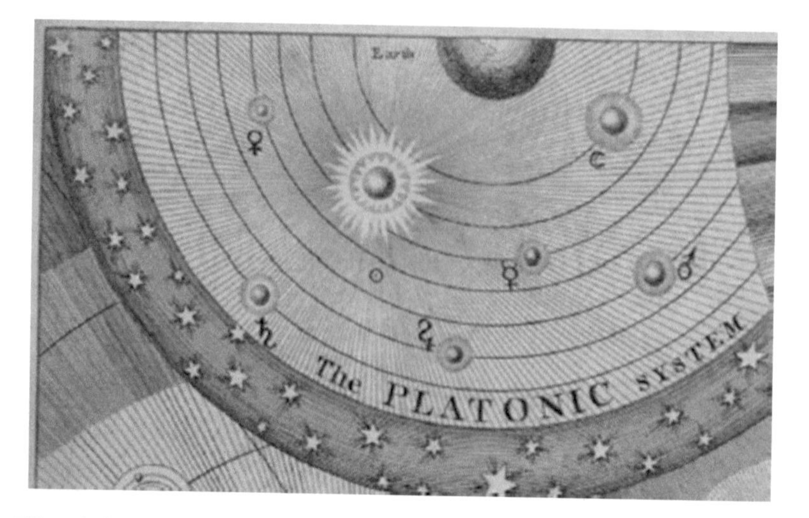


**Fig. 4.12** The "Platonic System" of nested spheres, described by Plato, Aristotle, and Ptolemy (image from Library of Congress from Thomas Wright, "A Synopsis of the Universe, or, the Visible World Epitomiz'd " (1742))

The pre-Socratics each invented some of the "ingredients" of Aristotle's cosmology. Pythagoras brought the notion of perfect eternal circular motion as being the only appropriate motions for the celestial objects. Empedocles brought the notion of fundamental elements of nature—earth, air, fire, and water—to which Aristotle added "ether," the fifth element, or quintessence, which comprised the celestial objects and the starry sphere surrounding the Earth. The notion of a spherical Earth, eclipses formed from shadows of the Earth and Moon, and the place of the Earth in the center of a concentric system had all been described from earlier philosophers. The Greek cosmologies after Plato were united in their emphasis on explaining planetary motions from combinations of pure circular motions. Such an explanation required a cosmos constructed of an elaborate system of nested spheres, in which each sphere would provide either the daily or annual motions and additional spheres were used to give slight wobbles and irregularities to the planetary motions.

The most basic of the planetary "phenomena" to explain was the retrograde loop, in which an outer planet appears to stop its normal "prograde" motion across the sky from east to west and then reverse course along a reverse or "retrograde" direction briefly before resuming its usual motion. We know now that these wobbles are simply due to the Earth passing an outer planet in its annual race around the Sun. For those who believed that planetary motions are uniform and circular, these "retrograde loops" were deeply disturbing. Fortunately, the Greeks were able to invent a system of nested spheres in which the combined motions of one forward moving and one backward moving sphere gives the observed looping behavior, known as an "epicycle" (Fig. 4.13).

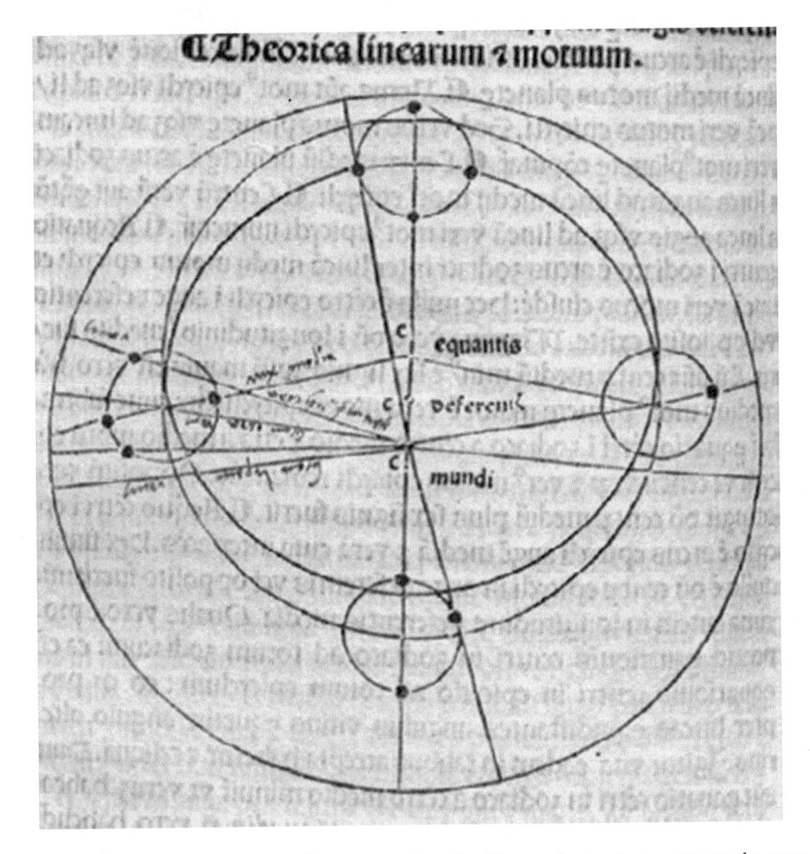


**Fig. 4.13** The apparent motions of an outer planet were described by ancient astronomers as the sum of a series of circular motions that include epicycles, equants, and a "deferent" which had the Earth as its center. This illustration from an early work shows the entire system for the planet Mars (from Regiomantanus, Johannes, "Sphaera mundi" (1482); image courtesy of Claremont Colleges Special Collections)

The philosopher Eudoxus developed the earliest of these celestial sphere and epicycle systems, in the period before 355 BCE. The Eudoxus cosmos described the motion of the Sun as the result of being attached to several connected spheres that surrounded the Earth. The motions of the Sun relative to the Earth could be explained as the combination of uniform circular motions produced by each sphere. One sphere, which produced the Sun's annual motion, was mounted with its axis offset by 23.5° from the Earth's pole and rotated once per year. A second sphere with an axis aligned with the Earth's north pole rotated daily to produce the Sun's apparent daily motion. A third solar sphere was added to account for some erroneous observational data indicating the non-regularity of the sunrise and sunset locations each year. Additional sets of spheres for the moon (3), each of the planets (4 per planet), and an outermost sphere for the stars (1) were added to give a total of approximately 26 spheres nested in a concentric system. Each set of outer spheres included pairs of spheres rotating in opposite directions, giving rise to the "figure 8" motions on the sky known by the Greeks as *a hippopede* or "horse fetter" (Evans 1998) (Fig. 4.14).

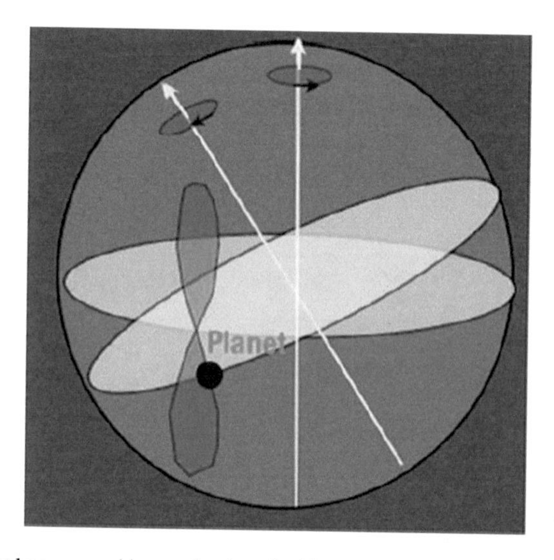


**Fig. 4.14** Motion of a planet known as a hippopede, described in the planetary system of Eudoxus. The hippopede is a "figure 8" type of motion resulting from combining two circular motions (original figure by the author)

The next refinement was provided by Aristotle's student and collaborator Callipus, who included extra spheres to make the planetary motions better match the improved naked eye astronomical observations of the time. Callipus included extra spheres for the Sun and Moon, five spheres for each of the inner planets and Mars, and added a set of four counter spheres for each planet to reverse these motions between one planet and the next. Callipus took the concept of nested spheres literally, and therefore had to reverse all of the motions from one planet to the next to give a "fresh start" to the spheres of the next outer planet. The total number of spheres for Callipus increased to 55 for the solar system and stars, but his system provided excellent agreement between the predicted motions and the planetary motions (Toulmin and Goodfield 1965).

## *Aristotle*

Aristotle (384–323 BCE) adopted the work of Callipus into his work *De Caelo* ("On the Heavens") to provide the definitive description of the cosmology of classical Greece. Aristotle also reasoned that the stars comprised the outermost sphere and that these stars can be proven to reside on the

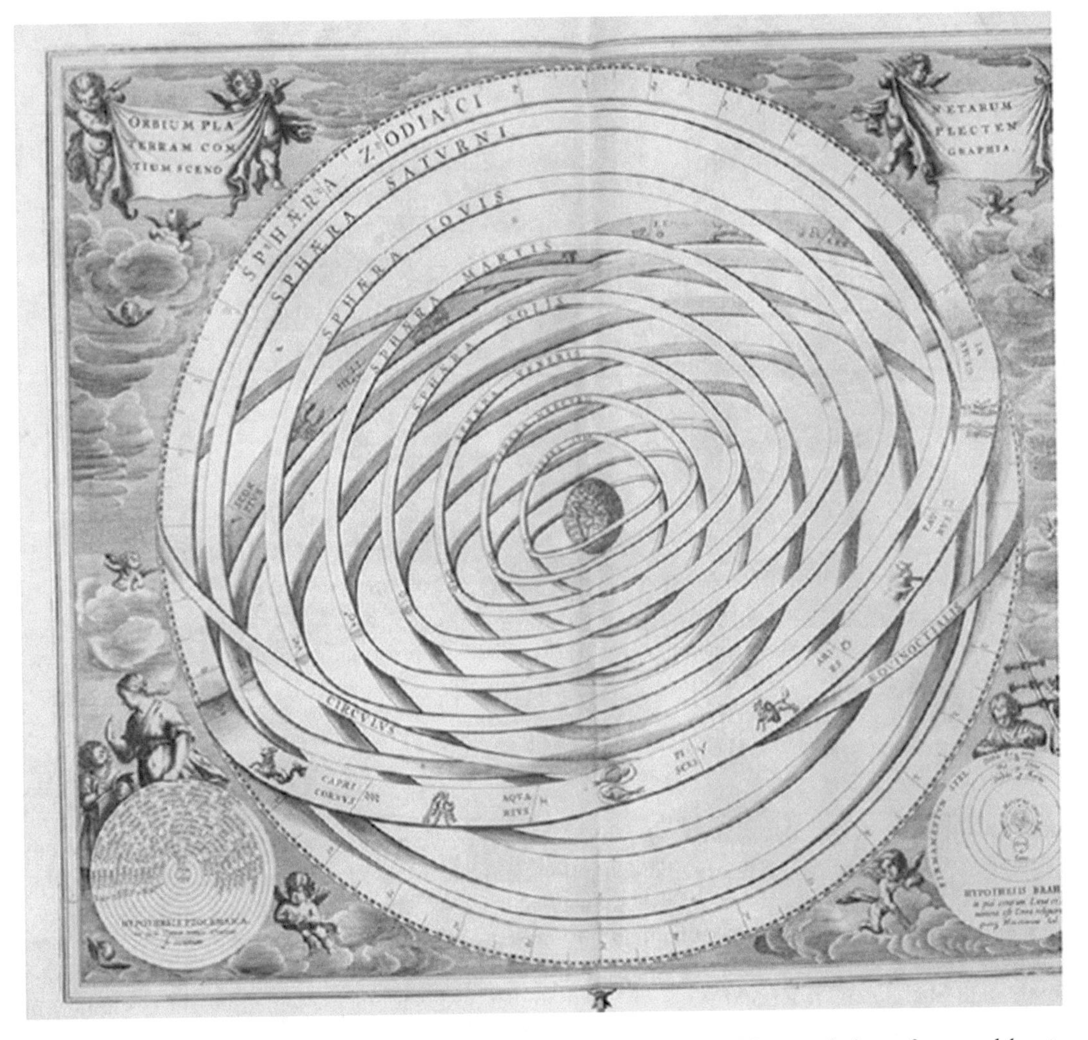


**Fig. 4.15** The cosmology of Aristotle, in which the nested spheres of the Sun, Moon, and planets form an elaborate system of perfect circular motion (image from Cellarius, 1500; image courtesy of the Huntington Library)

same surface, because they are not observed to move relative to one another (Fig. 4.15). In Aristotle's own words:

> It is established that the outermost revolution of the heavens is a simple movement and the swiftest of all, and that the movement of all other bodies is composite and relatively slow, for the reason that each is moving on its own circle with the reverse motion to that of the heavens… .
> We have next to show that the movement of the heaven is regular and not irregular. … circular movement, having no beginning or limit or middle in the direct sense of the words, has neither whence nor whither nor middle: for in time it is eternal, and in length it returns upon itself without a break.
> (Aristotle and Guthrie 1960, p. 45)

The Aristotelian cosmos was appealing to ancient and medieval minds, as it built on the notion of eternal circular motion and was consistent with a number of precepts both from Aristotle's metaphysics and those of the later Christian Church. This alliance of metaphysics and religion prevented major changes in the Aristotelian system for nearly 2000 years (Fig. 4.16).

**Fig. 4.16** The Aristotelian universe, which featured nested crystalline spheres for each of the planets, Sun and Moon, and a spherical earth in the center surrounded by a "sublunary sphere" of fire, air, and water (original image by the author)

## *Ptolemy*

After nearly 450 years of astronomical observations from Alexandria and the Greek world, many discrepancies between Aristotle's predicted motions and the observed planetary positions were found. Ptolemy in AD 150 was able to remove these discrepancies with his work *Almagest*, and the *Planetary Hypothesis*, which included the refinements of an "equant" and "eccentric," and incorporated new original observations from earlier astronomers such as Appolonius, Hipparcos, and Aristarchus, along with additional details from 800 years of astronomical observations contained within the library of Alexandria (Fig. 4.17).

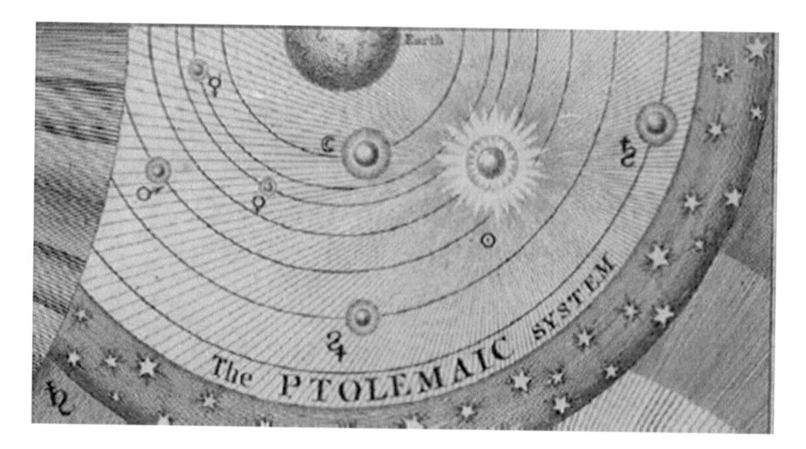


**Fig. 4.17** The Ptolemaic system, as illustrated by Wright (image from Library of Congress from Thomas Wright, "A Synopsis of the Universe, or, the Visible World Epitomiz'd" (1742))

Ptolemy kept uniform circular motions as a central theme, but stretched the hypothesis slightly by allowing the motions to be circular about a slightly non-concentric location (the eccentric) or to occur within a circular rotating center (the equant). By adding combinations of equants, epicycles, deferents, and eccentrics, Ptolemy could predict any conceivable motion of the planets as a sum of purely circular motions. With the freedom to add any number of these three elements, Ptolemy's system can produce arbitrarily complicated orbits—even a square orbit! (Toulmin and Goodfield 1965).

## Norse Cosmology

Like many ancient people, Norse imagined that a circular ocean surrounded the Earth. The creation song in Beowulf describes the Earth as "the bright plain which the water encircles" (Blacker et al. 1975), or in a different translation the Earth consists of "the fairest fields enfolded by water" (Gummere 1909).

The Norse also believed in a many-layered universe. The disk-shaped Earth and home for men was known as "Midgard" and contained a race of giants at its edge. Elves were thought to dwell within a magical place known as *Alfheim* which existed below the special enclave for gods and goddesses known as *Asgard*. Arching over all the world and connecting the various parts was the World Tree known as *Yggdrasil*, which contained near its roots a pair of magical springs known as Mimir's well (known as a source of wisdom) and a "well of fate." Below our home of Midgard were additional worlds known as *Svartalfheim* (home of the dark elves), *Hel* (land of the dead and origin of our word Hell), *Niffleheim* (home for dishonored dead), and *Muspelheim* (home of the fire demons). The entire structure of this universe is depicted in Fig. 4.18 and includes the looming Yggdrasil connecting and covering all the worlds.

**Fig. 4.18** The Norse cosmology describes the universe as a set of worlds connected by the World Tree Ygg Drasil. Ygg Drasil connects the Earth (Midgard) to a set of worlds through its roots and trunk. These worlds include the upper worlds of Asgard, Vanaheim, Alfheim, and Nidavellir, a ring world known as Jotunheim, and a set of lower worlds that include Muspelheim (a flame world), Niffleheim (an ice world), and the region of Hel, land of the dead (original figure by the author)

The world tree Yggdrasil provides an axis for the various levels of the Norse universe and connects them together with its roots, trunk, and branches. In some Norse poetry, the three roots of the great World Tree are said to reach into the worlds of Midgard, Jotunheim (home of the frost giants), and Hel. Some other poetry makes reference to "nine worlds" suggesting a vertical set of layers all connected by the World Tree, with Midgard, Hel, Nifleheim forming successively lower layers, and upper layers for two sets of gods—the Aesir gods (which include Thor, Odin, and Tyr) inhabiting Asgard and a separate set of gods known as the Vanir (which include Freyr, Freyja, and Njord,) inhabiting a layer known as Vanaheim (Blacker et al. 1975, p. 184). If one adds additional underworlds below Midgard to include Jotunheim, Svartalfheim, and Muspelheim, and the layer Alfheim above Midgard, the complete list of nine layers is obtained. The complete list of layers of the Norse universe is summarized in Table 4.1.

**Table 4.1** Summary of the many worlds in the Norse cosmology

| The Nine Norse Worlds |
|---|
| (1) Aesgard (home of the gods Odin, Thor, and Tyr) |
| (2) Vanaheim (home of fertility gods such as Freyr, Freyja, and Njord) |
| (3) Alfheim (home of elves) |
| (4) Midgard (home of human beings, and giants, surrounded by circular ocean) |
| (5) Jotunheim (home of the Frost Giants) |
| (6) Svatalheim (home of the dark elves) |
| (7) Hel (home of the dead, and ruled by its namesake god Hel) |
| (8) Nifleheim (home for the dishonored dead) |
| (9) Muspelheim (home of the fire demons) |

## Chinese Cosmology

The Chinese civilization brought to the world the first professional astronomical establishment, with bureaus recording positions of planets, comets, and supernovae for over 1000 years before European astronomers achieved comparable skill. The cosmology of the Chinese civilizations blended elements of Chinese Taoist and folk philosophy with accurate astronomical measurements of the Earth's diameter and the motions of the stars (Fig. 4.19).

Much of what we know about Chinese astronomical cosmology comes from the Zhou bi, a text from the Han dynasty (206–220 BC). This text gives a summary of three competing cosmological systems, which developed independently, and which had differing levels of validation from the "official" astronomers employed by the Emperor. This period is immediately after the "warring states" period and saw a reinstatement of a unified empire in 221 BC. This Warring States period, extending from the fifth century BC until the rise of the Qin Emperor in 221 BC, was an extremely dynamic period for Chinese philosophy, and helped catalyze the development of multiple models of cosmology (Cullen (1980); Cullen 2007)"Astronomy and Mathematics in Ancient China: The Zhou Bi Suan Jing. Cambridge Press). All of these cosmologies are less geometric than the Greek systems and align the emerging sciences of biology and physics to explain the sky—within a context of political power. For the Chinese the "Mandate of Heaven" gives a literal meaning to the "Power of Stars"—as the order or disorder within the skies confers legitimacy or cast doubts on the emperor's rule.

Three separate cosmologies competed during the period of the Han Dynasty, around 180 BC. Like the Greek cosmologists, the Chinese were keen observers of nature and explained the celestial sphere and motions of planets in terms of natural forces and elements free of the will of gods and other supernatural beings. The three cosmologies, known as the *kai thien* ("heavenly canopy"), the *hun thien* ("hen's egg), and the *hsuan yeh* ("infinite empty space"), are described separately below, in chronological order of their development.

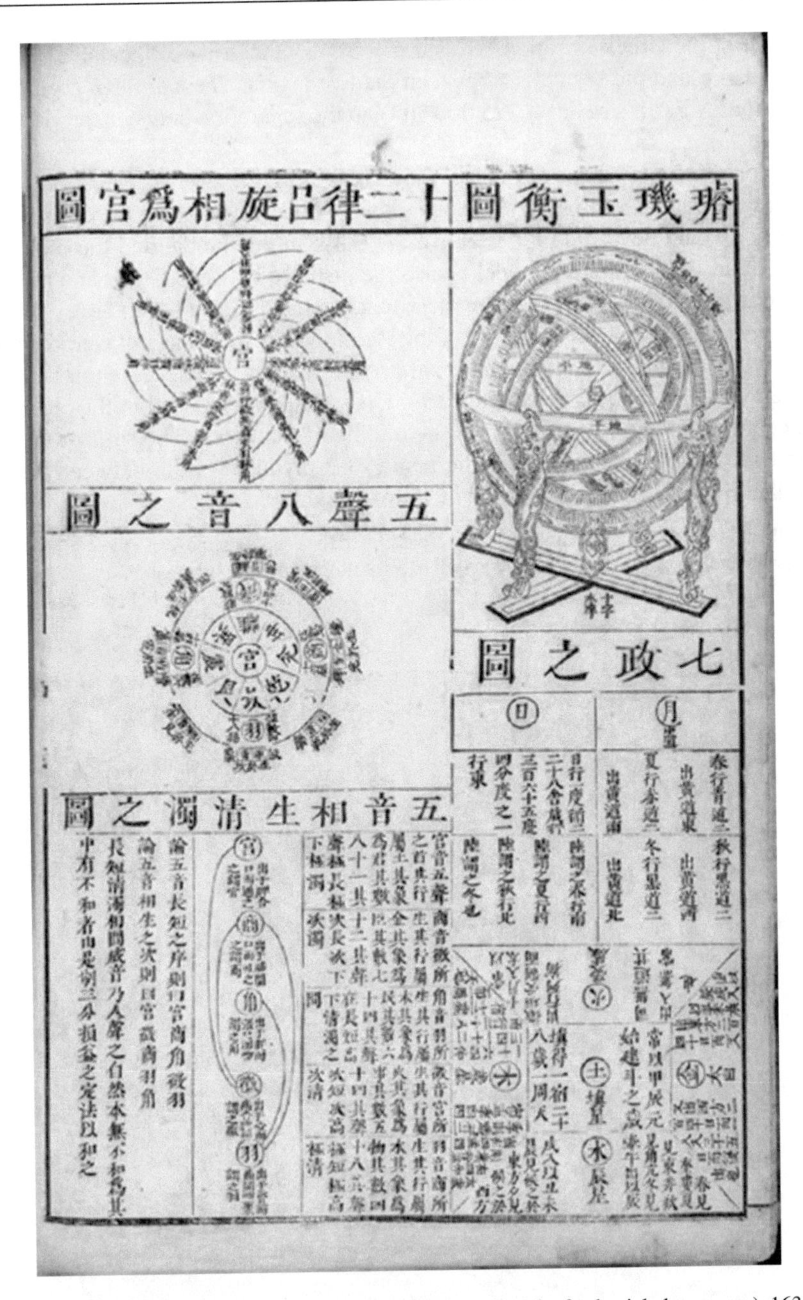


**Fig. 4.19** Image from Yu Tu Bei Koa Quan Shu (complete illustrated book of celestial phenomena), 1633, showing the five cardinal directions and five elements central to Chinese cosmology (image from Library of Congress, "Beginnings" online exhibit)

## Kai Thien ("Heavenly Canopy") Cosmology

The earliest system of Chinese cosmology described the universe as a set of nested bowls, with the heavens and earth forming concentric hemispheres, both resting on a circular ocean known as the "Great Trench" surrounding the Earth. The sky rotated about a celestial axis, much as the celestial sphere construct of the Greek astronomers, and precise distances were calculated for the diameters of the Earth and sky, and the separation between the two layers. The *Kai thien* cosmology included precise values for the Earth's radius—225,000 "li" and the separation between the Earth and Heaven, 80,000 "li."

If the commonly accepted value of the li of 0.57 km is used, these values are curiously large and at odds with more accurate measurements made possible by the data taken by the Chinese astronomical bureaus. The sky would be 27,000 miles above the Earth in this model and the radius of the Earth nearly 75,000 miles. Yet within the model is also the distance between China and the "edge" of the Earth, which was set at 6700 miles, a figure close to the actual radius of the Earth.

The *Kai thien* cosmology included the Taoist notion that the heavens were circular, and the earth square, which required the hemispherical earth to have a square base aligned with the four cardinal directions (Fig. 4.20). The constellation Ursa Major was placed in the center of the heavens and the country of China was placed in the middle of the Earth. The shift of the celestial axis by 36° from overhead was explained as the result of a mythical battle between Kung Kung and Chuan Hsu (the mythical fourth emperor of China), in which:

> Heaven's pillars broke, the bonds with the earth were ruptured, heaven leaned over to the north-west; hence the sun, moon, stars and planets were shifted, and earth became empty in the south-east.
>
> (Needham and Ronan 1980, p. 84)

**Fig. 4.20** The Kai Thien or Hemispherical Dome cosmology from China, in which the Earth sits atop a square base, around a watery sea, and in which the heavens enclose the Earth in a dome which contains the stars, and encloses the vapors of the atmosphere (original figure by the author)

This fanciful tale did include the actual observed tilt of the Earth's pole, which is responsible for seasons. However, the variation of weather within the seasons in ancient China was attributed to a combination of the arrival and departure of the *chi* from northern and southern skies, and also from motions of the celestial vault which were thought to move the pole of the sky further away from the Earth in summer than in winter.

## Hun Thien ("Hen's Egg") Cosmology

The *Hun thien* cosmology is associated with the Chinese philosopher Lo-his Hung, who died in 104 BC, and contains many similarities with the Greek notion of the celestial sphere. In the *Hun thien* cosmology, the Earth and sky are concentric spheres, with the Earth at the center "like the yoke of an egg." The ancient astronomical text "Commentary on the Armillary Sphere" dating from about AD 100 describes the basic features of the *Hun thien* model (Fig. 4.21):

> the heavens are like a hen's egg, and as round as a crossbow bullet. The earth is like the yolk of the egg and lies alone in the center. Inside the lower part of the heavens there is water. The heavens are supported by vapour, and the earth floats on the waters.
>
> (Needham and Ronan 1980, p. 89)

**Fig. 4.21** The *Hun thien* model, in which the Earth is "round like a crossbow bullet" and sits in the middle of celestial sphere "like the yolk in an egg" (original figure by the author)

The development of the *Hun thien* cosmology enabled the Chinese to construct actual models of the sky in the form of armillary spheres and to build sighting tubes with equatorial mountings. Within the cosmology is also a role for *qi* or *chi*, which can be translated as "vapor" but which also has the connotation of "energy." The *Hun thien* cosmology was the leading model for the Chinese astronomers for many centuries and helped provide a framework for interpreting astronomical observations, as well as new discoveries of physics and chemistry in early Chinese science.

## Hsuan Yeh ("Empty Space") Cosmology

Among the competing Chinese cosmologies, the *Hsuan yeh* cosmology presents a surprisingly modern conception of the universe in which space goes on forever, with the stars not arrayed in a crystalline sphere or canopy, but spreads throughout the dark and infinite space. From the commentary of the

Chin Shu (History of the Chin Dynasty) written by Ko Hung in AD 300, we can read some of the essential details of the *Hsuan Yeh* cosmology:

> The sun, the moon and the company of the stars float freely in empty space, moving or standing still, and all of them are nothing but condensed vapour. The seven luminaries sometimes appear and sometimes disappear, sometimes move forward and sometimes retrograde, seeming to follow each a different series of regularities... The speed of the luminaries depends on their individual natures, which shows they are not attached to anything, for if they were fastened to the body of heaven, this could not be so.
>
> (Blacker et al. 1975, p. 90)

Within all three of the Chinese cosmologies were elements of Chinese science and philosophy. The deep philosophical impact of Buddhism, Taoism, and Confucian thought played a role in developing these models of the universe. The concepts of yin and yang were assigned to the Earth and outer vapors, while the energy or *chi* of the universe was thought to produce a "Hard Wind" that moved the heavens, and the sky contained a "Great Emptiness" referred to by Taoist philosophers (Fig. 4.22). Within the Chinese cosmology was an implicit belief in balance, and therefore irregular motions of celestial objects or the appearance of new objects in the sky was interpreted to be a sign of imbalance within the universe. From the dedicated observations of the sky, and from the signs presented within the heavens, it was hoped that the emperor could maintain the "mandate of heaven" so central to the rule of law in the Chinese empire.

**Fig. 4.22** The Chinese "infinite empty space" cosmology described the Earth, the Moon, and "seven luminaries" suspended within space with the vapors or chi of the universe and provided a very modern conception of space nearly 2000 years ago (original figure by the author)

## Hindu and Buddhist Cosmology

The development of Hindu and Buddhist cosmology—as defined as a cosmological mapping of space and time—is intertwined and shares many common elements. Both Hindu and Buddhist cosmologies share the conception of vast expanses of time, expanding into timescales of millions and billions of years, and a notion of space that likewise is vast and perhaps infinite, containing multiple universes. The task of defining separate Hindu and Buddhist cosmologies is also complicated by the variations of both Hinduism and Buddhism across nations and regions. However there are some general notions that seem distinct within Buddhist and Hindu cosmologies, and common across most regions, which are outlined below.

## *Buddhist Cosmology*

The Buddhist temporal-spatial cosmology is inseparable from the emphasis within Buddhism on compassion and reincarnation. This necessarily creates a strong emphasis on the divide between the material and living universe. The material universe consists of the five elements (earth, water, fire, air, and space), which can be reassembled into living creatures, and form the vessels to contain life. The Buddhist terms for living beings (*sattva*) and the physical bodies of lifeforms (*bhajana*) refer to existence within the physical universe that we can see, but this universe in Buddhist cosmology is also only one of several multiverses, or world-systems (*cakravalas*), which can be "as numerous as the sands of the Ganges" (Ency. of Buddhism, p. 245). These physical worlds made of matter are suspended within space, which much like the ancient Greek conception of ether, is a separate element known as *akasa*. In addition to the vast numbers of physical worlds in the universe there are additional levels of the universe through which beings can travel with cycle of death and rebirth known as *samsara*. Typically the Buddhist cosmology recognizes six levels of the universe, and beings are able to move between these levels according to the *karma* or virtue that an individual expresses in each life. As such the symbol of *samsara*, the *bhavackra*, or "wheel of becoming" is often depicted with six parts that symbolize the six levels of the universe.

The Buddhist cosmology describes a multi-layered physical universe, in addition to the multiverse concept describe above. The word of humans (Manusya) sits between worlds above that the world of demigods (Asura), and the world of the gods (Deva). Below our world are a series of underworlds inhabited by demons and connected to the different vices. These underworlds include an animal realm (Tiryagyoni), a "hungry ghost" realm (Preta) and a region of unimaginable suffering which would be roughly equivalent to the Western notion of hell (Neraka). The demon Yama rules over these underworlds, and visual representations of the wheel of rebirth found in Tibetan art typically include images of Yama along with animals that symbolize the vices that would cause one to move into these underworlds. These include the pig, the snake, and the rooster, who are associated with the vices of ignorance, aversion, and attachment, respectively. Some depictions also include twelve embodiments of vice known as the *Nidanas*, which include lack of knowledge (avidya), grasping (upadana), and several others.

Buddhist art often depicts this cosmology to remind viewers of the dangers of the vices, and the redemptive power of the Buddha, often symbolized by the moon. The ultimate goal of breaking out of the cycle of birth, death, and rebirth through enlightenment is known as *nirvana*, which is a state of consciousness that transcends both time and space (Figs. 4.23 and 4.24).

## *Hindu Cosmology*

Assigning a single model of time and space to Hinduism is exceedingly difficult, as various texts provide different and often contradictory notions of the spatial nature of the universe. Like the Buddhist cosmology, the Hindu cosmology exists within a vast expanse of time and space, and is consistent with modern cosmology's notion of billions of years of cosmic time, and what a modern astrophysicist might call a multiverse. A single description of the Hindu cosmology is impossible as there are widely disparate accounts of the universe within Hindu literature, as well as a sense of multidimensionality to space and time that defies a simple explanation. In addition to the multidimensionality of space and time, the Hindu cosmology also speaks of repeated dissolutions of the universe, followed by renewals of the universe as the creator Brahma moves in and out of his "cosmic dream"—with the days and nights of Brahma providing each of cycles of creation and dissolution.

**Fig. 4.23** The *bhavacakra*, or wheel of becoming, representing the cycle of death and rebirth, along with the six partitions that correspond to the six levels of the Buddhist universe. The demo Yama is shown at top holding the wheel, and symbols of the Buddha along the outside represent the breaking of the cycle through enlightenment whereby one can attain *nirvana*

**Fig. 4.24** Hindu and Buddhist temples, with their central towers, often invoke a physical cosmology in which the universe is centered on Mount Meru, and in many cases are a microcosm of the Hindu universe. Here the central *vimanam* of the Hindu temple at Gangakhondapuram, Tamil Nadu (*left*), provides nested cornices and statues that provide recursive symmetries, while the Buddhist Borobadur temple in Indonesia (*right*) presents a model of the sacred mountain in which the devotee can climb to the top and recreate the journey of the Buddha, with scenes from Buddha's life represented in stone.

The earliest Vedic texts describe the universe as having three parts—the earth, the sky, and the heavens. Later Hindu texts, such as the Puranas, describe a physical structure of the universe—in which the locations of the various gods, demons, and spirits are specified in relation to each other. The Hindu cosmography typically includes three worlds or levels, in a system known as a *trilokas*. These worlds or *lokas* are each subdivided into seven sub-regions giving a 21 part universe. The seven-fold layering of the universe is described in the Bhagavata Purana, one of the 18 Puranas, which was compiled over 1200 years ago. The Bhagavata Purana describes this layered multiverse:

> Every universe is covered by seven layers — earth, water, fire, air, sky, the total energy and false ego — each ten times greater than the previous one. There are innumerable universes besides this one, and although they are unlimitedly large, they move about like atoms in You. Therefore You are called unlimited...
>
> Bhagavata Purana 6.16.37

The first universe or *loka* includes levels for the gods at the top, and various levels for the deities, semi-divine beings, sages and other celestial beings. Below these beings are layers for the atmosphere, the sun, moon and planets. In the middle is our *loka* which includes the earth, which sits above a series of underworlds. Below this *loka* are two other levels or *talas* that include serpents (*nagas*) and demons (*asuras*) on one level, and a bottom level that includes a form of hell known as *naraka* in which sinners are punished (Hindu Myths, p. 24).

In addition to this conception of a vertically layered universe, the Puranas provide a detailed description of a concentric universe that is largely horizontal, featuring seven concentric island continents (*dvipa*) separated by oceans (Matchett 2003). The central island continent of *Jambudvipa* is the earth, and beyond the earth is an ocean of milk. Additional worlds are separated from each other by oceans of sugar-cane juice, wine, clarified butter or ghee, curds, and freshwater. These worlds (in order of distance from the central world we inhabit) include *Plaksadvipa, Salmalidvipa, Kusadvipa, Krauneadvipa, Sakadvipa, and Puskaradvipa* (Vedic cosmography, p. 49). At the outer loop is a world known as *Lokaloka* which features forbidding mountains and marks the edge of the world of light and beyond which is darkness (Matchett 2003; in Flood, p. 140; chapter 7).

In the center of all is Mount Meru, and on its summit are the gods Brahma, Shiva and Vishnu, along with a retinue of minor gods, goddesses, asuras and demigods. Surrounding Mount Meru are guardians of the eight directions of space—that constitute the four cardinal directions and four intermediate directions. The Ganges river flows down from Mount Meru, and branches out to flow into our world near Bharat, India. In some accounts the entire assemblage of worlds is set within a cosmic egg, and various animals have mythological roles in either supporting the world or separating earth and sky. These mythological accounts include a one-footed goat named *Aja-ekapada* that holds earth and sky apart, a cosmic serpent named *Shesha* (sometimes referred to as a world snake or *Naga-padoha*) who supports the flat earth, resting upon the primordial tortoise known as *Akupara*, who in turn rests upon elephants, who arose from the early universe in the beginning of the universe (Dallapiccola 2003). Like Mount Meru, the base of the earth is supported by elephants arrayed in a pattern which supports the world at eight locations—one at each of the cardinal and semi-cardinal directions. Hindu and Buddhist temples both incorporate elements of this cosmography in their structure—with a large central mountain symbolizing Mount Meru, and guardian at each of the eight directions known as *Dikpalas*.

## Mayan Cosmology

The Ancient Maya believed that after the current universe was created, the role of human beings was to observe the stars, count the days, and honor the gods. While there are common cultural elements among all the Maya, a distinction does need to be made between the highland and lowland Maya, who lived in distinctly different environments. The highland or Quiche Maya inhabit present-day Guatemala and parts of El Salvador and Chiapas, Mexico, and their cosmology is described in the creation story

*Popul Vuh*. The lowland Maya, residing in present-day Yucatan peninsula of Mexico, Honduras, and Belize, constructed a similar cosmology and described some of their ritual practices in the book *Chilam Balam* or "the Book of the Jaguar Priest." We describe the main features of the Quiche Mayan cosmology below.

## The Quiche Classical Maya Cosmology

The spatial cosmology of the Maya is based on a layered model of the universe, somewhat reminiscent of the Mesopotamian universe, with a flat Earth surrounded by multiple "heavens" and multiple "underworlds" ruled by various gods and demons. Like the Native Americans, the Maya believed that the Earth was aligned at its edges with the four cardinal directions, with color symbolism for each of the directions—red, white, black, and yellow for east, north, west, and south, respectively. Each of the four corners was supported by gods known as *Bacabs*, while the sky was held up by four different species of trees at each corner. Through the middle of the Mayan universe rose a "world tree" which connected the nine layers of the underworld (known collectively as "Xibalba" the "place of awe"), the Earth, and the 13 layers of Heaven, and the various gods which ruled each layer (Coe 1999, p. 174) (Fig. 4.25).

**Fig. 4.25** The Mayan universe, in which sky gods and underworld gods surround a world in which the cycles of the universe over centuries and millennia require constant awareness and ritual. The figure includes underworld gods below playing the Mayan ball game, a world tree at center that ties together underworlds, the Earth and the sky, and a series of sky gods that helped inspire the Maya to practice advanced forms of astronomy and timekeeping (figure from National Geographic, reprinted with permission)

Number symbolism pervaded the Mayan cosmology, as the Maya were experts at mathematics. For this reason the universe was thought to be grouped into the visible earth and two invisible layers (the underworld Xibalba and a sky world). The upper or celestial world was thought to be ruled by the 13 Gods of the Upper world or *Oxlahuntiku* (based on the Mayan word oxlahun for 13, and ku for "god"). The underworld known as Xibalba had nine layers, over each ruled one of the nine Underworld gods known as the *Bolontiku* (Sharer and Traxler 2006, p. 534). The layers of the universe were also the home of the souls of the departed; those who suffer violent deaths were placed in one of the upper levels depending on the exact nature of their death, while those who suffered a peaceful death on the Earth moved to one of the nine lower layers. The nine lower levels of the Xibalba underworld were mirrored in nearly all of the Mayan step pyramids, which usually contained nine large steps, to create a representation in stone of the Mayan cosmology (Littleton 2002).

## Incan Cosmology

The Inca creation story tells the story of the descent of the ruler or Inti from the god Viracocha, who was the son of the Sun God himself. The creation of the first people was believed to have occurred in the region of Lake Titicaca in present-day Bolivia. The Sun God gave instructions on how to divide the city of Cuzco into two parts corresponding to a celestial male half and an earthly female half (Aveni 1989). Further subdivisions of the city into ceremonial shrines or *huacas* give an invisible landscape of sacred lines known as *ceques* which mirror the structure of the universe (Fig. 4.26).

**Fig. 4.26** The Inca universe, showing the center of the universe in the Inca capital of Cuzco, with the radiating lines of the ceque system, surrounded by the mountains of the Andes, bisected by the Vilcanota river, which extends both across the Earth and sky (figure from National Geographic Magazine, reprinted with permission)

Regular sacrifices of food and regular rituals at the shrines or huacas were thought to nourish the life forces of the universe and also linked people together in a web of ritual devotion which incorporated the Incan concepts of both the structure of time and space. Some authors have written that the network of lines and shrines is similar to the structure of the Incan *quipu*, a woven mesh device used for recording information, in which the position and pattern of knots convey numbers and words.

The intermeshing of time and space for the Inca is also found in their view of the night sky. The Milky Way appears above the Earth, as a celestial counterpart to the Inca's own Vilcanota River. The river on the Earth carried the water to the end of the Earth, where it was brought to the sky to complete a cosmic circuit of water as it moved across the sky in the Milky Way. Within the celestial river of the Milky Way appears several of the most significant Incan constellations, the Llama, Tinamou, and others, who appear in the dark clouds of the Southern Milky Way. The fertility of the land was embodied in these dark clouds, as the arrival of the life-giving rain coincided with the appearance of particular dark clouds and constellations. The Milky Way also served much as the hour hand of a cosmic clock in its annual rotation within the sky and helped divide the Incan year into four parts, much as the ceque lines divided the Inca cities into four parts.

The Inca practice of using the stars for weather prediction still survives in contemporary communities near Cuzco, as recounted by Father Jorge Lira, living in a contemporary Quechua village:

> If the stars of the sky appear bright and beautiful, everything will be good, materially and spiritually. If, in the Milky Way, there is an accentuation of the dark areas… it will be a year of pestilence and death.
>
> (Urton 1981, p. 173)

Observations of the star cluster of the Pleiades, known as *Collca* (the "storehouse"), were used to predict the future, since it was believed that the number of stars and their sharpness in the sky will increase during a good harvest year. A study of the appearance of the cluster was also used to reveal the best time to plant. Like the Chinese, Inca believed strongly that the sky can reveal the future and that the forces of nature reveal themselves in the transient variations of the night sky.

## Native American Cosmology

### *Hopi and Pueblo Cosmology*

The structure of the Hopi universe includes a succession of worlds, where people have emerged to the present world. The Hopi cosmology is suggestive of a "multiverse" in which our present world is the fourth. The first world was known as *Tokpela*, or Endless Space, and was associated with the color yellow, the mineral gold, and the westerly direction. In this world, people lived with the ant people until fire and volcanoes destroyed the first world. The second world was known as *Tokpa*, or Dark Midnight, and was associated with the color blue, the mineral silver, and the southern direction. The second world was destroyed when the two gods at the poles were ordered to leave their posts and enabled the Earth to spin out of control and freeze into solid ice. The third world was known as *Kuskurza*, an ancient word with no good translation, and was associated with the color red, the mineral of copper, and the easterly direction. A flood destroyed the third world, and the Creator led the people away to safety in rafts. The fourth world was known as *Tuwaqachi*, or World Complete, which was associated with the color of yellowish/white, the mineral *sikyapala*, a yellowish compound (perhaps sulfur or uranium), and the northern direction. This was the final world of the Hopi and their legends also describe their migrations within this world (Waters and Fredericks 1969).

Within the Hopi cosmology, the fundamental elements of the universe are enumerated. The first act of creation involved bringing together earth, water, and air, is reminiscent of some of the pre-Socratic thinking in its stress on fundamental elements being the primary reality of the universe. Within each world additional elements, *gold, silver, copper, and sikyapala* appear, along with the ordering principles of harmony, and named gods to hold the world's spin in place. The Hopi cosmology brings together these elements and forces to maintain a steady balance or equilibrium against destruction from fire, water, and cold.

## Navajo Cosmology

The cosmology of the Navajo centers on the creation of the universe by Black God, who placed First Man and First Woman together and taught them the way of life for the *Dineh* (*Dineh* literally means "the people" in the Navajo language). The creation story *Dine Bahanme* gives details of the gods and features of the Navajo universe. Within the story is the pattern of order for the universe, which is mirrored in the Hogan, the home of a Navajo family, and in the ritual sand painting and healing ceremonies that recreate the larger order of the universe in a smaller scale (Fig. 4.27).

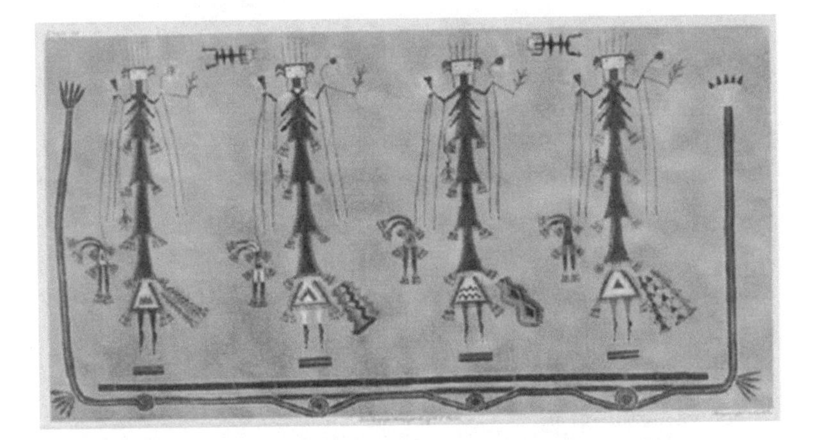

**Fig. 4.27** Navajo sand painting, showing four spirits used in a 9-day healing ceremonies. The emphasis on "fourness" is a fundamental aspect of Navajo cosmology (image source: http://commons.wikimedia.org/wiki/File:Navajo_sand-painting.jpg)

One prime ordering principle for the Navajo, as for other Native American groups, is the concept of fourness, which for Navajo was auspicious as it represented the balance of two pairs of opposites. In the Navajo conception of the universe there are four worlds, of which ours is the last, bounded by four sacred peaks (the four volcanic peaks surrounding the Navajo territory, Hesperus, Blanca, Taylor, and San Francisco peak). The world is believed to be bounded by four sacred beings (Evening Twilight Woman, Darkness Woman, Dawn Man, and Horizontal Blue Man; some accounts mention other groups of four), all associated with the four colors, white, black, blue, and red; four sacred plants (corn, beans, squash, and tobacco) (Fig. 4.28).

> As described by Navajo elder Harry Walters: "Each of the Sacred Mountains and the four directions have their own Holy Person who is either male or female. They are each paired. The Holy Person of the East is Talking God and is male and is paired with Hogan God who is female and represents the West. Born-for-Water is the Holy Person of the South, and is female and is paired with Monster Slayer, the Holy Person of the North, who is male. There are pairs of Mountains within the four Sacred Mountains, and they face each other and keep each other in check. There is a word which talks about all this: *alchisila*, which means 'balance.'
>
> (Griffin-Pierce 1992, p. 72)

The Navajo believed that the sacred Hogan was composed of four elements (white shell, abalone, turquoise, and obsidian). The Navajo described the sky as having four layers—the Sun, Moon, stars, and an opening to the sky beyond known as the sky hole (*yayohokaa*) that leads beyond to the stars. Time was also broken into four: four parts of the day (dawn, blue sky, evening twilight, and darkness) and four seasons of life (babyhood, adolescence, adulthood, and old age). Interestingly, the Navajo recognized two seasons only, *Hai*, or winter, and *Shi*, or summer, but these seasons are divided into groups of months and primary constellations, themselves grouped into four male and four female constellations (Griffin-Pierce 1992, pp. 67–75) (Fig. 4.28).

**Fig. 4.28** The Navajo cosmos. The image illustrates the combination of landmark peaks, the four colored gods of the horizons, and the sky gods from the Navajo people. Note the four primary colors in the sky gods that includes sun (*yellow*), moon (*blue*), two of the four primary beings at the edge of the earth, as well as the four sacred peaks that appear in the cardinal directions. Above the earth is a cosmic rainbow and father sky, with mother earth below (image from National Geographic; reprinted with permission)

## Chumash Native American Cosmology

The physical world of the Chumash (sometimes denoted as the *'antap*) is thought to be the middle of three worlds, which is interconnected to the lower world through springs and marshy areas and to the upper world through mountains. In the lower world, or *C'oyinahsup*, live snakes, frogs, and salamanders, including several giant snakes that support the *'antap* on their backs. The world we live in trembles with earthquakes when these primordial snakes writhe, a useful part of a cosmology in earthquake-prone southern California! Water creatures in our world were thought to be in contact with the powers of the lower world and were often depicted in rock art perhaps to bring more water to the Chumash or to appease underworld spirits at times of hunger or disease.

The world we live on can also be separated into two places, the *'antap*, where spirits of this world live and where shamans can travel in vision quests, and the *Itiashap*, the home of the first people.

The upper world of the Chumash cosmology was known as the *Alapay* and was inhabited by a cast of "sky people" who played important roles in the health of the people. The principal sky figures include the Sun, the Moon, Lizard, Sky Coyote, and Eagle. Pairs of these figures engage in intense gambling games throughout the year, with the Sun and Lizard on one team and Sky Coyote and Eagle on the other. The Moon keeps score, and at the end of the year, if Sky Coyote wins, rain and abundant food bless the mankind. If the Sun wins, however, drought, famine, and disease reign on the earth for the year (Blackburn 1975; Hudson and Underhay 1978).

The Chumash cosmology reflects some ambiguity about the all-powerful and sometimes overbearing southern California Sun. The Sun is seen as the source of all life and the principal rituals involve solstice dances which attempt to "pull back" the Sun through choreographed chanting and dancing at the winter solstice. Sun is also seen as a force for disease and death, and the sky person of Sun is described as a gruff, naked, old man who walks across the sky on a cord over the Earth with his blazing torch. While the Sun walks over the Earth he is always on the lookout for neglectful children or old people that he snatches and feeds to his two daughters. The Sun leaves his imprint in the radiant pattern of the sand dollar, as he is thought to rest in them on some of his journeys on the Earth.

The world system of the Chumash is shown below (Fig. 4.29) in a figure that includes most of the essential elements of their universe, which is a rich depiction of the Southern California natural world. Two key locations are the sacred site of Mt. Pinos, which was thought to be a hub of all three worlds, and Pt. Conception on the coast of Southern California, where some believed that people's spirits would rise to the upper world after death to become part of the "Journey of the Pinon Gatherers" or *Suyapo'osh*, who travel the road of the Milky Way to the land of the Dead or *Similaqsa* (Blackburn 1975, p. 33). The Chumash incorporated local landmarks, animals, and seasonal patterns into their model of the universe to help provide a more meaningful system of explaining the universe.

## Inuit Cosmology

The Inuit believed the Earth was a large and flat disk. This disk was necessary to keep water from spilling off, according to legend. It also enabled the chase of the Sun and Moon to occur overhead. However, additional layers were believed to exist both above and below the Earth. Within the sky were realms for those who have died, and in some accounts different types of deaths would result in different locations in the sky. In one account, most people when they died went to a location near the Moon, but victims of drowning went to the underworld, and others who died would rest between the Earth and Moon.

The sky in Inuit cosmology known as "Qilak" and is a round canopy made of hard materials. In a cosmology that is reminiscent of early Greek ideas, some Inuit believed that stars shined with light from beyond the holes in the sky. One account describes the journey of a shaman into this sky world in a dream. His dream allowed himself to see the holes clearly and put his head through this hole to see the world beyond:

> Raising himself up, he put his head through the nearest star hole and saw another sky with many stars shining above the first one. As he looked the sky sank slowly down until he could put his head through one of the star holes in it, and above this were shining the stars of still another sky. This too, sank slowly down, and standing up he found himself breast high above the third sky, and close by was a kashim surrounded by a village like the one in which he lived.
>
> (MacDonald et al. 1998, p. 32)

In one Inuit story, the sky of one world is the ground of another. In this conception, snow would be imbedded in the ground of the world beyond and could be shaken loose from those walking above, causing snowfall in the lower world of Earth. Below the stars and sky of the *Qilak* is the region known to the Inuit as *Sila* which includes the Earth, the air, the clouds and weather. *Sila* translates roughly as

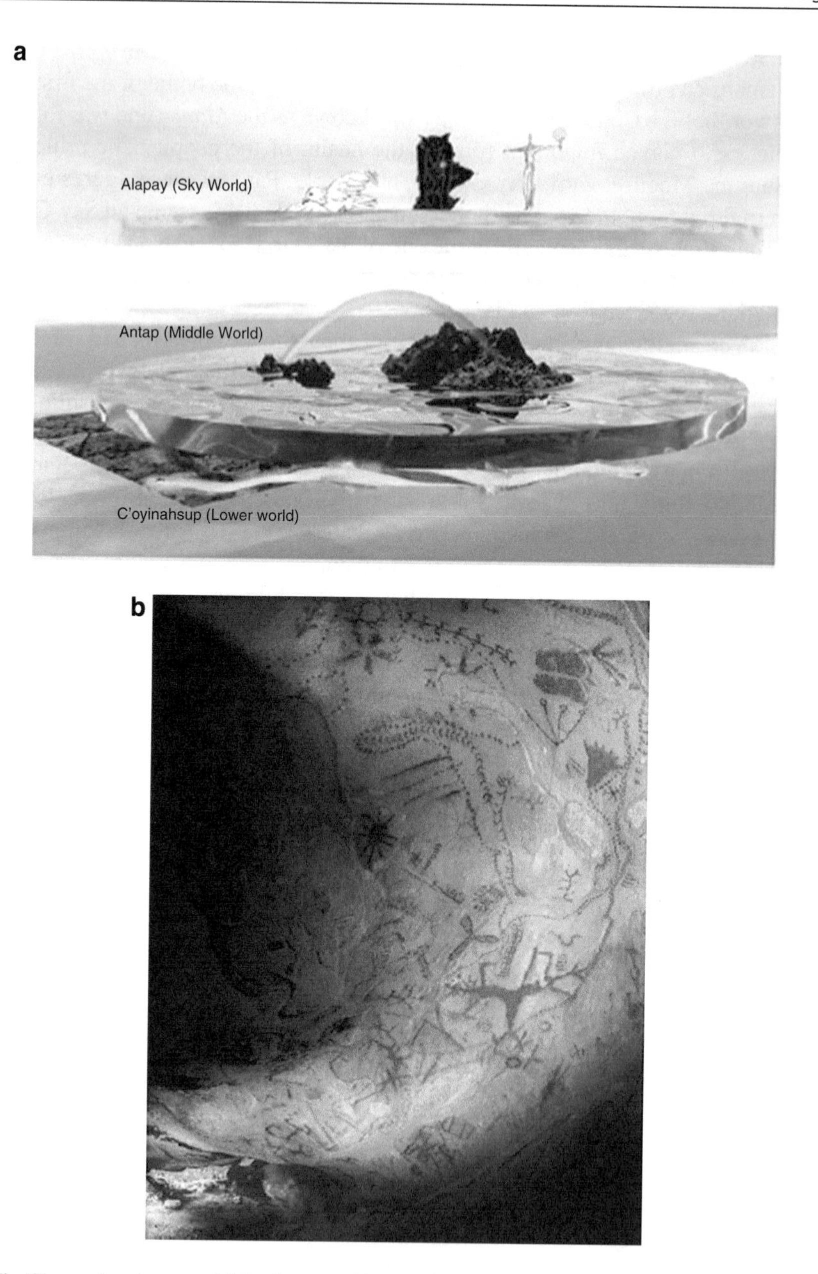


**Fig. 4.29** (**a**) The Chumash universe, with its three worlds—the sky world above, containing the "Sky People" of Sun (an old man carrying a torch across the sky), Sky Coyote, Eagle, and Moon, the 'antap or middle world, in which the islands are connected to the mainland by a rainbow bridge, and a lower world on which the world rests on the backs of snakes, filled with water underworld creatures (original image by the author). (**b**) Chumash painted cave in Kern County. The site has recently been suggested to have connections with a mythical struggle between the Sun and Polaris (Sky Coyote), key figures in the Chumash cosmology. The photograph itself is suggestive of the three levels of the Chumash universe, which included an underworld, a middle world (the Earth), and a sky world (Saing-Onge, Rex, John Johnson, and Joseph Talaugon, 2009, "Archaeological Implications of a Northern Chumash Arborglyph," Journal of California and Great Basin Anthropology, Volume 29, #1, Lynn Gamble, editor. Photograph by Rick Bury; reprinted with permission)

"environment" and is an abstract concept that includes both the physical world and the presence of an all-pervading spirit in the world. This spirit can be personified as weather, and as such would reflect moods and influences of natural as well as supernatural forces. The *Sila* is to be respected and watched carefully to enable survival. The Danish explorer Rasmussen described the *Sila* as a "great, dangerous and divine spirit" that "lives somewhere ... between sky and sea... and threatens mankind through the might powers of nature, wind and sea, fog, rain and snowstorm" (MacDonald et al. 1998, p. 36). The

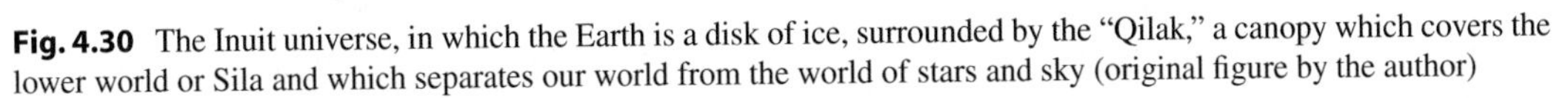

**Fig. 4.30** The Inuit universe, in which the Earth is a disk of ice, surrounded by the "Qilak," a canopy which covers the lower world or Sila and which separates our world from the world of stars and sky (original figure by the author)

Inuit survival depended on careful balance and observation of their environment, and their cosmology placed primary importance on harmony within the *Sila* (Fig. 4.30).

## African Cosmologies

The diverse range of cultures from Africa invented a wide range of cosmologies, which were transmitted through a vibrant oral tradition. Many of these take the form of star tales, which give insight into the larger universe as viewed from various African cultures. From the Khoikhoi and the San we learn that many of stars that commemorate famous people and events, as in other cultures. According to legend, the stars of Orion's belt were placed there when a small girl with magical powers turned a group of threatening lions into stars. The Milky Way was formed when an angry child threw some of the roots which her mother was roasting up into the air. The Milky Way also has been called the "Backbone of Night" by some of the San. According to the Tswana, "the sky is stone, and the Earth is flat. Water is beneath the Earth and above the sky." The Southern Cross according to a group known as the Credo Mutwa is the "Tree of Life" while the Xhosa believed that "the world ended with the sea, which concealed a vast pit filled with new moons ready for use" (SAAO 1998).

Some of the African stories of the Pleiades reveal deeper beliefs about the sky and provide some practical means for foretelling the future. The Pleiades were used both to predict rain and to help guard small children. For others the Pleiades were known as "digging stars" and their appearance was a sign to begin the planting. One tribe known as the Namaqua believed that the Pleiades were daughters of the sky god. The other stars in the same part of the sky form a family tableau where the star Aldebaran is the husband who is hunting three zebras (the stars of Orion's belt), with the star Betelgeuse in Orion's shoulder a lion watching the proceedings.

One interesting cosmology, which is in the form of a creation tale, tells of the formation of the universe from a primordial tree and a termite mound. The Tanzanian tribe known as the Wapangwa describes the first life in the universe:

> The sky was large, white and very clear. It was empty; there were no stars and no moon; only a tree stood in the air and there was wind. This tree fed on the atmosphere, and ants lived on it. Wind, tree ants, and atmosphere were controlled by the power of the Word. But the Word was not something that could be seen. It was a force that enabled one thing to create another… The Wind, it seems, was annoyed with the tree, which stood in the way, so it blew until a branch snapped, carrying away with it the colony of white ants. When the branch finally came to rest, the ants ate all the leaves, except one large one upon which they left their excrement. This excrement grew into a large heap and eventually into a mountain that approached the top of the tree of origin. Now in contact again with the primordial tree, the ants had more food to consume and waste to eliminate, from which they fashioned a huge object, the earth, with mountains and valleys, all of which touched the top of the tree of origin.
>
> (Ford 1999, p. 178)

The cosmology continues with a description the emergence of life, which is soon followed by a war in which the animals fight over food, causing them to shake the earth and start fires. At this point bits of the Earth break loose and form the Sun, the Moon, and stars.

## Dogon Cosmology

In the influential work "Conversations with Ogotemmeli," Marcel Griaule reports on many aspects of the African Dogon culture through the verbal traditions of one of its distinguished elders named Ogotemmeli. Ogotemmeli describes the universe as being formed by Amma, a creator god. "The stars came from pellets of earth flung out into space," and the Sun is "a pot raised once for all to white head and surrounded by a spiral of red copper with eight turns" (Griaule 1975, p. 16). Within the Dogon tradition the universe is seen as a living extension of the god *Amma*, who made the world from a lump of clay and gave the earth many human attributes, including body parts, in the form of a flat slab with both male and female attributes. After intercourse with the "Earth wife," the creator god gave rise to a pair of half human, half animal twins. These twins were known as the *Nummo* who in turn gave birth to a snake whose actions gave rise to the present-day form of the human body. The early snake was killed and its head buried underground. A man named Lebe descended into the Earth and the snake was resurrected, and when it came to life the snake consumed Lebe, but later regurgitated parts of Lebe in a shape which was that of the human body. The human body was then formed and became the template for the form of Dogon villages and society and also mirrored the form of the Earth and larger cosmos.

A village for the Dogon is laid out in a pattern that is thought to mirror the shape of the human body, with locations for "organs" of various types in analogous positions to those of the Earth. Dogon granaries in particular symbolize the structure of the universe with eight parts for the different organs of the world and of people, and roofs that mirror the shape of the sky above. An eight-part Dogon cosmology is also described in which directions, languages, drums, body parts, and human activity can all be divided into eight (Griaule 1975; Ford 1999, p. 182).

## References


Aristotle, and W.K.C. Guthrie. 1960. *On the heavens. With an English translation by W. K. C. Guthrie*. Cambridge: Harvard University Press.

Aristotle, and H. Lawson-Tancred. 1986. *De anima = On the soul. Harmondsworth, Middlesex, England*. New York, NY: Penguin Books.

Aveni, A.F. 1989. *Empires of time: Calendars, clocks, and cultures*. New York: Basic Books.

Blackburn, T.C. 1975. *December's child: A book of Chumash oral narratives*. Berkeley, CA: University of California Press.

Blacker, C., et al. 1975. *Ancient cosmologies*. London: Allen and Unwin.

Coe, M.D. 1999. *The Maya*. New York: Thames and Hudson.

Cullen, C. 1980. Joseph Needham on Chinese astronomy. *Past Present* 87: 39.

——. 2007. *Astronomy and mathematics in ancient China: the Zhou Bi Suan Jing (Needham Research Institute Studies)*. London: Cambridge University Press.

Dallapiccola, A.L. 2003. *Hindu Myths*. Austin, TX: University of Texas Press.

Evans, J. 1998. *The history and practice of ancient astronomy*. New York: Oxford University Press.

Ford, C.W. 1999. *The hero with an African face: Mythic wisdom of traditional Africa*. New York: Bantam Books.

Griaule, M. 1975. *Conversations with Ogotemmêli: An introduction to Dogon religious ideas*. New York: Published for the International African Institute by the Oxford University Press.

Griffin-Pierce, T. 1992. *Earth is my mother, sky is my father: Space, time, and astronomy in Navajo sandpainting*. Albuquerque, NM: University of New Mexico Press.

Gummere, F.B. 1909. *The oldest English epic, Beowulf, Finnsburg, Waldere, Deor, Widsith, and the German Hildebrand*. New York: Macmillan.

Hudson, T., and E. Underhay. 1978. *Crystals in the sky: An intellectual odyssey involving Chumash astronomy, cosmology, and rock art*. Socorro, NM: Ballena Press.

Littleton, C.S. 2002. *Mythology: The illustrated anthology of world myth & storytelling*. San Diego, CA: Thunder Bay Press.

Lloyd, G.E.R. 1971. *Early Greek science: Thales to Aristotle*. New York: Norton.

MacDonald, J., et al. 1998. *The Arctic sky: Inuit astronomy, star lore, and legend*. Toronto, ON: Royal Ontario Museum/ Nunavut Research Institute.

Matchett, F. 2003. In *"The Puranas" in the blackwell companion to Hinduism*, ed. Gavin Flood. Oxford: Blackwell Publishing.

Needham, J., and C.A. Ronan. 1980. *The shorter science and civilization in China: An Abridgement of Joseph Needham's original text*. New York: Cambridge University Press.

North, J.D. 2008. *Cosmos: An illustrated history of astronomy and cosmology*. Chicago, IL: University of Chicago Press.

SAAO (1998). African Starlore. From http://www.saao.ac.za/public-info/sun-moon-stars/african-starlore

Sambursky, S. 1963. *The physical world of the Greeks*. London: Routledge and Paul.

Sharer, R.J., and L.P. Traxler. 2006. *The ancient Maya*. Stanford, CA: Stanford University Press.

Toulmin, S.E., and J. Goodfield. 1965. *The fabric of the heavens: The development of astronomy and dynamics*. New York: Harper.

Urton, G. 1981. *At the crossroads of the earth and the sky: An Andean cosmology*. Austin: University of Texas Press.

Waters, F., and O.W.B. Fredericks. 1969. *Book of the Hopi*. New York: Ballantine Books.

Wheelwright, P. 1966. *The presocratics*. New York: Odyssey Press.

Wilson, A.M. 1995. *The infinite in the finite*. Oxford: Oxford University Press.

# Chapter 5

# Stars that Bind: Civilization, Calendars, and the Sky

> *And an astronomer said, Master, what of Time?*
>
> *And he answered: You would measure time the measureless and the immeasurable ... Of time you would make a stream upon whose bank you would sit and watch its flowing. Yet the timeless within you is aware of life's timelessness, and knows that yesterday is but today's memory and tomorrow is today's dream. And that which sings and contemplates in you is still dwelling in the bounds of the first moment which scattered the stars into space ...*
>
> —Excerpt from "The Prophet" (Gibran 1952)

The measurement of time catalyzed the development of numbers, mathematics, and astronomy as the first humans began counting days, hours, minutes, and seconds. The sophistication of ancient civilizations in noting the cycles of the Sun, the Moon, planets, and stars enabled them to unify their people and to coordinate the activities of their people with both the cycles of nature and with the cycles of their gods. For the study of time, the planets, the Sun, and the Moon set the beat by which human civilization developed.

Many of the earliest samples of writing in many cultures are related to astronomical record keeping that enabled the development of calendars. These early writings include notches on bones used in neolithic day counts of a lunar cycle and cuneiform tablets recording the positions and motions of the Sun, the Moon, and planets. Mayan hieroglyphic carvings give the dates in Mayan calendar of the creation of the universe and the dates of the heroic rise to power of the Mayan emperor, while Chinese Shang dynasty bones carved with dates of the appearance of a comet or supernova 5000 years ago, as part of an effort to divine the meaning of the events.

For some cultures, time and space are woven together. These cultures include both the Aboriginal Australian and ancient Mayan, as well as modern physicists, who make use of Einstein's theories to describe space and time as a four-dimensional continuum.

Most cultures realize in their count of days a reconciliation between the solar and lunar cycles. In some of the calendars, individual days, months, or even years are considered auspicious or dangerous, depending on their particular omens. Each culture develops original methods to bind together human activity with the periods of the Moon, the Sun, and stars, as well as planets such as Venus and Jupiter.

In this chapter, we will explore the process by which civilizations from around the world embark on their exploration and charting of the cycles of time. The concepts of time are as diverse and creative as are the concepts of spatial cosmology and include a wide range of computation, metaphor, and poetry that can inspire us, even after so many centuries.


B.E. Penprase, *The Power of Stars*, DOI 10.1007/978-3-319-52597-6_5

## Lunar Calendars and Timekeeping

The most obvious celestial timekeeper is the Moon. Any groups of people larger than just a few quickly find that they need to use the Moon to coordinate hunting activities and to coordinate migrations through the environment. The first calendars were probably therefore lunar calendars, and many examples still exist of the lunar calendar in present-day cultures.

### *Neolithic Calendars*

The Dordogne River valley in France is home to some of the most exquisite ancient cave paintings and neolithic artifacts from groups of people much like those depicted in Jean M. Auel's novel "Clan of the Cave Bear" (Auel 1980). From these groups came the nuclei of the first civilizations in Europe, as well as vivid depictions of hunting scenes and animals in the 15,000-year-old cave paintings of Lascaux. Even older artifacts from this region contain what may be markings corresponding to days in a lunar calendar, such as those seen in the 33,000-year-old bone fragment, found in the Abri Blanchard cave, and interpreted by Alexandar Marshack to depict notches in the shape of a waxing and waning Moon. The shapes and numbers of the notches are consistent with the 29 days of the lunar month (Fig. 5.1). Additional authors have found sets of 27–29 dots within cave paintings that could also relate to calendars based on lunar months. The practice of recording days of a lunar calendar appears to be common throughout the world, with evidence of lunar scribes being found in Native American sun sticks, Mayan wooden calendars that predate classical Mayan civilization, Nicobar Island calendar sticks, and calendar sticks of Siberian hunters (Marshack and D'Errico 1989).

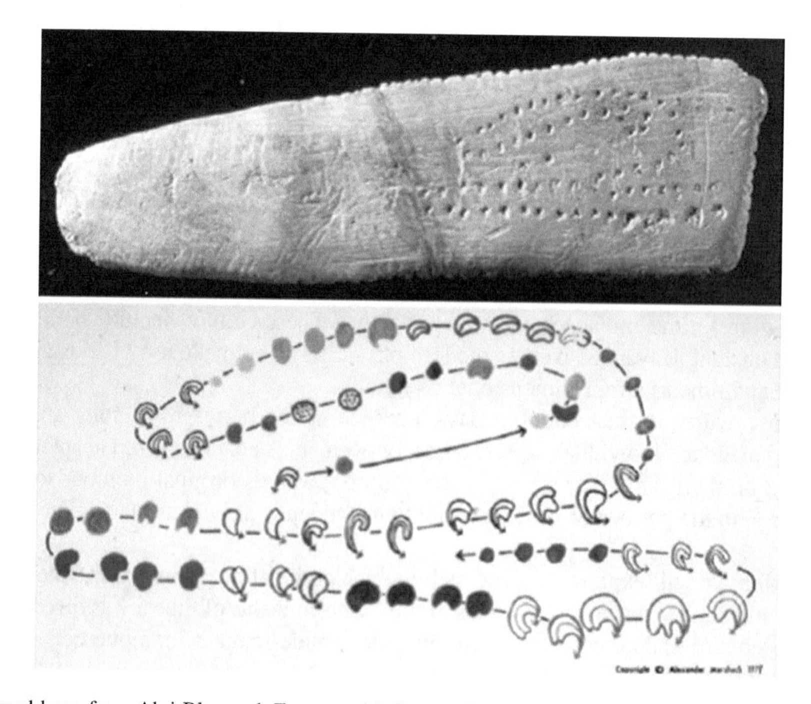

**Fig. 5.1** Carved bone from Abri Blancard, France, with figures that are grouped into what may be a neolithic lunar calendar from 15,000 BC (Marshack, 1991). The overlay (*below*) shows the outline of the figures, which appear in groups of 29 and have shapes suggestive of lunar phases. (c) President and fellows of Harvard College, Peabody Museum of Archaeology and Ethnology, 2005.16.318.38. (c) President and fellows of Harvard College, Peabody Museum of Archaeology and Ethnology, 2005.16.19.1

One particular site known as Presa de la Mula, near Monterrey Mexico, contains a fascinating set of lines in a huge petroglyph 10 feet across and 3-feet wide. The panel contains several sequences of 29, 27, 29, 28, 27, 7, 30, and 28 lines, which is close to the 207 days of seven lunar months. Discrepancies from the exact number are explained to result from the practical difficulties of observing the Moon every night (Ruggles 2005, p. 352). Additional sites in Mexico, such as the Boca de Potrerillos site, also show additional examples of panels with 205–207 markings, suggesting that the practice of observing lunar months was widespread in prehistoric Mexico.

## Solar/Lunar Calendars

With the advent of agriculture, the solar year becomes primary to help coordinate the planting of crops with seasonal weather patterns. Many civilizations also invented calendars to match the lunar cycles with the longer solar year.

The problem with matching both lunar and solar cycles is the simple fact that the numbers of days in the lunar month and solar year do not divide evenly. Since these numbers themselves are determined by the physical motions of our planet Earth in its orbit around the Sun, this lack of agreement should not come as a surprise. Reconciling the number of days in the lunar month (29.5333) and the number of days in the solar year (365.2422) was a great challenge for ancient civilizations and has been solved independently with many solar/lunar calendars, such as the Babylonian, Hebrew, and Chinese lunar calendars, as well as the Hawaiian and Hopi solar lunar calendars, which are described below.

### *The Babylonian Calendar*

The origins of our 24-hour day and 7-day week come from Babylonian timekeeping, as described in Chap. 1. Mathematical advances in Babylonia allowed their astronomers to factor and divide large numbers, which in turn allowed them to ascertain whole number multiples of the lunar months and the solar years, such as the 235 lunar month Metonic cycle, which provides a whole number of lunar months (235) and solar years (19), to enable the lunar and solar calendars to match every 6940 days.

The Babylonians, like many contemporary civilizations, used sets of 12 and 13 lunar month years to reach the 235 month completion of their cycle. These months were named after the various seasons within the Mesopotamian year and a month began with the visibility of the first crescent moon. The 12 months for Babylonian astronomers predate Babylon itself, as they were intoned in the first parts of the Babylonian creation story Enuma Elish, as Marduk is ordering the heavens:

> He constructed stations for the great gods,
> Fixing their astral likenesses as constellations
> He determined the year by designating the zones:
> He set up three constellations for each of the twelve months
> After defining the days of the year (by means) of (heavenly) figures,
> He founded the station of Nebiru to determine their heavenly bands …
> (Kelley and Milone 2005, p. 217)

These 12 Babylonian months are listed in Fig. 5.4, along with corresponding Babylonian cuneiform numbers. A thirteenth "intercalary" month was added as needed and usually given the name of "second Ululu" or "second Addaru" since it would follow the month of Ululu or Addaru, respectively, to give a complete set of lunar months to a year. Babylonian numbers were written in sexigesimal (or base 60) cuneiform, one of the earliest systems of number notation. We still use such sexigesimal numbers in our notation for time (HH:MM:SS), where the colons denote separation between hours,

**Fig. 5.2** Detail of cuneiform tablet from 2039 BC, in which scribe marks on clay tablets provided one of the earliest writing systems to preserve records of accounts, legends, and the cycles of the sky (*image source*: http://www.loc.gov/exhibits/world/images/s146p1.jpg)

**Fig. 5.3** Transition from 59 to 60 in Babylonian cuneiform. The number 59 (*left*) and the number 60 (*right*), which are a one symbol with a blank spot to the *right* to indicate place, as zero was not yet invented

minutes, and seconds (Fig. 5.2). The Babylonian cuneiform numbers were based on multiple marks to show small numbers such as two, which in cuneiform would appear 𒈫. Larger numbers would combine a symbol for 10 with the associated ones, such as the number 11, which would appear with a 10 and a 1 beside it (𒌋 𒁹).

The grouping of tens and ones would continue until 59 was reached (since their numbers were based on 60 or sexigesimal). Writing the number 60 caused the transition to a 1 symbol 𒁹 and a blank spot, just as we transition from the number 9 to 10 in a decimal system. Figure 5.3 shows how this transition would be accomplished. Incidentally, the Babylonians did not invent the number 0 (a technological innovation which would require many additional centuries) so in going from 59 to 60, some confusion might arise with the blank region of the tablet (Fig. 5.4).

## Egyptian Timekeeping

The Egyptian civilization perhaps had the longest baseline of time from which to develop an advanced calendar, with over 2700 years of history to chronicle! The Egyptian calendar was surprisingly simple, however, with an emphasis on preserving the 360 + 5 day Egyptian solar year and keeping track of the three main seasons of Egypt, which were tied to the agricultural practices of the Nile River valley. The three seasons were named "inundation," "cultivation," and "harvest," and divided the year into three parts of 120 days each. Within the three seasons were 30-day months, each named after gods and goddesses of Egypt. (Figure 5.4 shows the month names of the 365-day calendar, along with the corresponding Egyptian numerals.) Finally, night and day were each subdivided into 12 equal parts, giving a 24-"hour" day, but with hours of different duration depending on the seasons. The word "hour" itself derives from Egyptian timekeeping, since "hour" comes from the Greek word *hora*, which itself derives from the Egyptian word *hor*, which means "day" or "sun's path" (Lennox-Boyd 2005, p. 22).

**Fig. 5.4** Babylonian month names, along with corresponding cuneiform numbers (Kelley and Milone 2005, p. 219)

The Egyptians observed a ritual calendar year of 360 days and appended a five-day festival period to honor the chief gods of the Egyptian pantheon (Osiris, Isis, Seth, Nephthys, and Horus). Without any correction, the Egyptian calendar year (365 days) would drift with respect to the actual period of the solar year (365.2422 days) to give complete rotations of the seasons over the course of the thousands of years of the Egyptian civilization. A solution to this problem came from the stars. By watching the morning twilight for the first appearance of the star Sirius (or the *heliacal rising*), the Egyptians noticed that Sirius' heliacal rising coincided in time with the flooding of the Nile. By the later part of the Egyptian empire (according to a papyrus known as Carlsberg 9 dated from AD 144) the Egyptians routinely added an additional month of Thoth whenever the heliacal rising of Sirius happened within the last 11 days of the last lunar month (Krupp 1980).

The star Sirius, known in Greek as "Sothis," was associated with the goddess Isis and her life-giving flood of the Nile River. By beginning the Egyptian New Year in summer at the time of the heliacal rising of Sirius or Sothis, the Egyptians were able to develop one of the first lunar and solar calendars, known as the *Sothic year*. Egyptian priests noticed the drift of the Sothic calendar over centuries and were able to improve the match between the days of the year with the motions of the stars by means of the "decans" system. The decans were a group of 36 sets of stars which could be sighted in the early morning twilight in each interval of 10 days known as a "decan." In this way, ancient Egypt had two parallel calendars—one ritual calendar of 365 days subject to some drifting and a more accurate star calendar of the decan stars (Fig. 5.5).

For nearly 2000 years, the "Sothic year" was the primary calendar of Egypt and much of the ancient world. However, some attempt to fix the drift of the year was made late in the history of Egypt, in 286 BC, when the Canopus decree of king Ptolemy III added an additional day every fourth year. This Egyptian "leap year" predates the more famous Julian calendar by centuries and helped keep the Sothic year aligned with the decan star sightings by the Egyptian astronomer priests.

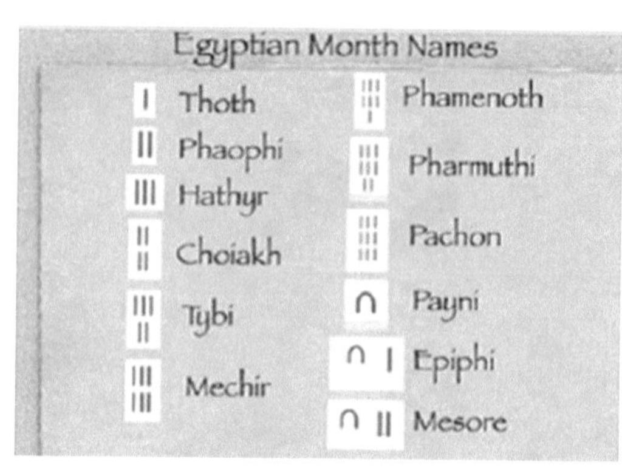


**Fig. 5.5** Egyptian month names along with Egyptian numerals for each month. Egyptian months were named after various gods in the Egyptian pantheon, and the year began with the heliacal rising of Sirius (Sothis) near the time of the Nile River flooding. The "Sothic Year" was named after Sirius and is based on the Egyptian calendar (Kelley and Milone 2005, p. 263)

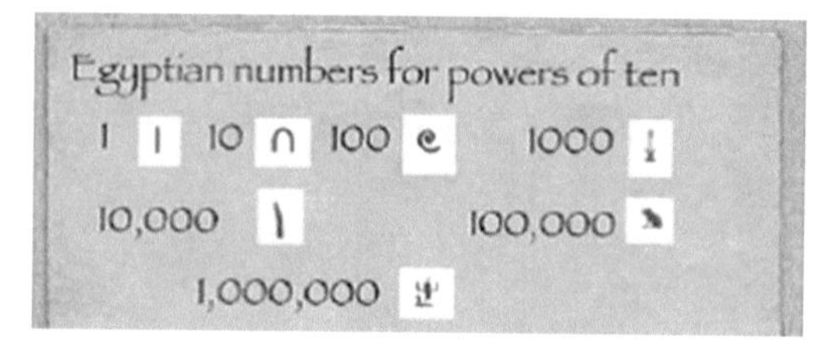


**Fig. 5.6** Hieroglyphic representations for various powers of 10. Hieroglyphics for larger numbers were distinct figures; a finger for 10,000, a frog for 100,000, and a kneeling person for 1,000,000 (figure generated with the hieroglyphics program at http://www.psinvention.com/zoetic/tr_egypt.htm)

The "decan stars" provided a celestial alarm clock that corrected the Egyptian calendar to match with the eternal cycles of the starry sky above. Ceremonial tables of decan stars are preserved on coffin lids, both as an aid to the departed Pharaoh in their effort to navigate the afterlife, and to preserve the timely functioning of the Pharaoh's activities in eternity. Lists of decan stars guided both the living priests and departed Pharaohs and kept Egyptian society in synch with the stars (Fig. 5.8).

In keeping track of time, and in constructing extremely large works (such as the pyramids), the Egyptians were able to develop a sophisticated base 10 number system. Some of these numbers are shown next to the Egyptian month names in Fig. 5.6. The Egyptian system of numbers allowed construction of very large base 10 numbers using the separate hieroglyphics for each power of 10 (Figs. 5.6, 5.7). Below are the signs for these larger decimal numbers. Having separate symbols for each power of 10 also solved the problem of not having invented the numeral zero, since it was easy to distinguish each of the symbols.

One wonders if the figure in the Egyptian hieroglyphic for 1,000,000 (see Fig. 5.7) is showing some signs reaching the limits of Egyptian mathematics in his gesture! One million was the largest number with a known hieroglyphic, comparable to the number of blocks in the Great Pyramid of Khufu (estimated at approximately 2.3 million blocks ranging in weight from 1.5 to 80 tons).

## *The Ancient Hebrew Calendar*

The development of the Hebrew calendar paralleled that of the Babylonian calendar, perhaps due to the Babylonian exile period between 450 and 419 BC. In some respects, the present-day Jewish calendar offers a "time capsule" of the ancient Mesopotamian lunar calendar. From the first month of Nissannu

**Fig. 5.7** Egyptian hieroglyphic representation for 2,300,000, the number of blocks estimated in the Great Pyramid of Cheops in Giza (figure generated with the hieroglyphics program at http://www.psinvention.com/zoetic/tr_egypt.htm)

**Fig. 5.8** Detail of star ceiling from the tomb of Seti I, who ruled from 1294 to 1279 BCE, showing a table that lists the deities associated with each of the Decans stars from the Egyptian calendar (*top panel*) (image by Jean-Pierre Dalbera)

in Babylonian calendar (Nisan in Hebrew) to the last of Addaru (Adar in Hebrew), the Hebrew calendar features month names nearly identical to the Babylonian, and like the Metonic system inserts extra lunar months over a 19-year cycle. The Jewish calendar is in fact a hybrid of the earlier Canaanite calendar, which started the year closer to the fall equinox, and the Babylonian, which began the year in spring, soon after the vernal equinox. This important difference makes the Jewish calendar 6 months out of phase with the Babylonian calendar (Kelley and Milone 2005, p. 219) (Table 5.1).

One additional feature shared by the Hebrew and Babylonian calendars is the 7-day week. The book of Genesis includes a description of a 7-day week, with 6 days of labor in which god created the world, and a seventh day of rest known as the "Sabbath." The word for Sabbath also shares roots in both Babylonian and Hebrew; in Hebrew the word is *Shabbat* which means "rest," while in Babylonian

**Table 5.1** Babylonian and Hebrew month names

| | Babylonian | | Hebrew | | Babylonian | | Hebrew |
|---|---|---|---|---|---|---|---|
| 1 | Nisannu | 7 | Nisan | 7 | Tasritu | 1 | Tishri |
| 2 | Ajjaru | 8 | Iyyar | 8 | Arahsamnu | 2 | Chesvan |
| 3 | Simanu | 9 | Sivan | 9 | Kislimu | 3 | Kislev |
| 4 | Duzu | 10 | Tammuz | 10 | Tebetu | 4 | Tebeth |
| 5 | Abu | 11 | Ab | 11 | Sabatu | 5 | Shebat |
| 6 | Ululu | 12 | Elul | 12 | Addaru | 6 | Adar |

The Hebrew calendar preserves a snapshot of the Mesopotamian calendar of 430 BC, with very similar month names (but shifted by 6 months), and the same Metonic-type lunar calendar as the Babylonian system (Kelley and Milone 2005)

the word is *sabattu* or "seventh." For the Babylonians, the *sabattu,* or seventh day, had ominous associations, as were several other days of the lunar month based on both theology and numerology. For this reason, the seventh day was reserved for rites of purification, and this tradition appears to have been shared by the Greeks and Hebrews (Steel 2000, p. 76).

Like many lunar calendars, the Hebrew calendar relies on sightings of a crescent moon to begin the month. Like the earlier Babylonian calendar, it uses a system of 12 and 13 lunar month years alternating in the 19-year cycle (Rich 2008), with extra intercalary months added as needed to match the solar year.

Additional months in the Hebrew calendar are given the name *Adar Sheini* or *Adar Beit* and are added to make for 13-month years, a practice which happens seven times in 19 years. Despite the fact that the main features of the calendar were set in place 2500 years ago, the first year of the Hebrew calendar is set over 5700 years ago, in the Gregorian year of 3760 BCE, which is thought by some to be the beginning of the universe as described in Genesis. Each year begins with the first of *Tishri*, when the festival of Rosh Hashanah ("head of the year") commemorates the creation of the universe and marks a time at which the Creator judges the universe. Additional holidays include Yom Kippur, the Day of Atonement, on the 10th day of *Tishri*, and Hanukah, which always occurs on 25 *Kislev* in the Hebrew calendar, even though its date will shift each year within the Gregorian calendar. Although the new year (and universe) is recognized to begin on 1 *Tishri*, the numbers of the Hebrew months begin with Nisan, as does the Babylonian calendar.

## *The Islamic Calendar*

The Islamic world still sets its sacred holidays with a calendar including 12 lunar synodic months of 29 or 30 days, for a total of 354 days. The beginning of a lunar month is set by the first visibility of the crescent moon, giving rise to the crescent moon as the international symbol of Islam. The beginning of the year is known as *Al-Hijra* and is the first day of the first month known as *Muharram*. The Islamic calendar begins its count of years in the year when Mohammed moved from Mecca to Medina and is given the name the Hijri calendar after this migration, known as the *hijra*. Year 1 of the Islamic calendar corresponds to the day Friday, July 16, 622 CE, making the year 2016 CE in the Gregorian calendar equal to year AH 1437.

The gap between the Islamic 12 lunar month calendar (354 days) and the solar year (365.25 days) means that the Islamic New Year shifts 11 days earlier each solar year, causing important dates to rotate through the solar year over time. Unlike many other lunar calendars, the Islamic calendar does not allow extra months to reconcile solar and lunar years, since the extra 13th intercalary month is forbidden by the Quran. All of the Islamic holidays will therefore rotate through the solar year. For example, the Islamic New Year date closest to the Gregorian New Year of 2017 begins on October 3, 2016 (AH 1438), and subsequent years begin 11 days earlier each year: September 22, 2017 (AH 1439), September 11, 2018 (AH 1440), August 31, 2019 (AH 1441), and so on, continuing 11 days earlier each year. Islamic holidays include the new year or *Al-Hijra*, and *Laylat al-Qadr (The Night of Power)*,

**Table 5.2** Major Islamic holidays during the year AH1438 and their significance along with the corresponding Gregorian dates in 2016/17 (later years have dates 11 days earlier each year)

| Holiday | Date in Islamic calendar | Meaning | Gregorian date in 2016/17 |
|---|---|---|---|
| Al-Hijra | 1 Muharram | Date of Mohammed's migration to Mecca (Hijra) | October 3, 2016 |
| Isra al-Miraj | 27 Rajab (7th month) | Ascent of the Prophet Mohammed | April 24, 2017 |
| Ramadan | 1 Ramadan (9th month) | Month during which Mohammed received revelation of the Quran | May 27, 2017 |
| Laylat al-Qadr | 21–29 Ramadan (varies by country) | "Night of Power" —date when Mohammed received first revelation | June 22, 2017 |
| Eid al-Fitr | 1 Shawwal (10th month) | "Breaking of Fasting" and end of the month of Ramadan | June 26, 2017 |
| Eid al-Adha | 10 Dhu'l-Hijja (12th month) | Date of Ibrahim's willingness to sacrifice Isma'il; end of Hajj | September 2, 2017 |

which is believed to be the day when Mohammed first received his revelation that began the Quran. It is believed that the Quran was revealed to Mohammed during *Ramadan,* the ninth month of the Islamic calendar and so the entire month is observed as a month of fasting during daylight hours.

The exact date of Laylat al-Qadr varies by country, and ranges between the 21st, 23rd, 25th, 27th, or 29th day of the month. During this night, prayers are made and the Quran is studied. The end of the month of fasting occurs on the first day of the following month of *Shawwal* and is the holiday known as Eid al-Fitr, celebrated by communal prayers followed by extensive family gatherings and feasting along with an exchange of gifts.

Other important holidays include *Isra al-Mi'raj*, which celebrates the ascent of the Prophet Mohammed (the 27th day of the seventh month of *Rajab*) and the *Eid al-Adha* (approximately 70 days after the end of Ramadan, on the 10th day of the 12th month of *Dhu'l-Hijja*). The *Eid al-Adha* is a time of prayer and sacrifice of a lamb or other animal, often a goat, and celebrates the willingness of Ibrahim (Abraham) to sacrifice his son Isma'il. This holiday occurs at the end of the hajj (the pilgrimage to Mecca). A summary of the Islamic holidays with their dates and meanings is given in Table 5.2.

## The Hawaiian Calendar

The ancient Hawaiians developed an original lunar calendar that made use of the lunar months and the position of the Pleiades star cluster to create a stellar–lunar calendar. The year was known as the *makahiki*, and it was divided into two equal sections for summer and winter. The key for maintaining accuracy of the lunar calendar was the synching effect of the Pleiades star cluster, known to the native Hawaiians as *Makali'i* (Johnson and Mahelona 1975).

The beginning of the first Hawaiian season and the Hawaiian year is in late October or November, and is marked by the rising of the Pleiades in the east just after sunset, known as *makali'i-hiki*. The first month begins when the crescent moon is first sighted in the evening (the day known as *hilo*), and sets of 30 nights are counted from this point. Each night had particular associations with agriculture (planting of certain crops were best on particular nights) and religion (including worship of specific gods during the *tabu* or holy nights). The second season of summer began in May when the Pleiades were first sighted in the pre-dawn sky, during their heliacal rising. The twelve 30-day lunar months of the Hawaiian calendar, with their 30-named nights, kept track of the year, but occasionally some months would need to be added to reconcile the annual cycle of the Pleiades with the lunar calendar, much as in other lunar calendars.

The Hawaiian New Year season was a period of celebration of the gifts of the land, closely associated with the god *Lono,* a god of fertility, agriculture, and medicine. A successful harvest and rebirth of the land was credited to the annual return of Lono, and the festivals and celebrations gave thanks to him. By coincidence, Captain Cook's final landing in Hawaii in February 1779 came just after the Hawaiian New Year period, and the Hawaiians may have interpreted Cook's arrival to be an inauspicious premature return of the god Lono, which required a ritual sacrifice to return order to the universe (Ruggles 2005, p. 179). This mistake in timing resulted in a misunderstanding between Cook and the Hawaiians which ultimately resulted in Cook's death—perhaps a cautionary tale on the importance of learning the calendrical customs of other cultures!

## The Zuni Calendar

Most of the pueblo Native American tribes, such as the Zuni, used rituals to help mark months and years. One key yearlong Zuni ritual reconciled and celebrated both solar and lunar calendars. The ritual is known as Sayatasha's Night Chant, in which an actor portraying the masked god Sayatasha keeps track of the days and months. The beginning of the recitation is timed with the sighting of the first full moon after the winter solstice ceremony. The actor plants prayer sticks every full moon throughout the year until October for 10 lunar months. After the 10th month, he begins counting the days before the beginning of an intense end of year dance known as *Shalako*, by untying a string which contains 49 knots, one for each day. At the end of the 48th day, Shayatasa's actor begins the long talk, and he and a set of other masked gods (the *kachinas*) participate in the Shalako dance until dawn. During the final night, the actor recites the precise poem intoned by hundreds of generations of past Sayatasha impersonators, reciting word for word a "Night Chant" which describes the motions of the Sun and the Moon, the rituals of the people, and their origins (Figs. 5.9 and 5.10). The experience of recreating the exact ritual of their ancestors binds the Zuni together and unites their activities with the larger celestial clockwork (Williamson 1984, p. 34).

## Native American and Tribal Seasonal Calendars

Many other Native American groups used changes in vegetation and animal behaviors to set their calendars and named their months based on these annual cycles. Like the Hawaiian and Zuni people, observations of the Moon, the Sun, and stars would give the "master clock" to reset the

**Fig. 5.9** A rare photograph from 1897 showing the procession of the Shalako, as he crosses the bridge toward Zuni pueblo to plant prayer sticks (image by Ben Wittick, 1845–1903, from Library of Congress; digital ID http://hdl.loc.gov/loc.pnp/cph.3c15456)

**Fig. 5.10** The Shalako and kachina dancers at Zuni pueblo, 1897 (photograph by Ben Wittick, from the Library of Congress. Digital ID http://hdl.loc.gov/loc.pnp/cph.3c01156)

calendar and start a new year. The stellar observations could include noting the position of the Big Dipper relative to the horizon at the beginning and middle of the night, as is the practice of the Mescalaro Apache, who know the Big Dipper as "nahakus" (pivot point) (Farrer 1991, p. 48). The Havasupai tribe began their new year with the heliacal rising of the stars known as "the Hoop" (Corona Borealis), which for marked the new year in mid-November. The Maricopa and Cocopa tribes of Arizona believed the stars were literally a hand of the creator, and lines of stars were the imprints of his fingers. These stars also worked like the hands of a clock, indicating seasons when plants are ripe such as the cactus which produces fruit in the summer when the Pleiades (*Xitca'*) appear on the eastern dawn horizon (Miller 1997, p. 199).

Across the Earth, African tribal cultures observed some of the same types of seasonal calendars. The Nuer, a tribe from Sudan, still practice traditions that divide their year into the seasons of *tot* (wet) and *mai* (dry), with subdivisions of the year based on their cycles of migration, and lunar months named after specific activities during each season. The Mursi people from Ethiopia also kept a lunar calendar based on the wet and dry seasons, with months named in a 12 lunar month cycle known as the *bergu*. The Trobriand people, living in small islands off the coast of New Guinea, marked time with cues from the nearby marine life and using a 10-month lunar calendar. During the full moon night between October 15 and November 15, a worm known as the annelid undergoes a massive spawning during this one night of the year. In this case, the astronomical observations to set the new year for the tribe are left to a small marine invertebrate! The month that starts the year is named *Milamala* in honor of this creature and is celebrated across the region. The exact date of the new year varies for each of the islands near the Trobriand, depending on exactly which full moon triggers the spawning (Aveni 1989, pp. 171–175).

## The Chumash Seasonal Calendar

The Chumash tribe from southern California used a combination of weather indicators, plants, and precise observations of the Sun to set their calendar. Each year a multi-day solstice ceremony took place near the winter solstice, in which a shaman would lead the people in a ritual that was believed to give new life to the Sun and return the Sun back to a northerly path. The ceremony involved using a number of feathered poles to mark important shrines in the territory, and a crucial sun dance in which a sun stick was used to "pull" the Sun back into its correct course along with precisely timed dances, chants, and songs.

The central instrument in the ceremony was a portable sun stick that included a carved and decorated stone head that was inclined precisely toward altitude of the winter solstice sun and was decorated with ray markings and shell beads in the shapes of constellations. Before the ceremony, solar solstice observations were taken from caves that had carefully arranged panels of rock art, in which a sliver of light would cross through a sun figure on the date just before winter solstice. Several of these "solar observatories" exist in southern California, and include the Burrow Flat cave, Condor cave, and Window cave (see Figs. 5.11 and 5.12 below). The sometimes harsh and dry climate of Southern

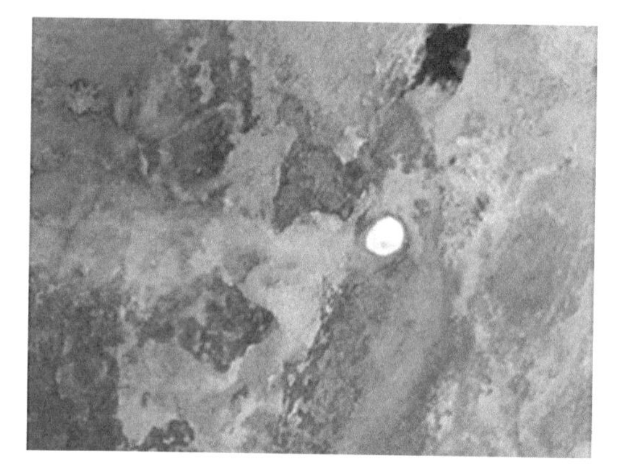

**Fig. 5.11** A hole carved into the side of "Condor cave" used for winter solstice observations by the Chumash. Chumash shamans would often perform a vigil near the winter solstice to help time the annual winter solstice ritual that included using caves which included small portholes that would light up a sun figure at the time of the winter solstice (photo courtesy of Jennifer Perry)

**Fig. 5.12** Two other Chumash solar observatories. Burrow Flat cave (*left*), near Chatsworth, CA, in which the winter solstice sun projects a blade of sunlight inside the cave to bisect the concentric circle sun figure inside. Window Cave (*right*), at Vandenberg Air Force Base, near Lompoc, CA, has a small window that beams the light of the winter solstice onto a small sun figure inside (photos taken by the author during Pomona College field trips)

California required careful monitoring of the Sun to help determine the best time for seasonal migration, fishing, and gathering of crops. As mentioned earlier, the Sun as a "Sky Person" was also a character to be reckoned with, and by timing their rituals precisely, the Chumash believed they could help influence the fate of their world in the coming year.

After the winter solstice, the Chumash would then begin their lunar calendar, and named each month for the changes in the plants and weather in their Southern California environment, described in Table 5.3. Each month also had an omen for people born in these seasons, which is included in the rightmost column of Table 5.3.

The Chumash believed that not only their solstice ceremony helped reverse the southward path of the Sun, but that the renewal of the Earth by the Goddess Hutash was aided by a fall festival named after this goddess. During the Hutash ceremony, which was usually held in early fall (August or September), the Chumash would celebrate for 5 or 6 days at each of the provincial capitals of the Chumash territory. A ritual enclosure was built for ceremonies of the 'antap cult, and the entire community would join in singing, dancing, and games. At the end of the week in selected years a mourning ceremony would conclude the *Hutash* festival, to commemorate the departed in the community. For the mourning ceremony, three concentric circles of the community were formed, with those men in mourning in the center, women in a circle outside of them, dancing counterclockwise, and a larger group of spectators watching an the outermost circle, as two elders led them in singing. Then at the end, the elders clapped their hands to stop the ceremony, and the belongings of the departed were placed into a large bonfire that brought a dramatic end to the year's *Hutash* ceremony (Hudson and Underhay 1978, p. 46).

**Table 5.3** The Chumash calendar, with natural associations for each month (based on information in McCall 1990)

| Month | Chumash | Omen for month |
|---|---|---|
| January | "Month of Datura" | Named for the sacred plant of the Chumash, people born in this month will be very successful |
| February | "Month when things begin to grow" | This is a rainy month, which brings forth life. People born in this month can be unsure, unreliable, but resourceful |
| March | "Month of spring" | As many leaves come out this month, some weak and some strong, so it is that people born in this month can be sickly and sad |
| April | "Month when flowers are in bloom" | People born in this month are pleasant and cheerful, as are the flowers that bloom now |
| May | "Month when Carrizo is abundant" | This month is named after a plant used for arrows and tobacco tubes. People born in this month are handy at making things, and at medicine |
| June | "Month when things are divided in half" | This month is named for the midpoint of the solar year, and is a time when people gather food in many directions. People born in this month are careful and respected by others |
| July | "Month when everything blows away" | Here dry summer winds blow hard in Southern California, and it is believed that people born in this month are not at peace and stir things up |
| August | "Month of fiesta" | During this month the Chumash celebrate a harvest festival, and people born in this month are thrifty and generous |
| September | "Month when those that are dry come down" | Many creatures migrate from the high points to the coast in this month, and people born in this month were thought to be careful and good at avoiding dangers |
| October | "Month of the great canoe builder" | This month is when canoes stayed at home due to rough seas. People born in this month will be very rich and with abundant shell bead money |
| November | "Month when rain keeps us indoors" | One of the months with the most rain, which led to the belief that people born in this month were never satisfied |
| December | "Month when the Sun's brilliance begins" | The date of the winter solstice ceremony is set in this month, and people born here start weak and grow stronger, just as the Sun does after this month |

## The Inuit Seasonal Star Calendar

Half a world away, in the frozen northern reaches of the Arctic, the Inuit developed a set of seasonal star markers for measuring the duration of the long winter night, which can last for months. The Inuit had a lot of time to contemplate the northern winter constellations, and recognized that within these stars were clues to the natural world, and to the cycles of life (Fig. 5.13).

**Fig. 5.13** Mural of the seasons of the Tongva (*top*), a tribe from Southern California next to the Chumash. The months of the Tongva, like the Chumash, were locked into the cycles of nature. This image shows some of the key species and natural cycles that set the rhythms of Southern Californian Native American. Detail of the mural's *right side* depicting winter (*bottom*) (original artwork by Sheila Pinkel, reprinted by permission. The work is entitled "*Journey to Tovangar, The World of the Tongva*" and is a large mural in the library in Sherman Oaks, CA)

The stars Altair and Tarazed formed the Inuit constellation known as *Aagjuuk*, and these two stars first became visible in late spring and were used to indicate the end of the long winter night. The two top stars of Orion's "shoulders," Betelgeuse and Bellatrix, were known as *Akuttujuuk* and were used to mark the middle of the winter, known to the Inuit as the "dark period." According to the Inuit elder Ijangiaq:

> There are two stars known as *Akuttujuuk*. These stars will show up after twilight during the dark period … When you see them during the [evening twilight] it is an indication that the days are getting longer.
>
> —(MacDonald et al. 1998, p. 53)

The Pleiades were used to mark time by many different Arctic groups, and had the name *Sakiattiak* among the Igloolik, who also used them for navigation. The Kobuk River people called the Pleiades *Sakopsakkat*, which translates as "the ones who close their eyes," as these stars were seen later in the evening near bedtime. At Kodiak Island, the appearance of the Pleiades in August signaled the start of a new year.

The star Sirius, known as *Sinquuriq* ("flickering" in Inuit), appeared low on the horizon for the Arctic peoples, was used both to tell time and the weather. Its appearance would flicker and also the weather could change the colors of this star, with a reddish color apparent in the coldest nights, and a lack of twinkling indicating the arrival of milder weather.

The bright star Arcturus, known to the Inuit as *Sivullik*, or "leader," is visible in the evening sky, being the first star (or "leader") to be sighted in the early evening in the west. For the Inuit of Coronation Gulf, the transit of Arcturus signaled the beginning of the sealing season (MacDonald et al. 1998, p. 78).

The Inuit observed a set of 13 lunar months, and a simple system was used to determine whether a year had 12 or 13 lunar months. If the winter solstice coincided with the first day of the 13th month, it was skipped, giving a natural alternating pattern of 12 and 13 lunar months over the years. Like many cultures, the Inuit used both cues from nature and from the stars to set their months and the month names, shown in Table 5.4, reflect this awareness of the environment. The various month names of the Inuit calendar, and corresponding cues from the sky and animals are summarized in Table 5.4.

**Table 5.4** Inuit season names, with their celestial and biological cues (MacDonald et al. 1998)

| Inuit month | Western month | Meaning | Season | Celestial cues | Biological cues |
|---|---|---|---|---|---|
| Siqinnaarut | Jan/Feb | "Sun is possible" | Ukiuq (winter) | Aaguuk stars seen at dawn; The Sun leaves horizon | Formation of new ice along floe edge; seal hunting |
| Qangattaasan | Feb/Mar | "The Sun gets higher" | Ukiuq (winter) | The Sun rises higher | Walrus migrate toward the land-fast ice |
| Avunniit | Mar/Apr | "Premature birth of seal pups" | Upirngak-saaq (early spring) | Spring equinox; days and nights equal | Premature seals born; snow conditions fast for dog sleds |
| Nattian | Apr/May | "Seal pups" | Upirngak-saaq (early spring) | Lengthening days | More snow and seal pups are born |
| Tirigluit | May/June | "Bearded seal pups" | Upirngaaq ("spring") | The Sun continuously above horizon | Bearded seal pups are born; snow too soft for igloos |
| Nurrait | June | "Caribou calves" | Upirngaaq ("spring") | The Sun continuously above horizon; summer solstice | Caribou calving + bird migration arrival |
| Manniit | Jun/July | "Eggs" | Upirngaaq ("spring") | The Sun continuously above horizon | Migratory birds nest |
| Sagguruut | July/Aug | "Caribou hair sheds" | Aujaq ("summer") | The Sun sets again, sea ice breaks up | Caribou skin sheds |
| Akullirut | Aug/Sept | "Caribou hair thickens" | Aujaq ("summer") | Twilight returns + bright stars visible | Seeds uncoil on Malikkaan plant; caribou skins best for winter clothing |
| Amirailaut | Sept/Oct | "Velvet peels from caribou antlers" | Ukiaksajaaq ("early autumn") | Autumnal equinox; days and nights equal | Shedding of velvet from caribou antlers; morning dew is heavy. Ice forms in small ponds |
| Ukiulirut | Oct/Nov | "Winter starts" | Ukiaksaaq ("autumn") | Days shorten | Caribous mate, fish spawn, sea ice starts forming |
| Tusartuut | Nov/Dec | "Hearing (news from neighboring camps)" | Ukiaq ("early winter") | Sunlight only a few hours | Sea freezes over |
| Tauvikjuaq | Dec/Jan | "Great darkness" | Ukiaq ("winter") | The Sunless period; pre-dawn appearance of Aaguuk stars | |

## Development of the Greek Calendar

The early Greeks made many of the same observations as Native Americans, as they timed their activities to match both natural cycles and astronomical events. The list of cues from nature within the Greek calendar is described in the "works" section of Hesiod's Works and Days, one of the earliest surviving works of Greek culture, dating from approximately 700 BCE. Celestial observations included noting the heliacal rising of stars and star clusters such as the Pleiades, as described by Hesiod:

> When the Pleiades, daughters of Atlas are rising (in early May), begin your harvest, and your ploughing when they are going to set (in November). Forty nights and days they are hidden and appear again as the year moves round, when first you sharpen your sickle. This is the law of the plains, and of those who live near the sea, and who inhabit rich country, the glens and dingles far from the tossing sea,—strip to sow and strip to plough and strip to reap, if you wish to get in all Demeter's fruits in due season, and that each kind may grow in its season.
>
> —Hesiod et al. (1914)

The Greek year began in October and continued on through September. Each month of the early Greek calendar had a combination of astronomical and natural cues, which helped keep the timing of natural and human activities in synch and these cues are summarized in Table 5.5. The Greek days were numbered from 1 to 30, like the Hawaiian month, and included auspicious days for various activities such as "bringing brides home," "giving birth to boys," "planting," "shearing," and of course "worshipping gods." The days were set by the phase of the Moon, with the month beginning at the first sighting of the crescent moon.

Each of the different Greek city-states had different month names and separate priests would call the beginning of the lunar month and decide when an extra month was needed based on stellar observations. The month names in Athens, for example, were *Hekatombaion, Metageitnion, Boedromion, Pyanepsion, Maimakterion, Poseidon, Gamelion, Anthesterion, Elaphebolion, Mounychion, Thargelion, and Skirophorion* (Evans 1998, p. 182). Year names were set relative to the reign of a particular ruler or relative to key events in local history.

One technique used to provide one common reference for year names in ancient Greece was the Olympic games, which began in the year 776 BC. Any confusion about the Olympic game number was fixed by attaching to the date the name of the champion of the most prestigious Olympic event,

**Table 5.5** Greek calendrical months from Hesiod

| Month | Celestial cue | Natural cue | Human activity |
|---|---|---|---|
| October | Pleiades hel. set | Cranes migrate, leaves shed | Start plowing and planting |
| November | | Winter rains | Late plowing; haul ships to land |
| December | Arcturus hel. rise | First snows | Keep oxen in doors |
| January | | | |
| February | | Swallows appear | Make clothes |
| March | | Fig leafs in growth | Finish pruning and digging vines |
| April | Pleiades hidden 40 days | Cuckoo sings | Spring sailing |
| May | Pleiades hel. rise | | Harvest |
| June | | Artichoke in flower; zephyros winds | Winnow and store food; summer sailing |
| July | Orion hel. rise | | Bring in hay and fodder |
| August | Sirius hel. rise | | Harvest grapes |
| September | Arcturus hel. rise; Orion + Sirius on meridian | End of summer | Cut wood |

*Sources*: West (1978) and Aveni (1989)

**Fig. 5.14** The ancient Greeks celebrated athletics and used the Olympic games to provide the basis for a universal calendar among the different Greek city-states. This Attic Greek vase from 530 BCE shows a running race from the Pan-Hellenic games of the time (*image source*: http://commons.wikimedia.org/wiki/File:Greek_vase_with_runners_at_the_panathenaic_games_530_bC.jpg)

the footrace known as the *stadion* (Fig. 5.14). For example, the historian Diodoros cited a date corresponding to the year 420 BC in ancient literature as follows:

> The Greeks celebrated the 90th Olympiad, in which Hyperbios of Syracuse won the stadion.
> (Evans 1998, p. 184)

By using the Olympic games to anchor their calendar, the Greeks were able to extend the universal language of sport to also provide a widely recognized international basis for measurements of time.

## Roman Timekeeping

The Roman calendar survives to this day, at least in name, in the form of the Julian Day. The Julian Day is not directly related to the Julian calendar, since its origin is set well before the first glimmerings of the Roman Empire, back on the date of January 1, 4713 BC (which happens to be a Monday!). This ancient date was selected as a point where three of the cycles observed in Renaissance Europe were all at their beginnings. The three cycles include the "indiction cycle," a 15-year cycle used in documents within Europe, the "Metonic cycle" of 19 years for lunar months, and a cycle of 28 years in which the day name and leap year in the Julian repeat. The product of all three cycles (15 × 19 × 28) gives 7980 years and sets the origin of the Julian Day count. The Julian Day is still used in astronomy research where the complexities of orbital periods require the use of a uniform count of days.

The origins of Roman timekeeping began not in 4713 BC, but in the period of the Roman monarchy, from approximately 776 to 509 BC, when the Roman calendar had 10 named lunar months. Each lunar month was divided into named days; the "calends" were the first days of the month, the nones were the 5th day of the month, and the Ides (Idus) was the 13th day of the month (Fig. 5.15).

The "calends" marked the beginning of both the lunar month and a business cycle, in which debts were to be paid at the beginning of the month, before the "nones," with the ides marking the middle of the lunar month. The calendar at first counted only those days when business was being done, and featured a 10-month year which started at the vernal equinox in Martius (named after the god Mars), and continued with Aprilis, Maius, Iunius, Quintilis, Sextilis, September, October, November, and December. Many of the month names (unchanged to this day) were simply their numerical place in Latin corresponding to "fifth" (Quintilis), "sixth" (Sextilis), "seventh" (September), etc. The months in the Roman Republic were of unequal length—Martius, Maius, Quintilis, and October were all of

| | | KL | .IANVARIVS. | SOLIS CAPRI. | | LVNAE S.G. | | S.G. | |
|---|---|---|---|---|---|---|---|---|---|
| 1 | A | | Circucisio domini | 20 | 3 | 0 | 13 | 0 | 13 |
| 2 | b | 4 noñ | Octaua.s.Stephani | 21 | 4 | 0 | 26 | 0 | 26 |
| 3 | c | 3 noñ | Octaua.s.Ioannis | 22 | 6 | 1 | 10 | 1 | 9 |
| 4 | d | 2 noñ | Octaua.s.Innocentum | 23 | 7 | 1 | 23 | 1 | 22 |
| 5 | e | Noñ | Vigilia | 24 | 8 | 2 | 6 | 2 | 5 |
| 6 | f | 8 idꝰ | Epiphanię domini | 25 | 9 | 2 | 19 | 2 | 18 |
| 7 | g | 7 idꝰ | Clauis.lxx. | 26 | 11 | 3 | 2 | 3 | 1 |
| 8 | A | 6 idꝰ | Erhardi episcopi | 27 | 12 | 3 | 15 | 3 | 15 |
| 9 | b | 5 idꝰ | Iuliani & sociorum eius | 28 | 13 | 3 | 29 | 3 | 28 |
| 10 | c | 4 idꝰ | Pauli primi eremitę | 29 | 14 | 4 | 12 | 4 | 11 |
| 11 | d | 3 idꝰ | AQVARIVS | 0 | 16 | 4 | 25 | 4 | 24 |
| 12 | e | 2 idꝰ | | 1 | 17 | 5 | 8 | 5 | 7 |
| 13 | f | Idus | Octaua epiphanię | 2 | 18 | 5 | 21 | 5 | 20 |
| 14 | g | 19 kal' | Febru. Foelicis in pincis | 3 | 19 | 6 | 4 | 6 | 3 |
| 15 | A | 18 kal' | | 4 | 20 | 6 | 18 | 6 | 16 |
| 16 | b | 17 kal' | Marcelli papę | 5 | 21 | 7 | 1 | 6 | 29 |
| 17 | c | 16 kal' | Antonij monachi | 6 | 22 | 7 | 14 | 7 | 12 |
| 18 | d | 15 kal' | Priscę uirginis | 7 | 23 | 7 | 27 | 7 | 25 |
| 19 | e | 14 kal' | | 8 | 24 | 8 | 10 | 8 | 8 |
| 20 | f | 13 kal' | Fabiani & Sebastiani martyrum | 9 | 25 | 8 | 24 | 8 | 21 |
| 21 | g | 12 kal' | Agnetis uirginis | 10 | 26 | 9 | 7 | 9 | 4 |
| 22 | A | 11 kal' | Vincentij martyris | 11 | 27 | 9 | 20 | 9 | 17 |
| 23 | b | 10 kal' | | 12 | 28 | 10 | 3 | 10 | 0 |
| 24 | c | 9 kal' | Timothei apostoli | 13 | 29 | 10 | 16 | 10 | 14 |
| 25 | d | 8 kal' | Pauli couersio | 14 | 30 | 10 | 29 | 10 | 27 |
| 26 | e | 7 kal' | | 15 | 31 | 11 | 13 | 11 | 10 |
| 27 | f | 6 kal' | | 16 | 31 | 11 | 26 | 11 | 23 |
| 28 | g | 5 kal' | | 17 | 32 | 0 | 9 | 0 | 6 |
| 29 | A | 4 kal' | Clauis.xl. | 18 | 33 | 0 | 22 | 0 | 19 |
| 30 | b | 3 kal' | | 19 | 33 | 1 | 5 | 1 | 2 |
| 31 | c | 2 kal' | | 20 | 34 | 1 | 18 | 1 | 15 |

**Fig. 5.15** Page from the medieval book *Calendarium*, by Joannes Regiomontanus (Venice, 1476), showing the Roman calendar days. The third column shows the "nones" (non), the "ides" (id), and the calends (kal) for the month of January, with Christian feast days appearing in *red*, leading to the expression "red letter days" (image courtesy of Claremont Colleges Special Collections)

31 days and Aprilis, Iunius, and Sextilis, September, November, and December were each of 29 days. After the end of the business and agricultural year, near the winter solstice, the early Romans simply stopped counting the days of the year until spring, which would signal the beginning of the new year in Martius.

In order to have a complete annual calendar, the Romans later added two additional months (Januarius, with 29 days, and Februarius, with 28 days) in 509 BC to cover the entire solar year, adding up to 355 days with 12 lunar months. In order for the solar and lunar years to match evenly, the irregular lengths of the lunar months were supplemented with an intercalary month roughly every other year, known as "Intercalcaris." This system required significant effort to maintain, and often the lunar months in some years would slip significantly with respect to the solar year (Steel 2000, p. 55).

Julius Caesar, as part of his duties as emperor and chief pontiff of Rome, discarded the old lunar calendar in 46 BC. To reform the calendar, Caesar added 10 days to the year by lengthening the five shorter 29-day months by two extra days, with the exception of February, which was kept at 28 days with an extra "leap day" added every 4 years. By the time of Caesar's reforms, the calendar had drifted by over 80 days and so the year 46 BC is known as the "year of confusion" (*annus confusionus*) since it was 445 days long! The result of Caesar's reforms was the Julian calendar, which includes 365 days, with an extra 0.25 days from leap years to give an average solar year of 365.25 days (Steel 2000). The Julian calendar, with 365.25 days, provides a close match with the length of the solar year of 365.2422 days. Caesar's calendar remained in place for over 1000 years, with the minor adjustment of renaming of Quintilis to Julius, after Caesar's assassination in 44 BC, and the later renaming of Sextilis to Augustus in AD 8 after his grandnephew, the emperor Caesar Augustus.

The Julian calendar is still in use today, with the (minor) correction known as the Gregorian reform of 1582 (named after Pope Gregory XIII), which took back 3 of the leap years in centuries not divisible by 4 (1700, 1800, 1900), while retaining leap year in those divisible by 4 (2000, 2400, etc.). The result of the Gregorian reform is to set our modern calendar to have a year of 365.2425 days, which gives an error or "drift" of only 0.0001 days in a year. The Gregorian reform was adopted gradually throughout Europe and resulted in a 10–15 day shift in dates as each country adopted the new calendar. The Gregorian reform not only enabled Easter to appear closer to the vernal equinox, but also caused many "anni confusioni" in the different countries of Europe as it was phased in. For example, if one were alive in the US colonies during the year 1752 when the Gregorian reform was enacted, one would go to bed one night on September 2, 1752 and awaken on the September 14, 1752 some 12 days later! (Evans 1998, p. 168).

The wide variety of calendars in use even in our own civilization can lead to confusion. For example, January 1 on AD 2000 is actually December 19, 1999 in the Julian calendar (without the Gregorian correction) and corresponds to Julian Day number 2451544, and also the year 5760 (Teveth 23) in the Hebrew calendar. Additional dates for this same day include year 1420 in the Islamic calendar (Ramadan 24), and for those who still keep track, January 1, 2000 is Annee 208 in the French Republican calendar.

## Sidereal Lunar and Planetary Calendars

### *The Incan Sidereal Year*

Many aspects of the Incan civilization are unique among all empires. The Inca built a road network more extensive than the Roman Empire, but did not invent the wheel. The Inca controlled a territory including hundreds of language groups and climates ranging from rain forest to moon-like desert, but did not have writing. The Inca built stone buildings with an original form of masonry combining enormous blocks to within a razor's edge, but without metal tools.

The Inca were also original in their approach to time. For the Inca, time was measured in many ways, from the natural cues of the environment, from the Moon and Milky Way, and from the sightings of stars, planets, and the Pleiades in the night sky.

The Inca recognized two groups of stars which were of special importance in managing their civilization; the *ch'aoka* ("shaggy hair") stars, which included planets or very bright stars visible near dawn, and the *coyllur* stars, such as the Pleiades, which were used for predicting weather. The position of the Sun on the sky and on the horizon moved between landmarks used to mark time. The landmarks included the Pleiades, close to the Sun during June solstice sunrise, and the constellation known as *atoq* ("fox"), where the sun appears during the December solstice sunrise (Urton 1981). Lines connecting these landmarks, and a corresponding pair of sunset markers, formed two axes of space, known as the *atoq* and *collca* axes, which divided the sky into quadrants.

The motions of the Milky Way in the night sky from 1 month to the next also served as a large "hour hand" of night and provided a similar splitting of the year into four parts. The Milky Way was known to the Inca as the celestial counterpart of the Vilcanota River, and the various dark cloud constellations symbolized the diverse forms of life on the Earth. According to the Spanish Chronicler Juan Polo de Ondegardo (1571), "The Incas believed all animals and birds on the Earth had their likenesses in the sky whose responsibility was their protection and augmentation."

Because of the Inca belief in the mirroring of animal life in the sky and on the Earth, the coincidence of animal dark cloud constellations (such as *atoq*) and the Sun were associated with the birth and gestation periods of these same animals on the Earth. The Inca believed that the dark cloud constellations provided the rhythm for life on the Earth.

The Inca calendar was also set by heliacal risings of bright stars, star clusters, and dark cloud constellations. The heliacal rising of the brightest star in the Southern Cross (α Cru, associated with *Yutu*, a bird, and dark cloud constellation) marked the beginning of the agricultural season, while the rising and superior culmination of the stars α and β Centauri (associated with the "eyes of the Llama") coincided with the birth season and mating season of the llama. The positions of star clusters were noted both above in the night sky (at zenith or transit) and below under the feet and crops of the Inca (at nadir or inferior culmination). For example, the appearance of the Pleiades or *collca* at nadir, directly beneath the Earth, was thought to be a harbinger of the best time for planting (Urton 1981).

The Inca religion also included extensive daily rituals which required precise timekeeping. Religious worship took place at precise times and spatial locations, which created patterns that mirrored the larger organization of the universe. Worship sites rotated through the city to various shrines or *huacas* on a daily cycle, arranged along patterns of space that connected the shrines to the horizon and the stars beyond. The ritual calendar linked all of the 328 *huacas*, one for each of the named days in the 328-day Inca year.

The somewhat unusual number of days in the Inca year comes from the observation of the Incan lunar month, which is based on a unique lunar/stellar calendar. The Incan month was closer to the 27.3 day *sidereal* period for the Moon to return to the same position against the stars, as opposed to the more familiar 29.5 day lunar *synodic* month of most other cultures (the synodic month is the time between full moons and is what we usually call a "lunar month").

The Inca would carefully watch the location of the Moon in the sky relative to the stars to determine the beginning and ending of lunar months. Groups of 12 lunar sidereal months combined to create an Incan year of 328 days, not only giving the annual calendar, but also leaving behind a group of 37 days which were uncounted (Aveni 1997). These uncounted days, much like those in the early days of the Roman Republic, were a time of rest and were outside the business and religious calendar. The new year (and the "counted" days) began again when the Pleiades were observed again at dawn, near the June solstice. By combining the lunar and stellar observations, the Incan calendar was prevented from drifting, but the duration of the uncounted days would vary slightly from year to year due to the mismatch between the lunar cycle and the solar year (Fig. 5.16).

## Hindu Naksatras and Lunar Calendars

The Moon's position relative to the stars was also observed carefully in ancient India, where the practice of Hindu astrology required the use of the Moon and star coincidences known as *naksatras*. Like the Inca, the early Hindu astrologers would watch the position of the Moon against the background of a set of indicator stars, which set the lunar sidereal calendar. For determining auspicious dates of weddings and births, the Hindu astrologer would combine the lunar phase of the date with the lunar sidereal location or *naksatra* to get a more precise reading of the sky.

The *naksatra* system dates to before 1000 BC, in the era of the Rig Veda. A set of 27 stars were chosen near the ecliptic plane as references for gauging the position of the Moon as it moves through its 27.3-day lunar sidereal cycle. The stars, their corresponding names in Hindu and European astronomy, and the corresponding constellations are presented in Table 5.6.

**Fig. 5.16** The ruins of Maccu Piccu attest to the inspiration and ingenuity of the ancient Inca civilization, which created a civilization with a unique calendar, and blending of time and space in their ritual (image from Library of Congress online exhibit "1492: An Ongoing Voyage")

**Table 5.6** Hindu *naksatras*, used to mark the path of the Moon through the stars during the 27.3-day lunar sidereal month, with star and constellation associations, and lunar month names (Ojha 1996 + Yano (2003)

| Naksatra | Location on sky | Shape | Stars | Lunar month name |
|---|---|---|---|---|
| 1. Krittika | Pleiades + Alcyone | Curved knife | 6 | Kārtika |
| 2. Rohini | Aldebaran | Cart | 5 | |
| 3. Mrigashira | $\lambda$ and $\rho$ Orionis | Face of a deer | 3 | Mārgashīrsha |
| 4. Ardra | Betelgeuse | Jewel | 1 | |
| 5. Punarvasu | Castor and Pollux | House | 4 | |
| 6. Pushya | $\gamma$, $\delta$, and $\tau$ Cancrii | Arrow | 3 | Pausha |
| 7. Ashlesha | $\delta$, $\epsilon$, $\eta$, and $\sigma$-Hydrae | Circular | 5 | |
| 8. Magha | Regulus | House | 5 | Māgha |
| 9. Poorva-phalguni | $\delta$ and $\tau$ Leonis | Manchaka | 2 | |
| 10. Uttara-phalguni | Denebola | Small cot | 2 | Phālguna |
| 11. Hasta | Spica | Hánd | 5 | |
| 12. Chitra | $\alpha$ to $\epsilon$ Corvi | Pearl | 1 | Chaitra |
| 13. Swati | Arcturus | Coral | 1 | |
| 14. Vishakha | $\alpha$, $\beta$, $\gamma$, and $\iota$ Librae | Torana | 4 | Vaishākha |
| 15. Anuradha | $\beta$, $\delta$, and $\pi$ Scorpii | Vali | 4 | |
| 16. Jyeshtha | $\alpha$, $\sigma$, and $\pi$ Antares | Circular earring | 4 | Jyaishtha |
| 17. Moola | $\epsilon$, $\zeta$, $\eta$, $\tau$, $\iota$, $\kappa$, $\lambda$ Sco (Tail and stinger of Scorpius) | Tail of a lion | 3 | |
| 18. Purva-shadha | $\delta$ and $\epsilon$ Sagittari | Tusk of an elephant | 11 | Āshādha |
| 19. Uttara-shadha (Abhijit) | $\zeta$ and $\sigma$ Sagitari | Manchaka | 2 | |
| 20. Shravana | $\alpha$, $\beta$, and $\gamma$, Aquilae (Altair) | Aeroplane | 3 | Shrāvana |
| 21. Dhanishtha | Delphinis | Mridanga | 4 | |
| 22. Shatabhisha | $\gamma$ Aquarii | Elliptic | 100 | |
| 23. Purva-bhadra | $\alpha$ and $\beta$ Pegasi (Markeb) | Manchaka | 2 | Bhādrapada |
| 24. Uttara-bhadra | $\gamma$ Pegasi and $\alpha$ Andromadae | Twins | 2 | |
| 25. Revati | $\zeta$ Piscium | Mardala | 32 | |
| 26. Ashwini | $\beta$ and $\gamma$ Arietis (head of Aries) | Face of horse | 3 | Āshvina |
| 27. Bharani | 35, 39, and 41 Arietis | Yoni | 3 | |

**Table 5.7** The Hindu and traditional Indian calendar included both solar months and lunar months

| Solar months | Season | Gregorian months | Zodiac |
|---|---|---|---|
| Mesh | Vasant | April/May | Aries |
| Vrushabh | (spring) | May/June | Taurus |
| Mithun | Grishma | June/July | Gemini |
| Kark | (summer) | July/August | Cancer |
| Simha | Varsha | August/September | Leo |
| Kanya | (monsoon) | September/October | Virgo |
| Tula | Sharad | October/November | Libra |
| Vrushchik | (autumn) | November/December | Scorpius |
| Dhanu | Hemant | December/January | Sagittarius |
| Makar | (autumn–winter) | January/February | Capricornus |
| Kumbha | Shishir | February/March | Aquarius |
| Meen | (winter–spring) | March/April | Pisces |

*Source*: http://www.nationmaster.com/encyclopedia/Hindu-calendar#tithi

The Hindu astronomers divided the ecliptic into 27 parts (the naksatras, described above) and also divided the sky during any individual day into 27 parts, known as yogas. The yogas (analogous to the hours) were based on the angular distance along the ecliptic of the Sun and the Moon, which divided the sky into 27 regions 13.333° wide. Additional terms and concepts of interest to the Indian astronomers and astrologers include the *tithi*, which is defined as the angular distance between the Sun and the Moon (something akin to lunar phase), and the *karana*, the time required for the distance between the Sun and the Moon to increase by a distance of half of a *tithi*.

The Hindu solar calendar functioned much like many other calendars, with a new year beginning with the first lunation after the vernal equinox, in April. A set of 12 solar months, or *rashi*, was observed based on the Sun's position within the ecliptic. Each of the 12-named solar months included one of the 12-named lunar months. The named lunar months include (in order): Kārtika, Mārgashīrsha, Pausha, Māgha, Phālguna, Chaitra, Vaishākha, Jyaishtha, Āshādha, Shrāvana, Bhādrapada, and Āshvina. Each lunar month is matched with a solar month, and lunar events determine any corresponding festivals within that month. For example, when the Sun is Capricorn, during January, it is the 10th solar month (*rashi*), known as Makar, and the lunar month is known as Pausha. A festival (or *mela* in Bengali) in this month in the province would be known as *paush mela.* A summary of the month names within the Indian calendar is shown in Table 5.7.

The Hindu calendar, like many ancient calendars, included a 7-day week. Days of the 7-day week in the Hindu calendar, known as the *vaasara* in Sanskrit, are named after the seven celestial bodies known to the ancients, in the same order as in the Western calendar. The sequence of days (in order) includes days named after the Sun, the Moon, Mars, Mercury, Jupiter, Venus, and Saturn, which in Sanskrit are Ravi, Soma, Mangala, Budha, Guru, Shukra, and Shani, respectively. Therefore the Indian day names are Ravi vasara (Sunday), Soma vasara (Monday), Mangala vasara (Tuesday), Budha vasara (Wednesday), Guru vasara (Thursday), Shukra vasara (Friday), and Shani vasara (Saturday).

India as a modern country is as diverse as Western Europe, with 22 officially recognized languages and over a dozen separate historic cultures within its borders. Variations in the names of months, days, and timing of festivals across India that reflect the linguistic and religious diversity of the country. However, many of these basic calendrical practices, set in place by Hindu astrologers over 5000 years ago, are still observed throughout modern India. In order to provide some standardization of the calendar across India, the "Indian National Calendar" was adopted in 1957, and this calendar attempted to reconcile the many different ritual calendars of the nation. However, the Indian National calendar is far from perfect, and the effects of precession are causing many of the ritual holidays of India to lose their astronomical underpinnings, such as coincidence with solstice or equinox dates (Aslaksen and Rugulagedda 2001; Nath 2008).

## The Chinese Calendar and Zodiac

The scientific astronomy of China was part of an even longer tradition of astrological forecasting, which was firmly grounded in both astronomy and ancient philosophy. Some of the writings of ancient China on astronomy and philosophy predate the Greeks by centuries, and in some cases millennia! The earliest artifacts of Chinese civilization date from a neolithic tradition that produced small settlements and pottery between 5000 and 2000 BC. The first large-scale civilizations of China, the Xia or Hsia dynasties, date from 2180 to 1600 BC, and were followed by the Shang Dynasty, from 1600 to 1027 BC. These early dynasties produced many of the traditions the later unified China would follow, as well as some of the earliest bronze artifacts, writing, astronomical observations, and calendars known on the Earth.

Among the first innovators of ancient Chinese civilization were the legendary astronomer brothers, Xi and He, who were assigned to observe the heavens by the legendary emperor Yao. From an ancient Chinese book known as the *Shu jing* or "Book of Documents" we can read the commission to the astronomers given by Emperor Yao as the first astronomers began their duties:

> Thereupon he (Yao) ordered Xi and He to accord reverently with august Heaven, and its successive phenomona, wiht the sun, the moon and the stellar markers, and thus respectfully bestow the seasons on the people.
> (Cullen, 2007, p. 3)

Archeology confirms that the practice of astronomy in China began early. The first astronomical observations from China appear to be recorded as inscriptions on fragments of bone known as "oracle bones" for their use in fortune-telling. The practice included writing some observation or question on the bone, placing the bone in a fire, and observing from small cracks within the bone clues about the future. Oracle bones with some of the earliest Chinese characters discovered from the Shang Dynasty have been deciphered and include specific astronomical observational dating as far back as 1500 BC. Astronomical observations include mention of "guest stars" (supernovae), and some of the key seasonal stars such as the "Bird Star" (α Hya, a marker of the Spring), the star known as "Great Fire" (α Sco, or Antares, a marker of the summer), the star known as "Void" (β Aqr, a marker of Autumn), and the "Mao" (Pleiades, a marker of Winter) (Kelley and Milone 2005, p. 317). Inscribed pottery dated between 4500 and 2300 BC gives hints that the astronomical knowledge of the Chinese was helpful in predicting floods, and some of the oracle bones seem to record astronomical conjunctions visible only in the period around 2400 BC.

From the earliest times, the Chinese believed the sky mirrored the Earth and reflected the "Heavenly mandate" of the emperor. The Chinese recognized a set of five elements that describe all earthly and celestial phenomena. Like the early Greeks, the Chinese believed all things were made of these fundamental elements. The five-named elements included wood, fire, the Earth, metal, and water, which had celestial counterparts in the planets Jupiter, Mars, Saturn, Venus, and Mercury, respectively. The solar year was also divided into five seasons, one for each element. The cycles of time and motions of the heavenly objects tied into these numbered elements and connected the essence of these elements to life on the Earth. Motions of the planets, appearances of "guest stars," and "sweepers" (now known as supernovae and comets) gave a picture of the universe that reflected the "heavenly mandate" of the emperor on the Earth.

The Chinese also divided space into systems that included 4, 10, 12, and 28 parts, which arose from both ancient folk traditions and Chinese philosophy developed over thousands of years. The ancient Chinese believed that each of the four cardinal directions had a symbolic animal and color, much like many Native American peoples. The Chinese associated each cardinal direction with animals and with colors that included the Black Warrior/Tortoise for North, the Red Bird for South, the Blue Dragon for East, and the White Tiger for West. Early Shang writings refer to "Ten Heavenly Stems" and "Twelve Earthly Branches," a set of reference stars set the calendar, and the planet Jupiter, respectively. The "heavenly stems" provided a system similar to the decans stars of the Egyptians and helped synchronize the calendar with the seasons. The "Earthly Branches," mentioned earlier, were a timekeeping system in which the 24-hour day (and the sky) was divided into 12 segments, named after animals from an earlier folk tradition.

The Chinese calendar was the culmination of thousands of years of folk tradition and scientific measurement. The calendar also fulfills the second half of the commission given to the astronomers by the legendary emperor Yao in 400 BC:

> Oh you Xi and He! The period is of three hundreds of days, and six tens of days, and six days. Use intercalary months to fix the four seasons correctly, and to complete the year…
>
> (Cullen, 2007, p. 3)

The establishment of the accurate calendar was an urgent priority for the first astronomers of China, and they used all the tools available—the height of the sun was measured with gnomons constructed across the empire, and stars were used to map the locations of the planets and transient events. One of the most important markers was the planet Jupiter, which circles the sky in a 12-year cycle, the basis of the well-known "Chinese Zodiac." From Jupiter's motions come the 12-year cycle that form the basis of the familiar Chinese New Year names. These animals (rat, bull, tiger, hare, dragon, serpent, horse, sheep, monkey, rooster, dog, and pig) give the name to the new Chinese New Year, and also designate the station that Jupiter occupies in the sky in each year of its 12-year synodic period.

These longer timescales were mirrored in smaller units of time within the Chinese calendar. Units of 10 days, known as the *xun*, provided day names from the characters of the *tian gan* or "Heavenly Stems." These 10 days were paired with another set of 12 characters, the *di zhi* or "Earthly Branches." The pairing of names from these two groupings of 12 and 10 characters provided a unique combination 60 day names that would repeat six times over the course of a year. Groups of these day names would be used to locate a day within a solar year, a Jupiter cycle, and these day names would be marked with one of 60 characters to create a 60 solar year cycle of unique day names (Cullen, 2007).

**Table 1. Stems and branches**

| | Stems | Branches |
|---|---|---|
| 1 | 甲 *jia* | 子 *zi* |
| 2 | 乙 *yi* | 丑 *chou* |
| 3 | 丙 *bing* | 寅 *yin* |
| 4 | 丁 *ding* | 卯 *mao* |
| 5 | 戊 *wu* | 辰 *chen* |
| 6 | 己 *ji* | 巳 *si* |
| 7 | 庚 *geng* | 午 *wu* |
| 8 | 辛 *xin* | 未 *wei* |
| 9 | 壬 *ren* | 申 *shen* |
| 10 | 癸 *gui* | 酉 *you* |
| 11 | | 戌 *xu* |
| 12 | | 亥 *hai* |

Table Caption Chinese Heavenly Stems (*tian gan*) and Earthly Branches (*di zhi*) provided a set of characters that together defined a cycle of 60 days. The ancient astronomers noted days from this cycle, and placed the day names in the context of the 12-year Jupiter cycle, a 60 year cycle of names, and the date within the rule of particular emperors (table adapted from Cullen, 2007).

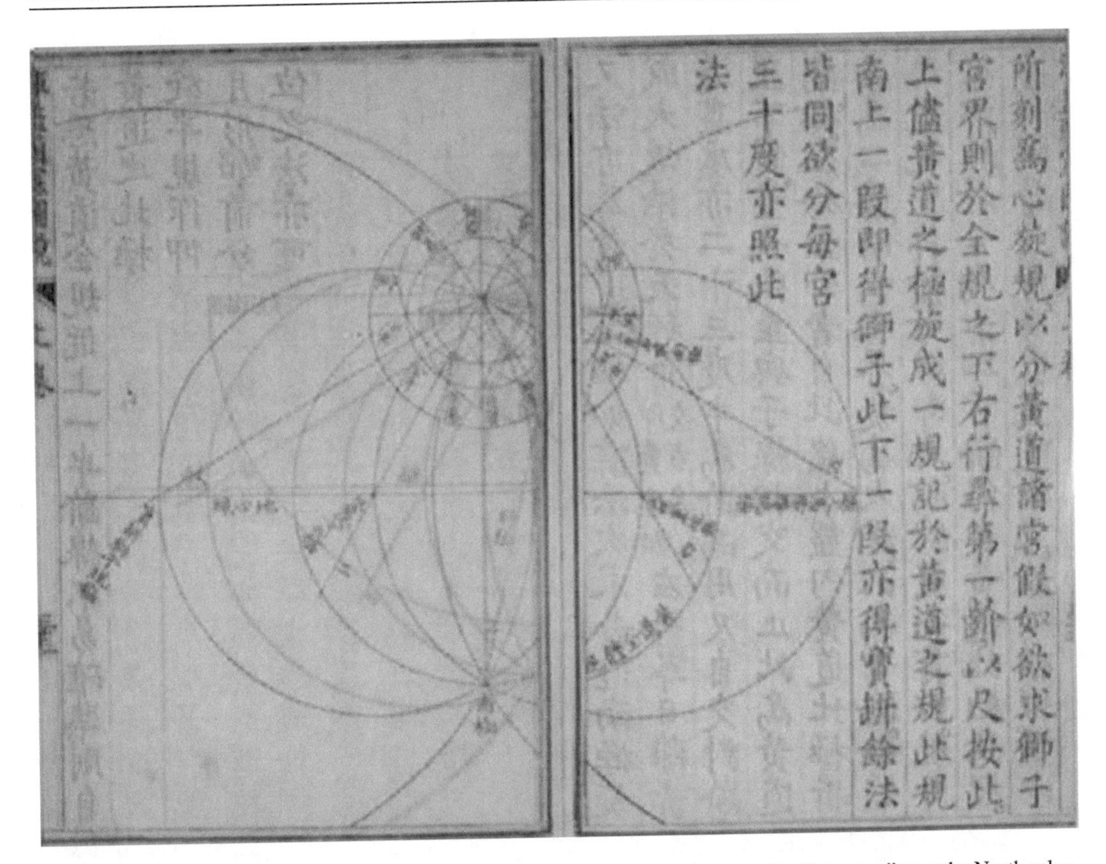


**Fig. 5.17** Illustration of the Chinese hsiu system, showing the connection between the "key stars" near the North celestial pole (*top*) to segments of the celestial sphere (*lines arcing below*). This chart was compiled by one of the first Westerners to make a systematic study of Chinese astronomy, Matteo Ricci, a Jesuit who became well known in the Chinese court during his extended stay in China. This chart was made between 1610 and 1620, based on a date of the ruler Wan-li (image from Matteo Ricci, "*Hun-kai t'ung-hsien t'u-shuo*", from Library of Congress Vatican exhibit "Rome Reborn")

Another key set of landmarks on the sky for setting the calendar were stars. Stars were grouped into 28 *hsiu* or "lunar mansions," and their first appearance at dawn or sunset provided an accurate solar date to reconcile the solar and lunar calendar (Figs. 5.17 and 5.18).

By the time of the Han Dynasty (206–220 BC), the Chinese had developed extensive astronomical bureaus dedicated to compiling records of all the celestial activity visible in each of the 28 lunar mansions. The duties of the astronomer during the Han Dynasty were enumerated in an early Chinese document from 200 BCE:

> He concerns himself with the twelve years (the time Jupiter takes to complete a circuit of the sun), the twelve months, the twelve (double) hours, the ten days, and the positions of the twenty-eight stars. He distinguishes them and orders them so that he can make a general plan of the state of the heavens. He takes observations of the sun at the winter and summer solstices, and the moon at the spring and autumn equinoxes, in order to determine the succession of the four seasons
>
> Needham and Ronan 1980, p. 76.

With the combination of the cycles of the Moon, the seasons, and the Sun, the Chinese astronomers were able to maintain a multilayer lunar, solar, and planetary calendar. The large number of cycles and philosophical associations also provided a rich foundation for a complex astrological system, which we will consider later.

**Fig. 5.18** Chinese astronomical instruments at the Nanjing Purple Mountain Observatory. (*Left*) An armillary sphere used for astronomical calculations, and (*right*) an abridged armilla, which is designed much as a modern equatorial telescope mount for visual observations of the positions of stars and planets. These bronze instruments date from 1437 AD, and provided the professional astronomical bureaus in ancient China the precise astronomical data that provide some of the most sophisticated astronomical records of any historical civilization (photographs by the author)

It is interesting to note that with all of the advanced technology developed for astronomy and technology, the Chinese tended to date events relative to the reign of particular emperors, making a difficult task of calculating the longer elapsed periods between eclipses and comets. One Chinese astronomical record from the Sung Dynasty describes the Great Crab supernova of AD 1054:

> In the fifth month of the first year of the Chih-Ho reign period, Yang Wei-Te (Chief Calendrical Computer) said, 'prostrating myself, I have observed the appearance of a guest-star; on the star there was a slightly iridescent yellow colour'. Respectfully, according to the disposition for emperors, I have prognosticated, and the result said, The guest-star does not infringe upon Aldebaran; this shows the Plentiful One is Lord, and the country has a Great Worthy.
>
> (Needham and Ronan 1980, p 207)

This extract gives very useful astronomical and cultural data, including the location in the sky in Taurus, the date relative to the emperor Chih-Ho, and a hint of the degree of specialization of astronomers in China ("Chief Calendrical Computer"), as well as the cultural interpretation of the supernova as a form of judgment on the emperor.

The professional Chinese astronomical records cover over 1600 years of observations, from approximately 200 BC to AD 1400, and provide rich material for dating ancient astronomical events. From the Chinese records come descriptions of Halley's comet going back to 240 BC, and also records of sunspots for over 500 years, taken by ancient Chinese astronomers looking at the Sun through thinned jade disks. Shen (1969) gives records for four separate supernova observed by Chinese astronomers, including those in AD 185, 396, 487, 902, and 1006. Below are samples of Chinese observations of ancient supernovae, translated from the original records:

> "A star appeared to the south of the Ti lunar mansion, east of Ku-Lou and west of Chi-Kuan. At one time it resembled the half moon, shown with pointed rays, and was so bright that objects could be seen by its light" (description of the Supernova of 1006)
>
> "…it was visible by day, like Venus; pointed rays show out from it on all sides; the color was reddish-white. Altogether it was visible in the daytime for 23 days" (description of the Crab Supernova of 1054)
>
> "The star was as big as a peach.…" (description of Supernova of 902 AD, which was also visible for a year).
>
> (Kiang 1969; Shen 1969)

In several cases, modern astrophysicists have been able to confirm the accuracy of the Chinese observations. One example is the supernova remnant known as G11.2–0.3 from an image taken by the Chandra X-ray telescope (Fig. 5.19).

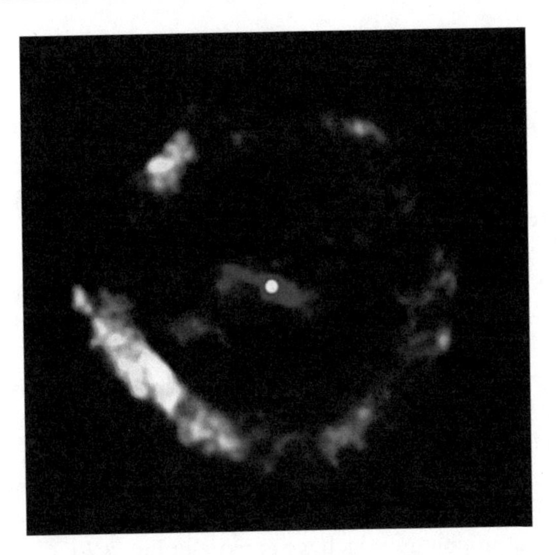

**Fig. 5.19** Supernova G11.2–0.3 observed in 2001 by the orbiting Chandra X-ray observatory and recorded during its original eruption by Chinese astronomers in AD 386 (*figure Credit*: NASA/McGill/V.Kaspi et al.; from http://chandra.harvard.edu/photo/2001/1227/index.html)

Over 1600 years earlier in AD 386, Chinese astronomers recorded a "guest star." This same object can be seen today with the orbiting Chandra X-ray observatory as an expanding nebula of hot plasma. By combining modern space-based observations and ancient Chinese records, astronomers are able to better determine the rate of expansion and cooling of supernovae. The patient writings of the Chinese astronomers 1700 years ago are helping modern astrophysicists make new discoveries, a tribute to the power of writing, and to the power of the stellar astronomy of the ancient Chinese!

## The Long Run: Cosmic Celestial Timekeeping of the Eons

### *Mayan Calendars and the Cycles of the Universe*

Perhaps the most intense and focused exploration of time by ancient civilizations came from the ancient Maya. The Maya were obsessed with time and kept detailed records and counts of each day, with accurate accounting for the positions of planets, the placement of the date both within recent cycles of Venus and the Sun, and within the larger cycles of creation and destruction of the universe. The cycles of time considered by the Maya boggle the mind and encompass thousands or even millions of years.

The interlocking calendars of the Maya contained two short-term calendars, the *tzolkin* and the *haab*, which were based on a 260-day cycle of 13 × 20 day months and a 360 + 5 day cycle of 12 × 30 day months. In addition to these two calendars, the Maya kept track of cycles of much larger eras or eons of time, by bundling the days into groups of 20 and 13, into what we might call centuries, millennia, and 10,000-year periods, which the Maya called tuns, katuns, and baktuns. The largest cycles were thought to be recurrent cycles of creation and destruction, and the Maya believed that each of the named days within their cycles contained a destiny which was relived in each of the many eras of a cyclic universe.

## The Kin

The day or *kin* for the Maya was the fundamental unit of time and the basis of their advanced mathematics for counting the days since the beginning of time. The Mayan creation myth, the Popul Vuh, states explicitly that one of the purposes for human in the present universe was "to keep the count of days" and the Maya were experts at counting.

The Mayan number system was developed to keep track of extremely large numbers and is unique in two aspects: (1) the use of a base 20 number system (common to Mesoamerican people, but perhaps invented by the Maya) and (2) the development, independent of Indian and Arabian mathematicians, of the number zero.

Examples of Mayan numerals are given below, with the actual Mayan notation from both written sources (known as codices; accordion-like books written on bark) and carved notation (known as glyphs). The counting in base 20 (perhaps derived from using both fingers and toes?) provides a different basis for grouping numbers than our base-10 system. In base 10, groups of 10 years (decades) or 100 years (centuries) seem natural. For the Maya, groupings of 20 days or years and 20 × 20 (400) years provide the equivalent groupings, which we explore in Fig. 5.20.

## The Tzolkin

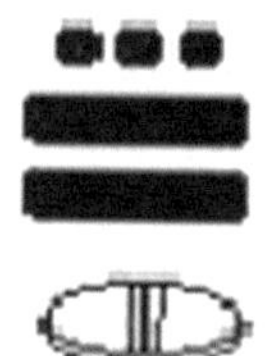

By grouping days into bundles of 20-named days, the Maya formed the basis of the "week" in their calendar, which when combined with 13 sacred numbers made a 260-day cycle known as the *tzolkin*. The word tzolkin or tzolk'in means "count of days" in Yukatek Maya and was the fundamental basis for the Mayan calendar. Each of 20-named days rounding through a cycle with 13 numbers made a cycle known as the *uinal*, or a Mayan "week," in which 20 days were grouped. Each day name would be paired with a number, until the *uinal* was complete. The pairings would continue until every possible pairing of 20-day names and 13 numbers had been made, after 260 days.

The combination of 20-day names and 13 numbers provided the fundamental cycle of time for the Maya. Some have likened the process of pairing names and numbers to interlocking wheels; one wheel would have 13 numbers and would rotate next to a wheel with 20 names. The imaginary assemblage would continue rotating until all 260 pairs had been made (Fig. 5.21).

The 260-day period of the tzolkin seems to lack any correlation with the other calendars we have encountered previously. A period of 260 days does not easily match with lunar months, the solar year,

**Fig. 5.20** Examples of Mayan base 20 (or *vigesimal*) numerals showing their representation of zero (developed independently from Indian and Arabian mathematicians!) and counting with groups of *dots* and *bars* representing ones and fives, respectively, until reaching 20, the base number for their number system (*source*: http://commons.wikimedia.org/wiki/Image:Maya.jpg)

**Fig. 5.21** Interlocking numbers and day names comprise the basis of the Mayan tzolkin, a 260-day calendar pairing 20-day names (represented by the *Jade gear*) and 13 numbers (represented by the *stone gear*). Each day the system advances by one cog, producing a day name and number (original figure by B. Penprase)

or even the motions of Jupiter. However, many authors have noted that 260 days are roughly half the period of the Venus "year"—the time Venus spends within either the Eastern or Western sky before disappearing in the glare of the Sun. The period between "maximum elongations" of Venus is 584 days, which divides into two 260-day periods for visibility and two 32-day periods in which Venus crosses either behind or in front of the Sun.

Since the principal god of the Maya, Quetzalcoatl or feathered serpent, is associated with Venus, it is possible and perhaps even likely that the Maya placed primary importance on the Venus year instead of the solar year. Others have also noted that 260 days is close to the 9-month gestation period for human birth and also close to the period of time needed for the growth of maize. Perhaps these additional appearances of the 260-day cycle heightened its significance for the Maya.

## *The Haab*

Since the Maya needed to maintain crops and time their activities with the stars, a second calendar was needed to fix dates within the solar year. The official count of solar days for the Maya was known as the *haab,* or "vague year," and took secondary importance to the tzolkin. The haab operated much like the tzolkin, but had its basis in 18-named months, with 20 numbered days, and operated somewhat more simply, with its names consisting of 1 of 18 month names, and 20 numbers, progressing along with day names such as "1 Pop," "2 Pop," "3 Pop," until 20 was reached and then the next month would be used. The 18-named months and 20 numbered days together gave a 360-day period, close to the solar year, known as the "tun." To keep the calendar from drifting, an extra unnamed (and very unlucky) 5 days were added to the end. The year-ending days were known as the "uayeb" to give a 365-day solar year, much as in the early Egyptian calendar, except the *uayeb* was a time of dread instead of a time of festivities. Groups of 20 of the haab years formed the grouping known as the *baktun*, which is analogous to a Mayan "decade."

## The Calendar Round

The combination of a *tzolkin* date and a *haab* date gave a unique number and name combination that uniquely specified a day within the Mayan calendar cycle. Some basic mathematics will show that the two cycles will both repeat after 73 × 260 = 18,980 or 4 × 13 × 365 = 18,980 days which corresponds to about 52 solar years. The pairing of these two cycles forms the basis of the shortest and most profound of the Mayan cycles of creation and destruction, the calendar round.

The calendar round was believed to be a cusp of a new era, and special effort was required to prevent the destruction of the universe during the critical day in which the cycles "roll over" to begin a new count of days. Every 18,980 days or 52 solar years, the universe would begin again unless ceremonies were performed to purify or expiate the gods. Both the Maya and Aztec civilizations believed in the critical nature of the 52-year "binding of the years" presented by their calendar and faced something like an ancient Y2K syndrome in which the turning of a purely mathematical count of time was thought to have profoundly dangerous possibilities. The intermeshing of these two calendars can also be visualized as a set of two imaginary wheels and when both wheels have turned through all combinations of tzolkin and haab names (after 18,980 days or 52 years) the calendar round would be complete (Fig. 5.22).

One interesting mathematical byproduct of the meshing of the two calendars is that only 4 of the 20-day names will begin a solar year. These 4 day names were known as the "year bearers" and provided additional possibilities for forecasting omens for the entire year. The year bearers for the Maya were the days *Kan*, *Muluc*, *Ix*, and *Caucac*, associated with the Chac Gods in the four cardinal directions of east, north, west, and south. The four different types of years required four different new year ceremonies, soon after the winter solstice. The new year ceremony typically involved setting up a ceramic idol of the year-bearer god in a prominent place toward the appropriate cardinal direction during the last 5 days of the previous year. Various offerings of corn, and sacrificial animals would be made, and a procession would bring a large number of people into town through the appropriate cardinal direction (Sharer and Traxler 2006, p. 550).

Most dreaded were the *Caucac* years, which were associated with dreadful calamities such as swarms of killer ants and flocks of crop-destroying birds. For these years a complete set of four gods were created, and more elaborate ceremonies were performed including extensive sacrifice and purification rituals.

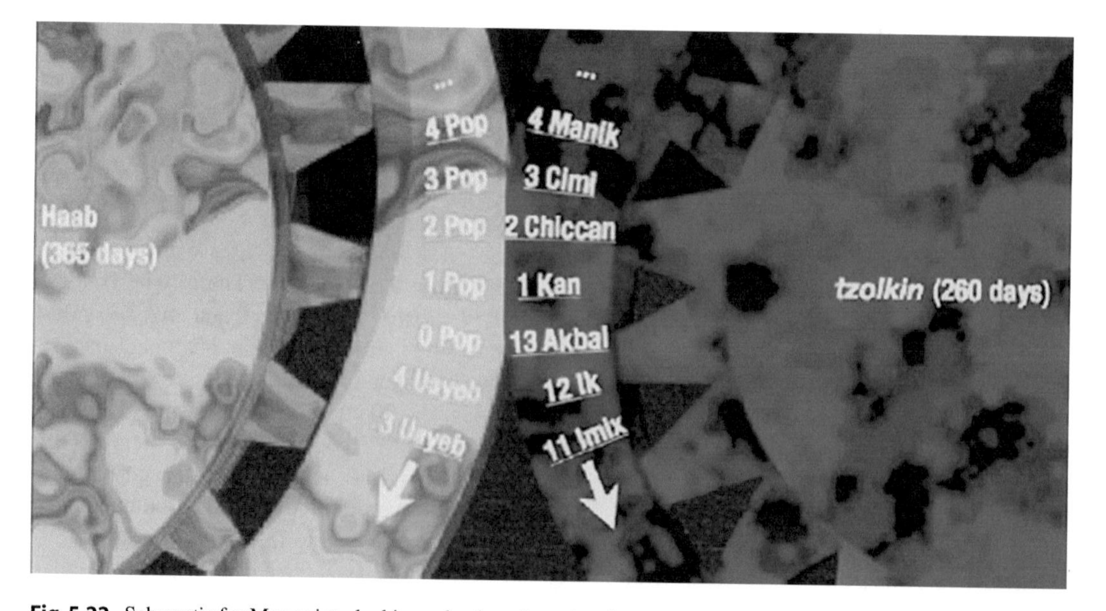


**Fig. 5.22** Schematic for Mayan interlocking calendars, featuring day names and numbers for the 260 day tzolkin (*left*) interlocking with the 365-day haab calendar (*right*). The cycles continue with unique combinations of day names for 52 years before repeating and this cycle is known as the calendrical round (original figure by the author)

After the completion of the 52-year cycle, many Mesoamerican civilizations believed that the universe had a chance for destruction and renewal. For the Aztecs, this time was dreaded and pregnant women were kept indoors, dishes in the house broken, and all fires were put out. A "New Fire Ceremony" was performed at the top of a temple, where astronomer priests would watch the Pleiades to see if it stopped its motion in the middle of the night, signifying the end of the present world. Upon observing that the Pleiades were not frozen on the sky, fires were lit, and chants were sung to help renew the world and preserve order in the cycles of the heavens "Echoes of the Ancient Skies"—in the bibliography this is Krupp (1983, p. 208). The presence of fires atop the mountains could be seen far away, signaling the renewal of a new cycle to the villages below.

The Maya had many similar rituals and beliefs to the Aztec, which perhaps included something like a New Fire ceremony (Fig. 5.23). The Maya also tied the calendar round into the more elaborate

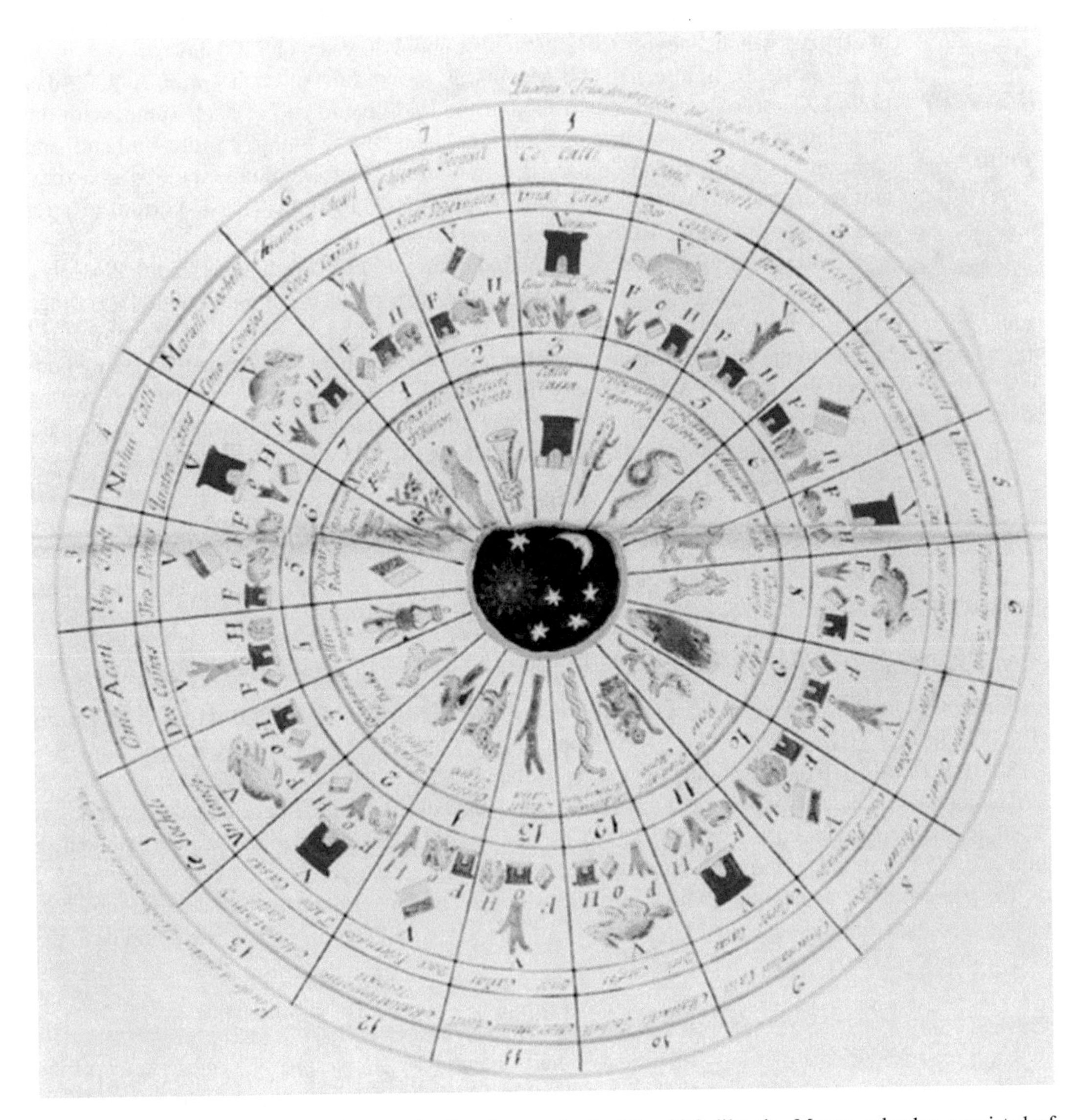

**Fig. 5.23** Representation of the Aztec calendar, the *tonalpohualli*, which like the Mayan calendar, consisted of 260 days, with 20 months of 13 days. The image shows 13 wedges for each of the days of a month, with icons showing the "year bearer" days (rabbit, reed, etc.), and the names of the different days in the Aztec language Nahuatl (from Echeverria, "*Historia del origen de las gentes que poblaron la America septentrional*," an early nineteenth century facsimile; image courtesy of the Library of Congress, from the online exhibit "1492: An Ongoing Voyage")

cycles of planetary motions within the Mayan Zodiac. For the Maya the stars of the Pleiades were known as *tzab*, which referred the rattles of a rattlesnake, and the stars in Gemini were known as *aac*, or “the tortoise.” Additional stars formed a complete set of Mayan zodiacal constellations, including the Mayan Scorpion, Tortoise, and Rattlesnake constellations (Sharer 1994, p. 580). The motions of planets within the zodiac were carefully observed and recorded by the Maya, and the results of some of these observations survive in the form of the Dresden and Paris codices.

## *The Katun and Baktun*

Larger groupings of time, analogous to our centuries and millennia, were also of great interest to the Maya. The first of these is the *katun*, mentioned earlier, which is a group of 20 tuns, which amounts to 7200 kins, or about 20 years or 7200 days.

Of particular interest to the Maya was the ending of each *katun* or 7200-day period. Ceremonies were performed at the middle and end of each katun, with the installation of a monument containing the dates of beginning, middle, and end, and some of the history of the era carved into stone. These monuments provide excellent records of the Mayan kings during the Classic and Late Classic periods (Sharer and Traxler 2006, p. 546).

Groupings of 20 *katuns* form a *baktun* which is 144,000 days, or about 394 years. These very long groupings of days form the basis of a Mayan long count date, which would record in each digit the number of baktuns, katuns, uinals, and days since the beginning of the current age within the long count. By adding together all of the numbers and the corresponding bundles of days, it is possible to place a date exactly relative to the beginning of the current age. A long count date would be completed by including the *tzolkin* and *haab* designation for the date at the end.

As an example, to specify a long count date one might see the following numbers and names:

$$9.15.10.0.103\ \textit{Ahau}\ \textit{Mol}$$

The long count date specifies in its five (base 20) digits the number of days since the beginning of the current age. The example above would include (starting from the left, which counts the largest grouping of days):

$$9\times20\times18\times20\times20+15\times20\times18\times20+10\times20\times18+0\times20+10\quad days=1,407,610\quad days$$

For this example, as we add up the days we include 9 *baktuns* (1,296,000 days) + 15 *katuns* (108,000 days) + 10 *tuns* (3600 days) + 10 kins (10 days) which places the Mayan long count date 9.15.10.0.10 1,407,610 days from the beginning of the universe.

The long count date also includes a tzolkin date and can be tied to our Gregorian calendar. For our example, the long count date above, 9.15.10.0.10, is named *3 Ahau 3 Mol* in the Mayan calendar or June 30, AD 741 in the Gregorian calendar (Sharer, p. 574).

The beginning of the universe, which forms the “origin” of the Mayan time-based cosmology corresponds to the Gregorian date of 13 August 3114 BC. On the first day of the age we would have the long count date of:

$$0.0.0.0.0.\ \textit{with Mayan day names}\ 4\ \textit{Ahau}\ 8\ \textit{Cumku}=13\ \textit{August}\ 3114\ \textit{BC}$$

The very ancient date of 3114 BC predates the Mayan civilization, since the oldest known long count date from Mayan archaeology is from Stela 2 at Chiapa de Corzo, which is listed as approximately 114 BC. Presumably, the beginning of the calendar in August of 3114 BC is simply a result of numerology and mathematics, much like the beginning of many other calendars we have seen. Other calendars with such beginning dates include the Julian dates, which begin in 4713 BC, the Jewish calendar, which begins in 3761 BC, the ancient Greek Olympiad-based calendar, which placed the first Olympics in 776 BC, and the Greek Orthodox calendar, which begins in 5509 BC (Sharer and Traxler 2006, p. 568).

The last date of the current age is when the leftmost zero (the *baktun* indicator) goes from 12 to 13, which corresponds to the last day of the current age.

$$12.19.17.19.19 \rightarrow 13.0.0.0.0 = 23\ \ December,\ 2012\ \ AD$$

The time between these dates, which would be the duration of the "age" we live in, is 13 *baktuns*, 1,872,000 days or 5200 years! The Mayan astrologers would expect that at this date we would have not only a complete renewal (and destruction) of our world, but also with the expectation that life and time would begin again, in an endless cycle of recurrent events. This numerical effect from the Mayan calendar is the basis of the "2012" hysteria which has emerged in popular culture.

## The Long Count and Cycles of the Universe

The largest cycles of days for the Maya come into play in the form of the long count described above, which defines "ages" of 5200 years. However, in some contexts (such as in monumental carved stoned stelae), long count dates can be specified in an even longer format, with two additional digits beyond the usual five places of the long count. Below is an example.

$$13.13.13.3.11.8.0.195\ \ Imix\ \ 9\ \ Kanka$$

These additional digits refer to even longer units of time, which would be groups of 20 × 400 years (a *pictun*), 20 × 20 × 400 years (a *calabtun*), 20 × 20 × 20 × 400 years (a *kinchiltun*), and 20 × 20 × 20 × 20 × 400 years (an *alautun*), corresponding to cycles of 8000, 160,000, 3,200,000 and, 64,000,000 of the Mayan *tuns*. The number of solar years within each of these units of time is slightly less than the number of *tuns*, since a *tun* is 360 days and the solar year is 365.2422 days. The largest of these time units recognized by the Maya is the *alautun*, which corresponds to about 63,000,000 solar years or 23 billion days!

To summarize the complete set of Mayan time units we provide the following listing of the many units within their reckoning of time below. Each line in the summary below corresponds to one position in a long count date.

**Summary of Mayan units of time**

1 kin or 1 day
20 kins = 1 uinal or 20 days
18 uinals = 1 tun or 360 days or ~ 1 solar year
20 tuns = 1 katun or 7,200 days ~ 19.7 solar years
20 katuns = 1 baktun or 144,000 days = 400 tuns ~ 394.24 solar years

---

20 baktuns = 1 pictun or 2,880,000 days = 8000 tuns ~ 7885.25 solar years
20 pictuns = 1 calabtun or 57,600,000 days = 160,000 tuns ~ 157703 solar years
20 calabtuns = 1 kinchiltun or 1,152,000,000 days = 3,200,000 tuns ~ 3,154,000 solar years
20 kinchiltuns = 1 alautun or 23,040,000,000 days = 64,000,000 tuns ~ 63,000,000 solar years.

Sharer and Traxler (2006)

The exact meaning and purpose of these longest Mayan time cycles are not well known, but their existence hints at a skill in higher mathematics of time that inspired and challenged the Mayan timekeepers and philosophers at an abstract level, well removed from any practical calendrical need.

## Hindu Ages and the Mahayugas

The ancient Maya are not the only civilization to contemplate the vast expanses of time. The Hindu civilization imagined that time as we experience it is only a transient flash in the eyes of the god Brahma, and that each age within our universe is a small part of a dream in Brahma. In the Hindu cosmology, time unfolded in a set of four ages, in a descent from pure order into decay, before destruction and renewal.

Hindu astronomers were masters of time and the cycles of astronomical motion—and kept track of intervals of time from minute divisions of a second to billions of years. Table 5.8 below describes the range of time units within Hindu calendrics and astronomy from timescales of a few microseconds to aeons of billions of years.

The first of the Hindu ages was known as *kritya* or *satya* and was symbolized by the bull of creation on all four legs. This first age lasted 4800 "divine years," which corresponded to 1,728,000 of our "human years." During this first age, the universe was in a state of purity and the laws of dharma were obeyed. The second age was known as *treta*, and in this age of 3600 divine years (1,296,000 human years) the first hints of decay are visible, and the cosmic bull is symbolized to have only three legs. The third age was known as Dvapara, and lasted 2400 divine years (864,000 human years). The fourth and final age, which is our current age—is known as the *Kali Yuga* or the age of Kali, and lasts for 1200 divine years (432,000 human years). After these four ages are complete, it is believed that a short period of twilight known as *Sandhya* would precede a cycle of destruction and re-creation to bring the universe back into its pure *kritya* state (Dallapiccola, 2003).

The enormous grouping of this vast expanse of time in the four ages comprised the length of time known as the *mahayuga*, or "great cycle." Each grouping of four ages amounts to approximately 4,320,000 years, and comprises a brief segment of a dream within the sleep of Brahma. One thousand of these cycles creates a length of time known as the *kalpa*, which amounts to about 4.3 billion years, close to the current scientific estimate of the age of the Earth, the Sun, and the solar system. This enormous expanse of time was also imagined to be the length of one full day in the life of Brahma. However, even longer lengths of time were implied by the life of Brahma, which if it extended to 100 of these "Brahma years" would amount to approximately 315 billion years, longer than the

**Table 5.8** The time-divisions within Hindu calendrics and astronomy (from "Knowledge Traditions and Practices of India—Astronomy in India" Central Board of Secondary Education, India. p. 22)

| |
|---|
| lotus-pricking time = 1 *truti* (about 9 $\mu$s) |
| 100 *trutis* = 1 *lava* |
| 100 lavas = 1 nimesa (twinkling of eye) |
| 4 1/2 nimesa = 1 long syllable |
| 4 long syllables = 1 kastha |
| 2 1/2 kasthas = 1 asu (respiration) = 4 s |
| 6 asus (pranas) = 1 sidereal pala (casaka, vinadi or vighatiku) |
| 60 palas = 1 ghatikas = 24 min |
| 60 ghatikas = 1 day |
| 30 days = 1 month |
| 12 months = 1 year |
| 4,320,000 years = 1 yuga |
| 72 yugas = 1 manu |
| 14 manus = 1 kalpa = 4,354,560,000 years |
| 2 kalpas = 1 day (and night) of Brahma |
| 30 days of Brahma = 1 month of Brahma |
| 12 months of Brahma = 1 year of Brahma |
| 1 year of Brahma = 72 × 14 × 2 × 30 × 12 yugas = 725,760 yugas = 3,135,283,200,000 years |
| 100 years of Brahma = Life of Brahma or mahakalpa |

current estimates for the age of the universe. Incidentally, a Brahma night and a Brahma day would amount to about 8.6 billion years, the current scientific estimate of the period within which most of the present-day stars and galaxies were formed. It is amazing to realize that Hindu scholars imagined accurate astrophysical timescales over 5000 years ago!

## References


Aslaksen, H. and A. Rugulagedda. 2001. Regional varieties of the indian calendars. From http://www.math.nus.edu.sg/aslaksen/calendar/indian_regional.html.

Auel, J.M. 1980. *The clan of the cave bear: A novel*. New York: Crown.

Aveni, A.F. 1989. *Empires of time: Calendars, clocks, and cultures*. New York: Basic Books.

———. 1997. *Stairways to the stars: Skywatching in three great ancient cultures*. New York: Wiley.

Cullen, C. 2007. *Astronomy and mathematics in ancient China—The 'Zhou Bi Suan Jing'*. London: Cambridge University Press.

Dallapiccola, A.L. 2003. *Hindu myths*. London: British Museum.

Evans, J. 1998. *The history and practice of ancient astronomy*. New York: Oxford University Press.

Farrer, C.R. 1991. *Living life's circle: Mescalero Apache cosmovision*. Albuquerque, NM: University of New Mexico Press.

Gibran, K. (1952). *The Prophet*. New York, Knopf.

Hesiod, Works and Days, translated by H.G. Evelyn-White. 1914. Loeb Classical Library #57.

Hudson, T., and E. Underhay. 1978. *Crystals in the sky: An intellectual odyssey involving Chumash astronomy, cosmology, and rock art*. Socorro, NM: Ballena Press.

Johnson, R.K., and J.K. Mahelona. 1975. *N*a inoa h*ok*u: A catalogue of Hawaiian and Pacific star names*. Honolulu: Topgallant.

Kelley, D.H., and E.F. Milone. 2005. *Exploring ancient skies: An encyclopedic survey of archaeoastronomy*. New York: Springer.

Kiang, T. 1969. Possible dates of birth of pulsars from ancient Chinese records. *Nature* 223(5206): 599–601.

Krupp, E.C. 1980. Egyptian astronomy: the roots of modern timekeeping. *New Scientist* 85(1980): 24–27.

Krupp, E.C. 1983. *Echoes of the ancient skies: The astronomy of lost civilizations*. New York: Harper & Row.

Lennox-Boyd, M. 2005. *Sundials: History, art, people, science*. London: Frances Lincoln.

MacDonald, J., et al. 1998. *The Arctic sky: Inuit astronomy, star lore, and legend*. Toronto, ON: Royal Ontario Museum/Nunavut Research Institute.

Marshack, A. 1991. *The roots of civilization: The cognitive beginnings of man's first art, symbol and notation*. Moyer Bell: Mount Kisco, NY.

Marshack, A., and F. D'Errico. 1989. On wishful thinking and lunar "Calendars". *Current Anthropology* 30(4): 491–500.

McCall, H. 1990. *Mesopotamian myths*. London: British Museum Publications.

Miller, D.S. 1997. *Stars of the first people: Native American star myths and constellations*. Pruett: Boulder, CO.

Nath, B. 2008. *Medieval mistake*. Chennai: Frontline.

Needham, J., and C.A. Ronan. 1980. *The shorter science and civilization in China: An Abridgement of Joseph Needham's original text*. New York: Cambridge University Press.

Ojha, G.K. 1996. *Predictive astrology of the Hindus*. D.B. Taraporevala: Bombay.

Polo de Ondegardo. 1571. Relación de los fundamentos acerca del notable daño que resulta de no guardar a los indios sus fueros, Madrid.

Regiomontanus, J. 1476. *Calendarium*. Venice: Bernhard Maler (Pictor), Erhard Ratdolt, Peter Löslein.

Rich, T.. 2008. Judiasm 101. From http://www.jewfaq.org/calendar.htm.

Ruggles, C.L.N. 2005. *Ancient astronomy: An encyclopedia of cosmologies and myth*. ABC-CLIO: Santa Barbara, CA.

Sharer, R.J. 1994. *The ancient maya*. 5th ed. Stanford University Press: Stanford, CA.

Sharer, R.J., and L.P. Traxler. 2006. *The ancient Maya*. Stanford, CA: Stanford University Press.

Shen, C.S. 1969. Pulsars and ancient Chinese records of supernova explosions. *Nature* 221(5185): 1039–1040.

Steel, D. 2000. *Marking time: the eqic quest to invent the perfect calendar*. New York, NY: Wiley.

Urton, G. 1981. *At the crossroads of the earth and the sky: An Andean cosmology*. Austin: University of Texas Press.

West, M.L. 1978. *Hesiod–works & days*. Oxford: Clarendon Press.

Wittick, B. 1897. Photograph from Library of Congress. http://www.loc.gov/pictures/resource/cph.3c15456/ (photo of the pueblo + river from 1897).

Williamson, R.A. 1984. *Living the sky: The cosmos of the American Indian*. Boston: Houghton Mifflin.

Yano, M. 2003. Calendar, astrology and astronomy. In *Blackwell companion to Hinduism*, ed. Gavin Flood. London: Blackwell.

# Chapter 6

# Modern Timekeeping and the Development of the Science of Time

*"Five hundred twenty-five thousand*
*Six hundred minutes,*
*Five hundred twenty-five thousand*
*Moments so dear.*
*Five hundred twenty-five thousand*
*Six hundred minutes*
*How do you measure, measure a year?"*

Seasons of Love, lyrics from the Broadway Musical "Rent"

Civilizations across the globe invented calendars and refined their observations of the Moon and planets to help synchronize their lives to these cycles. The modern world is distinctive for its advanced technology, for its material culture, and for its embrace of science. The development of all three of these was inseparable from the development of the science of time.

The modern world is distinctive for its advanced technology, for its material culture, and for its embrace of science. The development of all three of these was inseparable from the development of the science of time. With the advances of science we have accelerated the pace of our lives, and now we must synchronize our lives to much shorter units of time. The hectic modern person is more likely to be counting minutes and seconds, as measured by accurate satellites and atomic clocks. These devices in turn are able to divide time into microscopic intervals—modern GPS satellites and spacecraft contain atomic clocks accurate to within one ten-thousandth of a trillionth of a seconds.

Physics laboratories are able to split time into even smaller parts, using lasers and high-speed cameras to record the motions of atoms, and the flashes of reactions in particle accelerators. From studying the nature of time, modern science has given us a profound awareness of how time and space are inseparable, as revealed by Einstein's Theory of Relativity, and also how time and space arise together in the first instants of the Big Bang. This section will explore how we got to this point, through an analysis of the history of the science of time.

Major developments in timekeeping did not just happen in the last few centuries, but grew from the inventions of the Chinese, Greek, Arabic, and later European scientists as they developed water clocks, mechanical devices for measuring time between Olympics, accurate astrolabes, and exquisite mechanical clocks capable of measuring time for months at sea and to within seconds during astronomical observations. In this section, we will explore some of these key developments to help understand the origin of our modern system of time. Science advanced from these timekeeping technologies, up until the present, when we employ electrical and laser-optic devices for our clocks, all of them ultimately synchronized by astronomical events.


B.E. Penprase, *The Power of Stars*, DOI 10.1007/978-3-319-52597-6_6

## Shadow Clocks and Sundials

For the first few millennia of history, the measurement of time required the use of shadow clocks. The first shadow clocks, such as those of ancient Egypt, were capable of measuring the 12 divisions of the day through a T-shaped shadow and markings on the base of the device, which represented the state of the art in 1500 BC. Additional precision for measuring both the hours and date of the observation came from marking the *base* with horizontal curves, making the first sundials capable of accurate timekeeping throughout the year (Fig. 6.1).

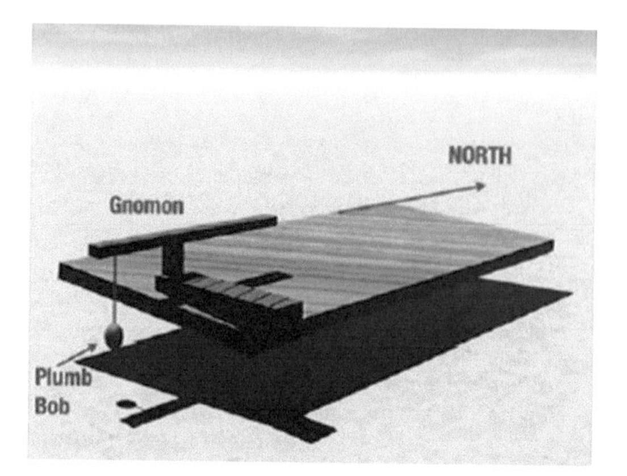


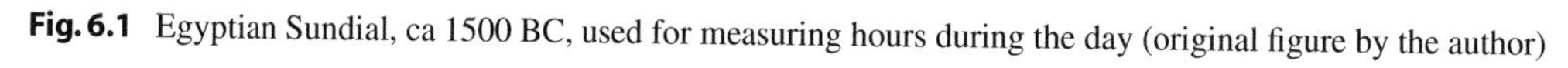

**Fig. 6.1** Egyptian Sundial, ca 1500 BC, used for measuring hours during the day (original figure by the author)

Sundials took many forms throughout China, India, the Arab world, and Europe and embodied the highest standards of artwork, materials, and craftsmanship throughout the world. Early Greek and Roman sundials were described by the Roman author Vitruvius, who in his work *De Architecturus* from 30 BC reported some 13 distinct types in use, including horizontal and vertical sundials, as well as devices using the *analemma* (the "figure eight" pattern of solar positions each day of the year) to account for variations in the sunrise and sunset from the orbit of the Sun.

Some of the most accurate ancient sundials were developed in China and trace their history to some of the first-known timekeeping devices from about 400 BC. Chinese devices include hybrid sundials with water clocks, featuring bronze mirrors carved with figures for noting the locations of shadows, as well as mechanical indicators to tell time from the water clock (Needham et al. 1986, p. 144). The Chinese solar observatory of Guo Shou Jing at Gao Cheng Zhen (near Dengfeng, in the Henan province) made use of an enormous 40-foot tower which cast a long shadow along an accurately ruled and water-leveled meridian scale, to measure the exact dates of solstices and the precise length of the solar year (Krupp 1984; Kelley and Milone 2005).

The Chinese also constructed portable sundials in several types, with ingenious arrangements for measuring time and location on the Earth. The simplest sundials made use of a stretched string to cast a shadow and a compass needle for alignment of the device. A second portable Chinese sundial made use of a pole-oriented gnomon which was adjustable for latitude, and which also contained a compass for orientation, and inscriptions noting the 24 fortnightly periods or *chhi* within the Chinese tradition. The two types of sundials appeared between 1000 and AD 1600 and the latter seems to be an original Chinese invention. A third type of Chinese sundial dates from about AD 1260 and appears to integrate technology from the Arab world, which includes inscribed ivory plates marking both the hours of the day and lines for different months, and containing three different plates for use in different latitudes (Needham and Ronan 1980, p. 149).

One of the most famous of all the large stationary solar observatories is the observatory of Ulugh Begh, in Samarkand, developed as part of the Mongol empire. Ulugh Begh was a Mongol prince who developed his observatory in present-day Uzbekistan in 1424, to replicate and surpass the technology that existed within Persian observatories of the time. The crown jewel of this royal observatory was the giant transit circle, which includes a huge subterranean stone semicircle, on which the light of the Sun would be cast from an elaborately decorated south-facing portal. The meridian circle was 40 m in diameter and could measure the exact moment of noon, and thereby determine the exact length of the day in combination with water clocks (Kelley and Milone 2005, p. 80). The design of this ancient observatory has inspired the modern Griffith Observatory in Los Angeles, California to install its "Transit Corridor" which similarly tracks the exact moment of noon, but with additional modern day electronics (solar cells) which light up an electronic display during the transit of the Sun (Fig. 6.2).

**Fig. 6.2** Solar observatories old and new. The transit circle of Ulugh Begh, in modern day Uzbekistan (*top*), used to measure the height of the sun and length of the day in 1424, and the modern Griffith Observatory "Transit Circle (*bottom*)," as demonstrated by director Dr. Edward Krupp. The Griffith "Transit Circle," delights modern visitors by precisely measuring the height of the noon sun and lighting up an adjacent display using photodiodes (*top image source*: http://commons.wikimedia.org/wiki/File:Ulugh_Beg_observatory.JPG; *bottom* two photos by the author)

The sophisticated mathematics of the Arabic and European astronomers combined with their abilities in metal work provided some of the world's most beautiful sundials, which also came in portable varieties, used by aristocratic or royal personages, and in stationary sundials that graced palace gardens, public buildings, and courtyards. Many of these devices have stunning paintings and engravings and are works of art in their own right. However, even the most modern dials use the same basic technology as the much older Egyptian, Greek, Roman, and Chinese counterparts and rely for their operations on the eternal motion of our nearest star—the Sun.

## The Earliest 24-h Clocks: The Clepsydra

While the sundial has been refined into a beautiful symbol of advanced culture, it has many drawbacks as an actual timekeeping device, most notably the impossibility of using a sundial to measure time on either cloudy days or at night! The measurement of time during the night becomes essential for astronomers, who track the motions of planets, the risings, and settings of stars, and who need to know the times of these events in order to apply mathematical calculations to determine the underlying theories of how the universe works.

The first 24-h clock came in the form of the water clock, or *clepsydra*, which was developed in Babylon, Greece, and China. The clepsydra uses a flow of water, usually through a set of reservoirs with successively smaller outlets dripping slowly to measure successively smaller units of time. The precisely controlled cascades of these drops were also sometimes coupled to springs and gears to indicate the times or even depict the motions of the Sun, the Moon, and planets.

In ancient Egypt the first water clocks were developed in the Middle Kingdom, by about 1500 BC, perhaps to help improve observations of the decans stars by Egyptian astronomer/priests, and both "inflow" and "outflow" types of water clocks were used in which time was measured as the device would fill or empty a container, respectively. The Babylonian astrologers used mainly the "outflow" type, in conjunction with their astronomical observations. The accuracy of these clocks enabled the Babylonian astronomers to measure exact times of solar eclipses, planet risings, and settings, and from these observations the Babylonians were able to obtain unprecedented precision in their astronomy, which was recorded in their cuneiform tablets.

By 100 BC, several types of Greek water clocks were in use. Some used a filling reservoir to move a float attached to an indicator, while others used the increasing weight of the water or its level in a marked container to measure the passing of the hours. By 50 BC, the most famous of the Greek water clocks, the Tower of the Winds, sometimes referred to as the "Horologion of Andronicos" was built. The Tower of Winds contained multiple sundials, a weather vane, a water clock, and also functioned as a "simulacra" to simulate the motions of the skies (Fig. 6.3).

Vitruvius, the Roman author on architecture, sundials, and other technologies, described a water clock from 30 BC known as the "anaphoric clock" which caused a float and weight to move with the water level of the clock, and turn a disk which indicated the locations of the Sun and stars relative to the horizon, and the position of different zodiacal constellations. Fragments of the bronze plates from this device survive, and show artistic renderings of constellation figures, which were attached to an axle that turned as the water level changed (Evans 1998, p. 156).

The most impressive ancient water clocks, however, would have to be those of the Chinese. Some examples include the tower of Su Sung/Chou Kung from AD 1088. Within this ingenious device were many completely new innovations in mechanical engineering, including "the oldest chain drive in any civilization" and a series of gears of unprecedented sophistication that coupled the energy of an 11-foot diameter waterwheel to an escapement, which initiated periodic motions to sound bells, gongs, or drums, at hourly and quarter-hour intervals (Needham 1981, p. 136). Additional wheels caused carved figures to appear at quarter-hour intervals with placards announcing the time. In addition, Su Sung's clock included an armillary sphere for recreating a realistic simulation of the motions of the stars in the sky (Goudsmit and Claiborne 1969; Needham et al. 1986) (Fig. 6.4).

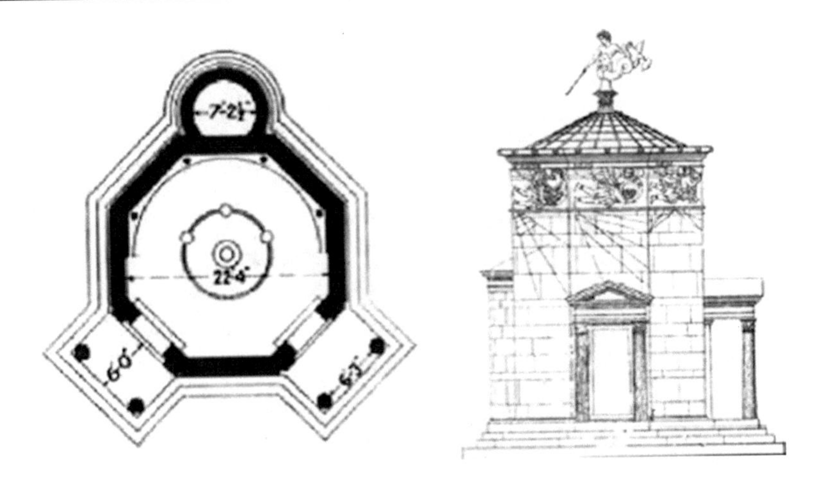

**Fig. 6.3** The Tower of the Winds, in Athens, Greece, featured a weather vane, several sundials, and a water clock, as well as a device from simulating the motions of the heavens. This clock was attributed to the Syrian architect Andronikos Kyrrestes and was constructed between 200 and 50 BCE (image from Stuart and Revett, Antiquities of Athens (1762))

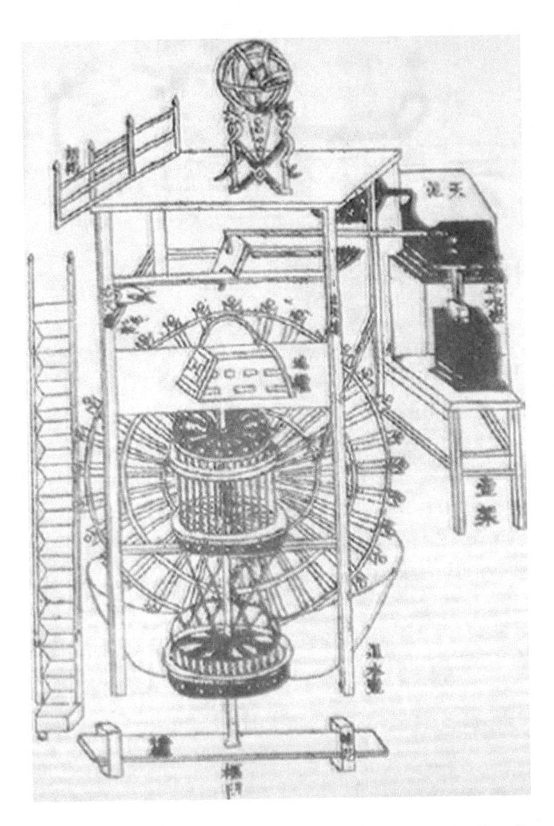

**Fig. 6.4** The Su Sung clockwork, from 1088 AD, showing the elaborate network of gearing, armillary spheres, and other devices, powered by water to replicate the motions of the sky (*image source*: http://commons.wikimedia.org/wiki/File:Clock_Tower_from_Su_Song%27 s_Book.JPG)

Centuries later, the knowledge of such advanced clockwork reached Europe (Fig. 6.5). The European clocks were inspired by Chinese, Indian, and Arabic devices, and by the 1500's evolved into more sophisticated mechanical designs that derived their power from weights or springs.

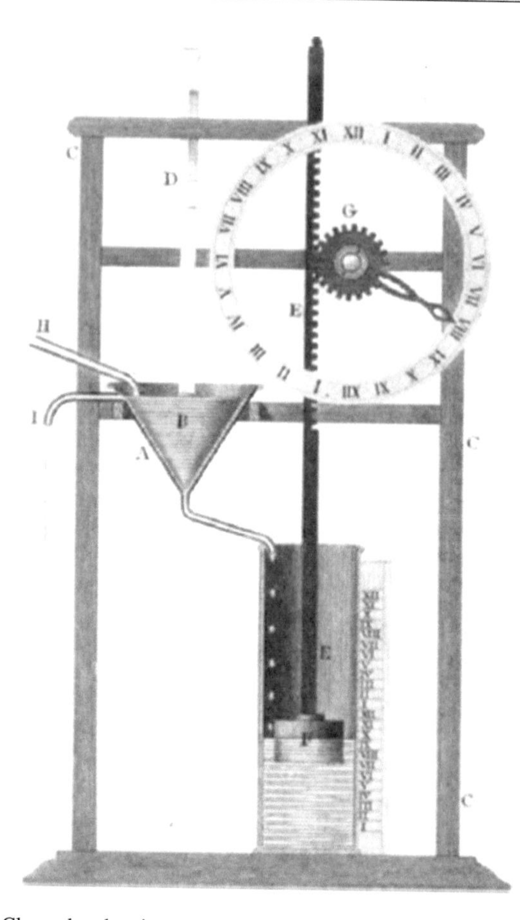


**Fig. 6.5** An early European Clepsydra showing a very simple design in which the rising water level raised a rod (E) which turned a gear attached to a timekeeping dial (Rees 1819)

As humans began to intensify the pace and geographical range of their activities, their clocks needed to keep up. Water clocks suffered from lack of portability—the largest clocks, such as the Su Sung water clock described above consumed 1000 pounds of water every 9 h! Portability demanded a purely mechanical clock, in which the source of energy came not from the Sun or water, but from falling weights or springs. A reliable and portable clock was also a vital tool for navigation, where the accuracy of the ship's clock could make the difference between a safe arrival on the shore of a newly discovered continent, or being lost at sea.

## The Antikythera Mechanism

Ironically, the earliest known purely mechanical clock appears to have been discovered on a shipwreck and was therefore lost at sea. Sponge divers discovered the device, known as the "Antikythera mechanism," in 1901 when they themselves were blown off course and anchored near the Antikythera Island, between Greece and Crete (Fig. 6.6). They discovered a shipwreck at the bottom, along with numerous bronze and marble artifacts. Four bronze items were brought to the surface that appeared to be fragments of an object that used a geared mechanism to move plates. Subsequent exploration of the underwater site, by none other than explorer Jacques Cousteau, enabled scientist to date the remnants of a ship to 87 BC. The Antikythera mechanism is at least as old as the shipwreck and possibly significantly more ancient (Brecher and Feirtag 1979, p. 52).

**Fig. 6.6** Main fragment of the Antikythera mechanism, recovered from an ancient Greek shipwreck, and analyzed with modern physics to reveal its purpose for timekeeping (*image source*: http://en.wikipedia.org/wiki/File:NAMA_Machine_d%27Anticythère_1.jpg)

The Antikythera mechanism has inspired creative speculation on its capabilities and origins. Some speculated that it was an ancient computer that simulated the nested spheres of ancient Greek cosmology or to aid in astrological forecasts. A fascinating union of ancient and modern technology revealed the actual function of the device, as the latest tools of modern physics were used to peer inside the works of the badly corroded artifact. Gamma-ray photographs were taken in 1974, and a series of atomic beam and X-ray images were used to reveal the mechanism's inner workings. Inside the device were a series of gears which appeared to be set to have ratios of 254 to 19, so that rotating one wheel 19 times produced 254 rotations of a second wheel, that in turn gave rise to a differential rotation of $254-19 = 235$ rotations. This last number coincides exactly with the 235 lunar months in a Metonic cycle of 19 years and led authors to conclude that the device was used to count lunar months (Brecher and Feirtag 1979).

Further analysis of the Antikythera mechanism with more advanced technologies has provided several recent discoveries, which were published in the prestigious journal *Nature*. An international team of researchers based in Wales and Greece used surface imaging and high-resolution X-ray tomography to reconstruct the function of the 30 precision, hand-cut bronze gears and to decipher some of the inscriptions inside the device. The team concluded that the device was able to predict lunar and solar eclipses based on Babylonian eclipse algorithms, as well as to display some of the planetary and lunar positions, and even had gears to simulate irregularities in the Moon's motion (Freeth 2006). Careful study of the inscriptions revealed by X-ray imaging enabled researchers to discover a set of month names, which were of Corinthian origin and were consistent with the use of the device over a 223 Metonic lunar month cycle, useful for predicting eclipses (Freeth et al. 2008). The researchers also concluded that the upper dial was not used to simulate epicycles of planets, but instead was used to track the 4-year cycle of the pan-Hellenic games, more commonly known as the Olympics.

More recent work suggests that the device contained a series of dials that pointed toward the Sun's position on the Zodiac, with individual dials for the five known planets (Mercury, Venus, Mars, Jupiter, and Saturn) that would indicate when they would begin the retrograde motion. An additional feature inferred from recent studies is a revolving black and white ball that would indicate the lunar phase (Marchant 2010). Credible reconstructions of the interior of the device have been drawn based on the X-ray imaging and combining years of detective work (Lin and Yan 2014; Lin 2011). The modern reconstructions show the layout of banks of gears that can recreate the known heavenly cycles of the time—the Calippic Cycle (a 76-year cycle with 940 lunar months), the Olympiad Cycle (with a 4-year cycle, like today), the Metonic Cycle (the famous 235 lunar month cycle, that coincides with 19 years), the Saros cycle (a 223 lunar month cycle associated with recurrent eclipses), and the Exiligmos Cycle (669 lunar months or three Saros cycles). A complete summary of the Antikythera Mechanism with its various gear subsystems drawn in exacting detail has been presented recently reveals the mechanical ingenuity of the ancient designers of this device (Lin and Yan 2016). The complexity of the Antikythera device was a triumph of Greek science and engineering and was not matched in the rest of Europe for over 1500 years (Fig. 6.7).

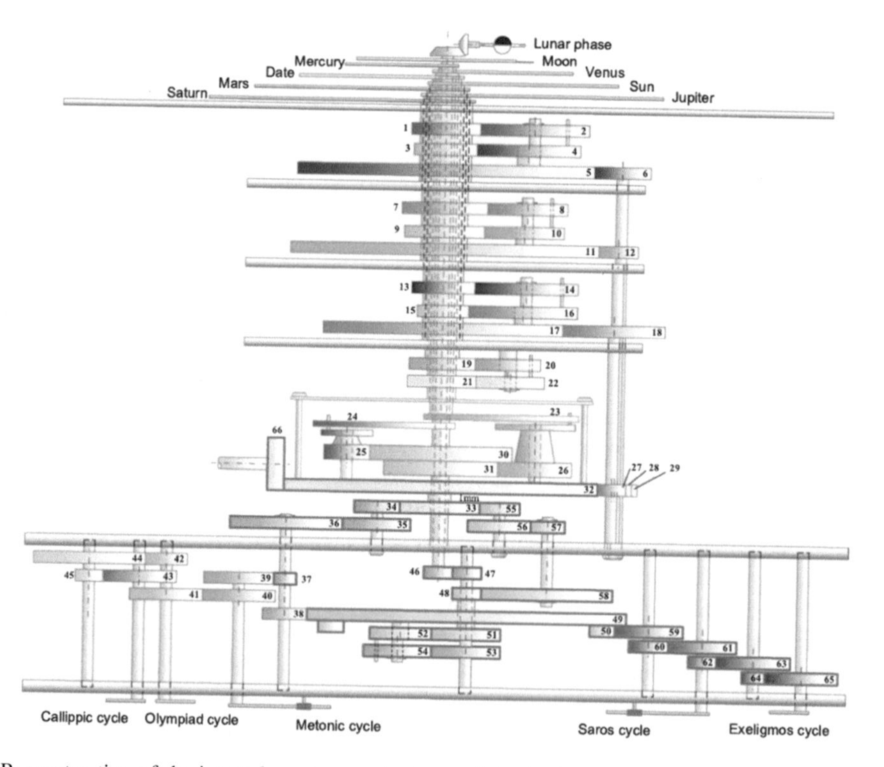


**Fig. 6.7** Reconstruction of the internal gearing mechanisms of the Antikythera mechanism, showing the incredibly complex machinery that recreated the known cycles of the universe at the time useful of predicting eclipses, motions of planets, lunar and solar months, and of course the Olympic games (from Lin and Yan 2016)

## Mechanical Clocks from Europe

Some of the first mechanical clocks in medieval Europe were of the falling weight type, which rang in the hours in monasteries and cathedrals. By the fourteenth century, clocks in England and Germany were powered by weights, which drove an "escapement" which caused a toothed wheel to regulate the falling drive train to move at a slow and even rate. One of the oldest mechanical clocks is in the cathedral tower at Salisbury, which by 1386 included massive gears and wheels to move clock hands for the first time in England.

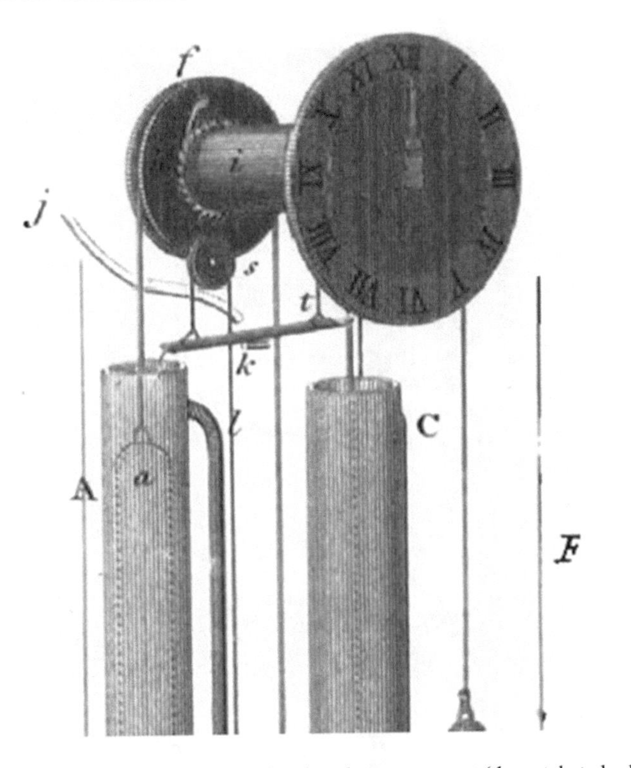


**Fig. 6.8** Schematic of an early falling weight clock showing the escapement (the ratcheted wheel) which metes out equal units of time as the weight falls downward (figure from Lockyer and Seabroke 1878)

The locksmith Peter Henlein in Nuremberg, Germany was able to make miniature versions of the Cathedral clocks, which used wound iron springs to power tiny clocks placed within the famous "Nuremberg eggs." The eggs were collected by wealthy Europeans and were more useful as decoration than as timekeepers, since the clocks slowed down as the spring played out, much as a child's windup toy. By connecting the spring to a conical-shaped "fusee" or gearing, the Czech clockmaker Joseph Zech was able to vary the gear ratio to match the slowing spring and provide a constant timekeeper. These clocks were also decorated with elaborate astronomical figures and often included additional dials for the lunar month and for the position of the Sun (Goudsmit and Claiborne 1969) (Fig. 6.8).

## Pendulum Clocks

Further innovations in timekeepers were made by the best minds of Europe in the seventeenth century and included contributions by the greatest minds of the time; inventors of clock mechanisms include the likes of Galileo, Huygens, and Hooke.

According to legend, Galileo Galilei observed that the period of the swinging lamp was constant by comparing it with his pulse during a long church ceremony, and then used this information to develop a new theory of falling bodies, as well as a purely practical device, the pendulum-driven clock. Within Galileo's clock design are all the elements of a modern grandfather clock—the pendulum, an escapement, and a set of gears very similar to those in modern clocks to provide an even motion for measuring time (Van Helden 1995).

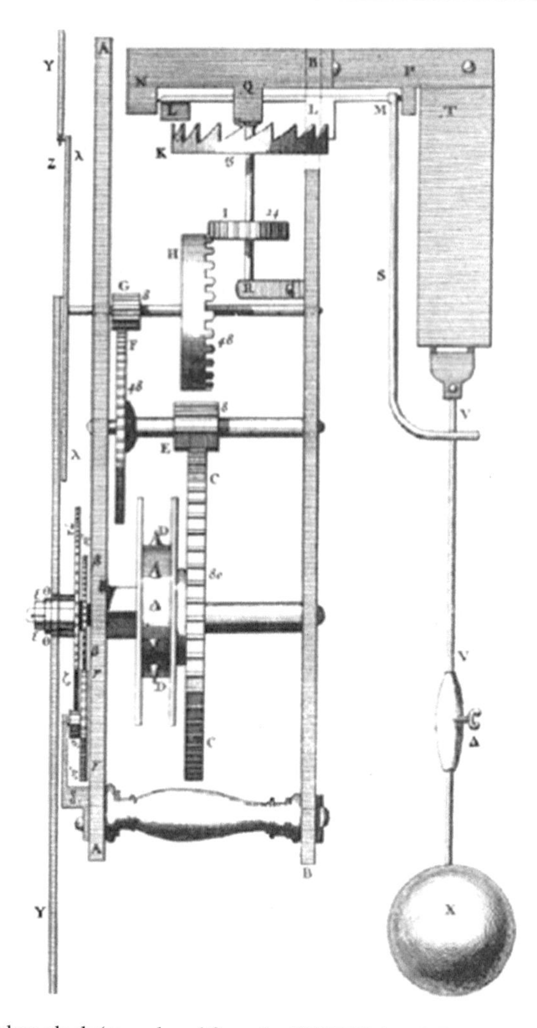


**Fig. 6.9** The Huygens pendulum clock (reproduced from the 1832 Edinburgh Encyclopedia)

Christian Huygens, working independently in Holland, made use of the older falling weight technology as his drive train, but added an ingenious pendulum regulator to release the falling weight at a slow and exact rate through a series of gears and an escapement. Huygens also developed a pure pendulum clock and wrote up his discoveries on timekeepers in his work *Horologium* in 1673 (Huygens 1673) (Fig. 6.9).

Robert Hooke, a contemporary and rival of Isaac Newton, developed even more advanced spring-driven clocks as part of his duties for the Royal Society (Goudsmit and Claiborne 1969). After traveling to the West Indies in 1662, Hooke observed both the variation of gravity across the surface of the Earth and the practical limitations of a pendulum powered clock at sea. Hooke solved these problems by use of an ingenious pair of spiral springs and balances that were able to even out variations of a rocking pendulum. These devices were all miniaturized into one of the first ever pocket watches, which was presented on February 20, 1668 to the Royal Society (Smith 2000).

## Clocks for Navigation

The most famous of the spring-driven mechanical clocks was developed in the eighteenth century, as a result of an Act of the British Parliament in the year 1714, which offered the reward of 10,000, 15,000, or 20,000 pounds sterling for any clock that could determine longitude at sea within 60, 40, or 30 mi., respectively. While a sextant can easily measure latitude at sea, only an accurate clock can determine longitude, by comparing the local time at sea with the time at the Prime Meridian. It is easy to see that as one moves west, time shifts earlier—the sunrise becomes successively earlier, and likewise the stars will rise and set at earlier times. This shift is the basis for "time zones" which were established early in the twentieth century. The local time at sea therefore depends on longitude and is determined by observing the positions of the Sun and stars. The time at the Prime Meridian, however, can only be determined with an accurate timekeeping device that is calibrated at port and keeps time accurately throughout the voyage.

The eighteenth century contest for determining longitude was reminiscent of our twenty-first century "X prize" for Space Exploration. One important difference is that measuring longitude was of even more practical use in the eighteenth century than space exploration in the twenty-first century. The British created a Longitude Board after the tragic loss in 1707 of a fleet of four ships and 2000 men as they crashed into the Scilly Islands in an accident caused by inaccurate longitude reckoning. The accident could have been prevented had a more accurate timekeeper been available. The captain of the fleet, the colorfully named Sir Cloudley Shovel, miscomputed his longitude, based on an inaccurate ship's clock, resulting in the disaster (Goudsmit and Claiborne 1969, p. 81).

An obscure clock maker named John Harrison worked for nearly his entire life to solve the problem and claim the prize. Harrison succeeded only after making five successively more exquisite and more compact prototypes. The details of Harrison's work, and his struggles to perfect his timekeeper, are recounted in the popular book "Longitude" by Dava Sobel (Sobel 1995). Harrison began in 1728 by presenting plans for such a device to Edmund Halley, the astronomer Royal. Halley sent him back to the shop and suggested he construct the device first before expecting any further assistance. In 1735, after 7 years work, he returned with his first device, which included ingenious temperature compensating devices, and other innovations for providing a more regular motion of the spring-driven works inside. After a test on a barge, and then a round trip voyage to Lisbon in 1736, Harrison succeeded in measuring longitude with his device to within 1.5°, winning him a 500-pound grant to fund further improvement. A second device, known as H2, was ready by 1742, but war with Spain delayed further tests. The third device, known as H3, included a bimetallic strip to compensate for temperature variations and new bearing technologies. A miniaturized version, H4, was sent on a transatlantic journey in 1761, and survived the trip to Jamaica, where it only had lapsed by 5 s, providing an amazing accuracy of longitude to within 1 nautical mile or 1/50 of a degree.

The Board of Longitude refused to award the prize without a second trial at sea, however, since during the voyage some errors in operating the device nearly caused the ship to miss the island of Madeira. Such an error, according to a contemporaneous account, "would have been inconvenient, as they were in want of beer" (Goudsmit and Claiborne 1969, p. 82) (Fig. 6.10).

**Fig. 6.10** Clocks designed and built by John Harrison, winner of the British Board of Longitude prize for maintaining accuracy in long sea voyages. The first four of his inventions are shown here, including the H1 (1735) *upper left*, the H2 (1741) *upper right*, the H3 (1757) *lower left*, and the H4 (1759) *lower right* (Photographs by the author)

The Longitude Board only relented after a new version (H5) proved itself on another voyage to sea and only after frequent pleading by John Harrison and his son, William. The final payment did not come until 1773, which was just 3 years before the death of John Harrison. While John Harrison did not have long to enjoy the fruits of his labor, the results of his innovative clock making transformed navigation, astronomy, and timekeeping for centuries afterward.

## The Modern Science of Time

As humans moved from ships to airplanes, spacecraft, and satellites, timekeeping also accelerated. During ancient times, precision of timekeeping within a day was sufficient to coordinate rituals, hunting, and agriculture. Navigation and astronomy increased the needed precision to within seconds during the seventeenth and eighteenth centuries, and further innovation and the demands of science provided mechanical clocks during the nineteenth century capable of measuring time to within seconds. The twentieth and twenty-first centuries brought humans to an entirely new level of precision, however. Time could be measured to within one part in a billion by 1960, and by 2000 the most accurate atomic clocks could measure time to one part in $10^{15}$, providing only 2 min of error during the entire amount of time elapsed in the history of the universe—the past 13.7 billion years!

At first glance, there seems to be little practical utility to measuring time accurately over spans of billions of years, but there are surprisingly many practical benefits of accurate timekeeping at the level of one millionth of a second. Most of these demands come from our reliance on electric power and electronic communications. To prevent power generators from working against each other, their phase of electricity needs to be accurately synchronized across the electric grid. Communication signals have to be synchronized within fractions of a microsecond for efficient radio transmission. The use of geosynchronous positional satellites (GPS) requires time measurements to one billionth of a second, as the GPS unit you may use compares the exact arrival time of beacons from orbiting satellites to the nearest billionths of seconds to determine your position. Even greater precision is needed in scientific research, where the reactions of molecules and the destruction of nuclei in particles take place on timescales of $10^9$ (one billionth of a second or a *nanosecond*) and 10–15 s (one thousandth of a trillionth of a second or a *femtosecond*), respectively.

The most accurate modern timekeeping devices use no springs, water, or gears. Instead, a modern clock uses the generation of billions of photons each second coming from transitions within atoms, to measure time to an accuracy of trillionths of a second. Much as the Sun, the Moon, water drops, or a pendulum can create a regular "beat" to move a clock hand, crystals, molecules, and atoms can give a signal to provide a simple and regular unit of time. The "escapement" for a modern clock is not a ratcheted toothed wheel, as in Galileo's and Huygen's time, but instead comes from the rapid oscillations of tiny quartz crystals or the wiggling of cesium or rubidium atoms!

### *Electronic and Atomic Clocks*

The evolution of modern timekeeping devices followed some of the same patterns of miniaturization seen in the evolution from water clocks, to mechanical clocks, to portable pocket watches. In the twentieth century, pocket watches gave way to wristwatches as spring-based mechanisms were miniaturized and sped up. Eventually an even smaller and faster "pendulum" was found in the form of a tuning fork, and watches with these devices became standard during the middle of the twentieth century. By 1928, the first experimental clocks using a solid state crystal as the oscillator was developed by W.A. Marrison of Bell Laboratories (Goudsmit and Claiborne 1969, p. 104). With space travel came miniaturized solid state circuits, which allowed the invention of an even smaller and faster oscillator, the quartz crystal, which could oscillate more than 32,000 times in a second. Such quartz oscillators have replaced pendulum and spring and tuning fork devices in most of the home clocks in use today.

Developing ever more accurate clocks requires faster and smaller oscillators. Individual molecules and atoms have become the units for ultimate miniaturization. But how do these atoms and molecules oscillate? By providing a source of light or radio waves, it is possible to get a sample of molecules or atoms to absorb radio waves or light at a very exact frequency. The trick to making a clock out of such a device is to provide feedback to help the molecules or atoms "lock in" the exact frequency.

To make a molecular clock, an electrical circuit makes a radio wave, which may oscillate at a billion cycles per second. The radio wave enters a chamber where molecules (such as ammonia) are kept at a very uniform temperature. The molecules absorb this signal perfectly when it is at just the right frequency, and an electronic circuit measures the rate of absorption and adjusts the frequency to maintain the exact frequency for strongest absorption. Such a "feedback loop" in the case of a molecular clock is known as an ammonia microwave maser molecular clock, which was the state of the art in the 1950s, and operates at 23,870 million cycles per second. This clock is accurate to one part in 100 million.

A similar feedback loop can be made with atoms by using an optical signal (i.e., a beam of light) absorbed by a cloud of atoms maintained at a precise temperature in a test chamber. Here instead of a pendulum, or a spring, the electrons in the atom provide the regular "beat" which sets the regularity of the clock. And unlike the larger mechanical clocks, individual atoms precisely obey quantum mechanics, and have nearly exact energies for their transitions, which can be used to set a frequency standard that marks out time. The pendulum in this case is oscillating many trillions of times per second, however, giving billions of times more precision than a mechanic clock!

Within the world of atomic clocks, rapid evolution of technology and new applications of lasers and ultra-precise control of temperature has steadily increased the accuracy of atomic clocks to precision that defies imagination. Most atomic clocks rely on ultra-cold atoms to provide maximum stability. Lasers are used to check on the frequency of the atoms—any drift in the frequency of the laser changes the amount of absorption and allows the atomic clock to adjust itself to maintain an exact frequency. One of the most successful designs for atomic clocks was developed at the US National Institute for Standards and Technology (or NIST) laboratory. The NIST clocks known as F1 and F2 make use of a "Cesium Fountain" to provide atoms of precise temperature for the frequency standard of the USA.

A description of the NIST Cesium fountain clock is mind boggling, especially when compared to earlier water clocks, sundials, springs, and mechanical clocks. To make an atomic clock, one needs to produce a cloud of cesium atoms and suspend them by lasers in a microwave cavity at a temperature a few millionths of a degree above absolute zero. A precise electronic circuit monitors the microwave cavity and adjusts the microwave frequency so it exactly matches the resonance frequency of the cesium atom at 9.192631770 trillion cycles per second. By using the most advanced electronics available, it is possible to lock in the circuit to the resonance frequency of the atoms. The NIST-F1 clock (state of the art in 2008) was able to maintain an accuracy of about one part in $10^{15}$, or one part in a thousand trillion (NIST 2008).

An improved atomic clock named the NIST-F2 has improved this accuracy by more than a factor of 10 (Heavner et al. 2014). With this increased precision, the NIST-F2 will not lose a second of time in accuracy until 316,000,000 years elapse! As astoundingly accurate as NIST-F2 is, newer designs are providing dramatically greater accuracy. NIST is testing a new model, known as a "strontium lattice optical clock" (Fig. 6.11) that keeps accurate time to within 1 s in 15 billion years. This new clock will not lose a single second over an expanse of time longer than the age of the universe! This new clock makes use of an "optical lattice" which is a pancake of strontium atoms held in place by lasers. The atoms "tick" at a rate of 430 trillion times per second, which provides a faster and more accurate frequency standard (Nicholson et al. 2015). With this level of precision, Einstein's notions of relativity become critical—time passes at different rates due to the force of gravity. New atomic clocks can measure these effects of General Relativity to such precision that they notice how time slows in as little as 2 cm of vertical motion.

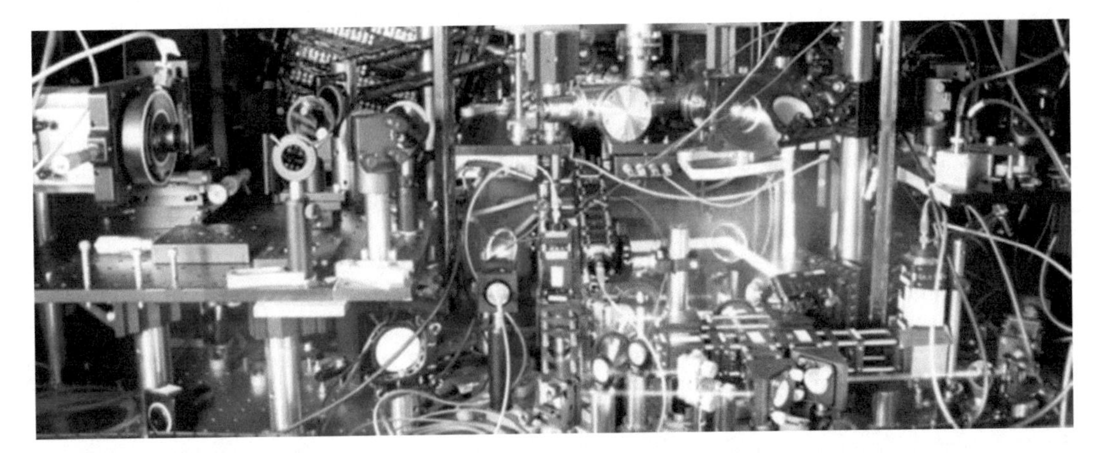

**Fig. 6.11** The current champion in atomic clock accuracy. This NIST strontium lattice optical clock is the modern alternative to the sundial, clepsydra, and the pendulum clock. This device levitates a pancake of ultra-cold strontium atoms at a precise temperature with lasers, and then uses the oscillations of those atoms as a frequency standard. The atoms then create an atomic metronome that beats 430 trillion times per second, with an accuracy to within 1 s over a timespan of over 15 billion years—more than the age of the universe! (figure from JILA 2016)

## *Solar Time or Atomic Time?*

Ironically, with this incredible precision, modern science has begun to separate humans from the most ancient of timekeepers, the Sun and the Moon. The units of time we use in our daily life—hours, minutes, and seconds—are all synchronized by the spin of our Earth. Our fundamental unit of time, the second, has not changed since its definition by French scientists in 1820–1/86,400 of a mean solar day. However, since the Earth itself is whirling like a giant top, it responds to the physical forces of the solar system. These external forces, such as the tug of the Sun and the Moon, give us observable effects, such as tides which slowly change the rate of the Earth's spin over centuries. The effects of libration, nutation, and deceleration, which come from wobbling of the Earth's spin axis from the tugs of the Moon and sun, slow down the spin of the Earth at a rate of 0.01 s per century or 0.27 μs per day (Goudsmit and Claiborne 1969, p. 105). While 100 μs of change in the length of the day per year does not seem like much, over a span of several years this difference becomes measurable, just as the 0.25 days between a 365-day Egyptian year and the 365.24 Julian year becomes measurable over centuries. Ten years include 31 million seconds, which is enough to cause a drift of 3 s due to the slowing of the Earth's spin!

The deceleration of the Earth's spin raises difficult questions about which time source should be the primary unit to synchronize human activity. How should humans set their clocks? Should we use the rotation of the Earth, which gives rise to the slowly lengthening day, or should we use the vibrations of atoms, which physics tells us remain the same at all locations and times in the universe?

## *Leap Seconds*

At the beginning of the twenty-first century, humans seem to have decided to keep both solar time and atomic time. Atomic clocks and the spinning Earth can be reconciled by use of "leap seconds." Just as the Julian and Gregorian reforms were able to improve our annual calendar by inserting additional days to account for the length of the year (which is 365.2422 days and not 365 days), the "leap second" is able to adjust individual days to account for the slow decrease of the Earth's spin rate. The official timekeeper in the USA is the US Naval Observatory, which features a set of departments with intriguingly obscure names such as the "Sub-bureau for Rapid Service and Predictions of Earth

Orientation Parameters" and the "International Earth Rotation Services (IERS)." According to these professional timekeepers, the Earth has slowed since 1820 by 2 ms (or 1.7 ms per century), meaning that the length of the mean solar day is presently 86,400.002 s, instead of 86,400 s, as defined. Leap seconds are used to keep us tied to the solar day, and since 1974, 27 leap seconds have been added to correct for the slowing of Earth's spin. Table 6.1 shows when the US Naval Observatory has introduced leap seconds since 1990. Further leap seconds will be announced on the US Naval Observatory web site—so you can keep your own atomic clock accurate!

The irregularity of leap seconds, which occur roughly once every 18 months, has, however, caused some hiccups in our advanced computer systems. Some computers have been programmed to include leap seconds, while others have not—both from limitations in software and from some computer programmers not being aware of the schedule for introducing leap seconds. Such complications, resulting in shifts of time of only 1 or 2 s, would seem insignificant, but our modern society requires computer systems to remain synchronized, and these tiny time shifts have had big consequences. Several computers, including one at a power station in Switzerland and one from Qantas Airlines, were caught out of synch from the leap second in 2012, and as a result flights were delayed, and some cities in Europe risked blackouts from the software error (Meyer 2015). These disruptions from leap seconds, and potential future problems, have caused some to advocate for abolishing the leap second. The result of abolishing leap seconds would be for solar time to shift slightly, and to disconnect our

**Table 6.1** Leap seconds for the years 1990–2016

| Year | 30-Jun | 31-Dec |
|---|---|---|
| 1990 | None | +1 s |
| 1991 | None | None |
| 1992 | +1 s | None |
| 1993 | +1 s | None |
| 1994 | +1 s | None |
| 1995 | None | +1 s |
| 1996 | None | None |
| 1997 | +1 s | None |
| 1998 | None | +1 s |
| 1999 | None | None |
| 2000 | None | None |
| 2001 | None | None |
| 2002 | None | None |
| 2003 | None | None |
| 2004 | None | None |
| 2005 | None | +1 s |
| 2006 | None | None |
| 2007 | None | None |
| 2008 | None | +1 s |
| 2009 | None | None |
| 2010 | None | None |
| 2011 | None | None |
| 2012 | +1 s | None |
| 2013 | None | None |
| 2014 | None | None |
| 2015 | +1 s | None |
| 2016 | None | +1 s |

The table shows that from 1990 until 2016, 12 leap seconds have been added (*Source*: http://en.wikipedia.org/wiki/Leap_second)

**Table 6.2** A summary of human timekeeping technology and its precision over the centuries; note that in recent years, atomic clocks have been improving their accuracy by a factor of ten every five years!

| Year | Place | Timekeeper | Precision (per day) |
|---|---|---|---|
| 1500 BC | Egypt | Sundial | 30 min |
| 700 BC | Babylon | Water clock (Clepsydra) | 1 min |
| 40 BC | Greece | Advanced water clock | 30 s |
| AD 1088 | China | Su Sung water clock | 10 s |
| AD 1358 | England | Salisbury geared clock | 1 min |
| AD 1500 | Germany | Spring-driven clock (egg) | 1 min |
| AD 1650 | Holland | Huygens clock | 10 s |
| AD 1720 | England | Harrison H1 | 0.1 s |
| AD 1761 | England | Harrison H4 | 0.04 s |
| AD 1800 | Germany | Pocket watch | 0.05 s |
| AD 1950 | USA | Maser ammonia clock | 1/millionth second |
| AD 1960 | Switzerland | Tuning fork-based watch | 0.001 s |
| AD 1970 | USA | Quartz watch | 0.0001 s |
| AD 1975 | USA | Cesium atomic clock | 1/10-billionth of a second |
| AD 2008 | USA | NIST atomic "Cesium Fountain" clock F1 | 1/100-trillionth of a second |
| AD 2010 | USA | NIST atomic "Cesium Fountain" clock F2 | 1/1000-trillionth of a second |
| AD 2015 | USA | NIST Strontium lattice optical clock | 1/10,000-trillionth of a second |

timekeeping from the sun and from astronomy. This disconnect over a long time period of 1000 years would cause the sun to be overhead at 1:00 PM instead of at noon!

Whether we decide as a species to break the bonds of our spinning earth in our timekeeping is something that will be decided from further debates in the coming year. Our timekeeping technology, inspired and developed from the eternal cycles of the sky, has brought us full circle and now has attained such accuracy that it may separate human time from the time of the Sun, the Moon, and stars (Table 6.2)

## Calendar Reform: Metric Time and the World Calendar

Innovations in clock making have taken us from sundials to atomic clocks, and yet our methods for counting hours and days seem to have changed little from the ancient Babylonian astronomy of 3000 years ago. The Babylonians gave us base 60 (sexigesimal) minutes and hours, and the ancient Egyptians divided the night and day into 12 segments, from which we get our 24-h day. Further complexities arise from the complicated mix of 30, 31, 29, and 28-day months from the Julian and the Gregorian calendar. Our breathtaking technological advances of the last century seem to have overlooked updating the calendar. Isn't there a way to simplify or rationalize the keeping of time?

Many thinkers have pondered the calendar in the past 200 years and have attempted to provide solutions. In the following sections we describe two attempts at calendar reform—the French Revolutionary calendar of the eighteenth century, and the twentieth century calendar known as the World Calendar.

### *The French Revolutionary Calendar*

During the French Revolution, the metric system was invented to rationalize weight, volume, and distance measurements. A similar rationalization was proposed for time measurement, in a sort of "temporal metric system" known as the French Revolutionary calendar. The French Revolutionary calendar consisted of 12 months of 30 days, with each month consisting of three equal 10-day weeks. Each day consisted of 10 h, each with 100 min, and each divided into 100 s. The result is a complete decimal time measurement system, grouped in periods of 36 weeks, to create 360 days. The extra 5 days of the year (6 in a leap year) were added at the end and were known as "jours complémentaires" or "complementary days."

The designation of five extra days at the end of the year was similar to the ancient Mayan and Egyptian calendars, which associated each of these extra days with gods and bad omens. The French system instead gave these "complementary days" names that were dedicated to the rational and revolutionary concepts of "virtue," "genius," "labor," "reason," and "reward." The sixth extra day for leap years was dedicated to the revolution itself.

The names of the months were derived based on "rational" associations with (Northern Hemisphere) seasons and were grouped into four seasons, each of 3 months. The year began on the Vernal equinox (September 21 in our calendar) during the year 1792. The months were given names by the poet Fabre d'Eglantine, and he appropriately began the year with "vintage month" (September), followed by months named for fog, sleet, snow, rain, and wind. By springtime, the improved weather inspired names for a seed month of *Germinal* (March), followed by a blossom month of *Floreal* (April), and a pasture month (May). The summer months were named for the harvest (June), for a heat month of *Thermidor* (July), and finally a fruit month *Fructidor* (August). Each of the month names for the four seasons were also chosen to rhyme, and to invoke the "sonority" of the season from the sound of the words. The French names for both the "extra days" and for each month of the Revolutionary calendar are presented in Tables 6.3 and 6.4.

**Table 6.3** Extra days, known as "complementary days" within French Revolutionary calendar, which were inserted after the 360-day year, as an end of the year festival

| Number | Day name | Translation |
|---|---|---|
| 1 | jour de la vertu | Virtue day |
| 2 | jour du génie | Genius day |
| 3 | our du labour | Labor day |
| 4 | jour de la raison | Reason day |
| 5 | jour de la recompense | Reward day |
| 6 | jour de la révolution | Revolution day |

**Table 6.4** Month names in the French Revolutionary calendar, with their translations and associations with nature

| Number | Revolutionary name | Translation | Gregorian month |
|---|---|---|---|
| 1 | Vendémiaire | Vintage month | September |
| 2 | Brumaire | Fog month | October |
| 3 | Frimaire | Sleet month | November |
| 4 | Nivôse | Snow month | December |
| 5 | Pluviôse | Rain month | January |
| 6 | Ventôse | Wind month | February |
| 7 | Germinal | Seed month | March |
| 8 | Floréal | Blossom month | April |
| 9 | Prairial | Pasture month | May |
| 10 | Messidor | Harvest month | June |
| 11 | Thermidor | Heat month | July |
| 12 | Fructidor | Fruit month | August |
| 13 | Sansculottides | Additional days | |

New clocks were designed showing both the French "metric" time units and the older non-revolutionary units—hours, minutes, and seconds. Many found the conversion to "metric time" difficult, as the 10-h day provided much longer hours (closer in length to the ancient Chinese and Egyptian hours) and unfamiliar decimal minutes and seconds. In a "decimal day" each revolutionary "hour" lasted 2.4 h, each revolutionary "minute" lasted 1.44 min, and each revolutionary "second" lasted about 0.86 s.

Adjustment to the longer 10-day week was also a challenge. Each of the 10 days of the week (known as a "decade") was named simply by rank order—the first day was called "first" (primidi), then came "second" (duodi), "third" (tridi), "fourth" (quartidi), "fifth" (quintidi), "sixth" (sextidi), "seventh" (septidi), "eighth" (octidi), "ninth" (nonidi), and tenth (décadi) (Goudsmit and Claiborne 1969). The last of these 10 days (décadi) was the day of rest, which came only every 10 days, and did not help the popularity of the system with workers, who preferred getting a day of rest every 7 days! Each of the days of the year was also associated with plants, animals, or tools, to replace the previous association with saints and ancient religions.

The French Revolutionary calendar, despite its success in rationalizing time units, went against the habits of people who had adjusted to the rhythms of the ancient months, hours, and days. This Revolutionary system did not survive and was cast off in 1806 by the emperor Napoleon I, who restored the calendar put in place by the much earlier emperor Julius Caesar, and modified by Pope Gregory. Despite achieving a simplifying "metric" calendar, with more "rational" units of time, the month names were keyed into seasons found mainly in Northern Europe, and the Revolutionary Calendar was at odds with the rituals and habits based on our more ancient 24 h, 7-day week system of counting days.

## The World Calendar

The American activist Elizabeth Achelis proposed a less extreme calendrical reform, called the "World Calendar," in 1930. The World Calendar built upon on earlier efforts by the Italian abbot Marco Mastrofini, who rearranged the lengths of months to enable the year to always begin on Sundays. The World Calendar was motivated by a desire to have all dates within a given year recur in every future year, and it was hoped that this system would help unify the world with a single rational system in all countries.

The World Calendar proposed a less radical adjustment of the calendar than the French Revolutionary calendar since it retained the same month names, the same day names, and same basic lengths of months, days, and hours. However, month lengths were adjusted so that the every year days would fall on the same dates, and the calendar was also divided into four equal quarters. Each year consisted of four 3-month blocks of 31-, 30-, and 30-day months (Fig. 6.12).

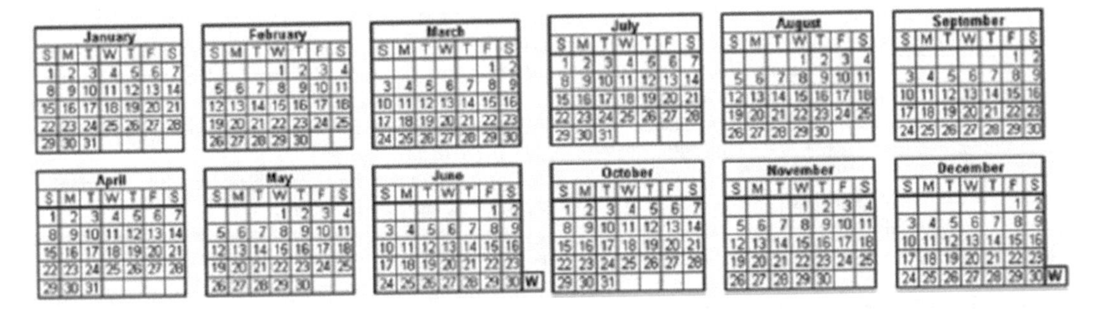

**Fig. 6.12** The World Calendar, which provides a perfectly repeating calendar divided into four equal quarters, with the two additional days added known as "World Days" to preserve leap year and provide an "uncounted" day each year. The simplicity of the calendar is that it could be reused forever (thereby strongly impacting the calendar industry!) but at the cost of rearrange our month lengths and having the extra "W" days (*image source*: http://en.wikipedia.org/wiki/World_Calendar)

In the World Calendar, January, April, July, and October had 31 days, while the other months all had 30 days. The combination of these new month assignments enabled each quarter to have exactly 91 days, giving a total of 364 days in four quarters. The remaining 365th day existed outside the quarters and was designated "World Day" instead of one of the seven traditional day names, and leap years would have two such "World Days."

By providing an additional "World Day" every fourth year (as a leap year) the World Calendar was able to preserve the repeating day names perpetually, at the expense of adding one or two mysterious holidays each year which would not be a Tuesday, Wednesday, or Sunday but instead a "World Day." The "World Day" in some ways is reminiscent of the ancient Roman and Incan calendars where there are periods of "uncounted time."

The benefits of this proposed calendar are that quarterly activities (such as school schedules and business reports) could use the exact same form each year, saving much time and difficulty. A date in the World Calendar (such as a birthday), which occurred on Friday, the 15th of March, would recur on this exact same date and day every year. The World Day would create some 8-day weeks, however, and since most of the world's religions are locked into a 7-day cycle of worship, these longer weeks would disrupt the pattern of religious worship in Christian, Judaic, and Muslim traditions.

Despite 25 years of advocacy by Ms. Achelis and her World Calendar Association, as well as two books and a *Journal of Calendar Reform*, resistance to the World Calendar was too strong to part with our ancient ways. Our civilization seems to prefer to keep its ancient system of named days, with its connections to earlier lunar calendars, Roman emperors, Babylonian base-60 mathematics, and ancient planetary gods for the foreseeable future.

## References

Brecher, K., and M. Feirtag. 1979. *Astronomy of the ancients*. Cambridge, MA: MIT Press.
Evans, J. 1998. *The history and practice of ancient astronomy*. New York: Oxford University Press.
Freeth, T., et al. 2006. Decoding the ancient Greek astronomical calendar Known as the Antikythera mechanism. *Nature* 444(7119): 587–591.
Freeth, T., A. Jones, J.M. Steele, and Y. Bitsakis. 2008. Calendars with Olympiad display and eclipse prediction on the Antikythera mechanism. *Nature* 454(7204): 614–617.
Goudsmit, S.A., and R. Claiborne. 1969. *Time*. New York: Time-Life Books.
Heavner, T.P., et al. 2014. First accuracy evaluation of NIST-F2. *Metrologia* 51: 174.
Huygens, C. 1673. *Christiani Hugenii Zulichemii, Const. f., Horologium oscillatorium, siue, De motu pendulorum ad horologia aptato demonstrationes geometricae*. Parisiis: Apud F. Muguet.
JILA. 2016. AMO physics and precision measurement. http://jila.colorado.edu/yelabs/research/ultracold-strontium. Accessed July 2016.
Kelley, D.H., and E.F. Milone. 2005. *Exploring ancient skies: An encyclopedic survey of archaeoastronomy*. New York: Springer.
Krupp, E.C. 1984. *Echoes of the ancient skies: The astronomy of lost civilizations*. New York: New American Library.
Lin, J.L., and H.S. Yan. 2014. Historical development of reconstruction designs of Antikythera mechanism. Presented in proceedings of the 9th international conference on history of mechanical technology and mechanical design, Tainan, 23–25 Mar 2014. (Applied mechanics and materials, vol 163, 1–62012.04.).
Lin, J.L.. 2011. Systematic reconstruction designs of Antikythera mechanism. Dissertation, Symposium on History of Machines and Mechanisms, VU University Amsterdam, The Netherlands.
Lin, J.L., and H.S. Yan. 2016. *Decoding the mechanisms of the Antikythera astronomical device*. New York: Springer.
Lockyer, N., and G.M. Seabroke. 1878. *Stargazing: Past and present*. London: Macmillan.
Marchant, J. 2010. Mechanical inspiration. *Nature* 468: 496.
Meyer, R., 2015, Clocks are too precise (and people don't know what to do about it). The Atlantic.
Needham, J. 1981. *Science in traditional China: A comparative perspective*. Cambridge, MA: Harvard University Press.
Needham, J., and C.A. Ronan. 1980. *The shorter science and civilization in China: An abridgement of Joseph Needham's original text*. New York: Cambridge University Press.

Needham, J., et al. 1986. *Heavenly clockwork: The great astronomical clocks of medieval China*. New York: Cambridge University Press.
Nicholson, T.L., S.L. Campbell, R.B. Hutson, G.E. Marti, B.J. Bloom, R.L. McNally, W. Zhang, M.D. Barrett, M.S. Safronova, G.F. Strouse, W.L. Tew and J. Ye., 2015, Systematic evaluation of an atomic clock at 2 × 10−18 total uncertainty. Nature Communications .
Sobel, D. 1995. *Longitude: The true story of a lone genius who solved the greatest scientific problem of his time*. New York: Walker.
NIST. 2008. NIST-F1 Cesium fountain atomic clock. From https://www.nist.gov/pml/time-and-frequency-division/primary-standard-nist-f1.
———. 2009. Ytterbium gains round in quest for next-generation atomic clocks. From https://www.nist.gov/pml/time-and-frequency-division/primary-standard-nist-f1.
Rees, A. 1819. *The cyclopædia; or, universal dictionary of arts, sciences, and literature*. London: Longman, Hurst, Rees.
Smith, M.. 2000. Robert Hooke—English father of microscopy. From http://www.microscopy-uk.org.uk/mag/artmar00/hooke4.html.
Van Helden, A.. 1995. The Galileo Project—Pendulum clock. From http://galileo.rice.edu/sci/instruments/pendulum.html.

# Chapter 7
# Celestial Architecture: Monuments of the Sky

*This grand show is eternal. It is always sunrise somewhere... Eternal sunrise, eternal sunset, eternal dawn and gloaming, on sea and continents and islands, each in its turn, as the round earth rolls."*

John Muir (Wolfe, 1979)

Humans viewed the stars, imagined the structure of the universe in their star tales and cosmologies, and then very quickly began to recreate this structure on the Earth in monumental buildings. These buildings took the forms of ziggurats, pyramids, cathedrals, temples, and kivas in various continents and centuries and served many purposes. One key element in making a structure part of the "celestial architecture" is to align the structure with the night sky or the morning or evening sunrise or sunset. In this chapter we explore the many ways in which humans have made "celestial architecture" that either mirrors cosmology, reflects the origins and beliefs of the people, or is aligned with morning sunrises, evening sunsets, or transits and standstills of the Sun, the Moon, and planets.

## Megalithic Alignments

### *Stonehenge*

The first and most widely discussed of the megalithic archaeoastronomy sites is Stonehenge, which helped define much of the field of archaeoastronomy over the course of the twentieth century. Stonehenge, with its five huge stone "trilithons" (each consisting of two upright and one horizontal lintel stones), and a circle of 30 gigantic vertical upright stones forming the famous "Sarsen Circle," presented an enigma to people attempting to decipher its purpose during the 5000 years since its construction. In some cases, modern interpretations reveal more about our own inclinations and natures than about the people who originally constructed Stonehenge!

Early during the Roman Empire, Stonehenge was already an ancient site. The conquering Romans, perhaps responding to the eerie mystery and the hold of Stonehenge on the Britons, attempted to destroy the monument and succeeded in removing some of the horizontal lintels within the stone circle (Castleden 1993). Later during the middle ages, Stonehenge's vast stones were thought to be evidence of magic by the wizard Merlin, while authors in the seventeenth and eighteenth centuries attributed Stonehenge to a mystical race known as Druids.

From archaeological evidence, we know that the task of creating Stonehenge was epic and occurred in three stages over the course of nearly 2000 years from 3000 BC until 1500 BC. Advanced analysis




B.E. Penprase, *The Power of Stars*, DOI 10.1007/978-3-319-52597-6_7

of minerals within the bluestones (sometimes known as "forensic geology") located their origin from a quarry in faraway Wales, nearly 200 km distant, while the Sarsen stones came from nearly 40 km away. The Aubrey Holes, a set of 56 small pits surrounding the stones of Sarsen Circle, have provided traces of bone, charred wood, and other organic material that enable dating of the phase of Stonehenge from 2500 BC. Additional evidence of a "Woodhenge" of posts surrounding the site suggests that there was continuous use of the site from 3500 BC until 2500 BC, with the site evolving into its present form between 2500 and 1500 BC. The Sarsen Circle, which consists of 30 stones of sizes ranging from 10 to 80 t, appears to have been put in place last, sometime around 1500 BC. Particular evidence from archaeological studies of the site include deer antlers dated to 3810 BC, charcoal from 2305 BC in the Aubrey holes, and a skeleton found within the ditch surrounding the site dated to 1765 BC, giving strong evidence of continuous use in ritual for over 2000 years.

The stunning grandeur and mystery of Stonehenge has challenged scientists to explain both the method by which it was built and also to question the motives of the long-lost ancient civilization that built Stonehenge. Was Stonehenge a site of ritual sacrifice? An astronomical observatory? A burial site? An outdoor chapel? Interestingly the answer seems to be yes–Stonehenge served all of these functions. And as technology advanced during the course of the late twentieth century and now in the twenty-first century, we are finally understanding more about how Stonehenge might have been used during ancient times.

Stonehenge certainly seems to have been a site of sacrifice—evidence of animal parts within the Sarsen Circle and buried within the Aubrey holes seems indisputable. Stonehenge also seems to have functioned as an astronomical observatory. The main axis of Stonehenge defined by the heel stone and a long avenue connecting the interior of the Sarsen Circle with the outside circle aligns nearly perfectly with the summer solstice sunrise. Additional alignments between lintels and the lunar standstills have been proposed, and while less definite, there appears to be a basic alignment between Stonehenge and the extreme horizon locations of the Sun and Moon.

One of the first astronomical interpretations of Stonehenge dates from 1740, when William Stukely noted the alignment of the "heel stone" with summer solstice sunrise. Sir J. Norman Lockyear in 1901

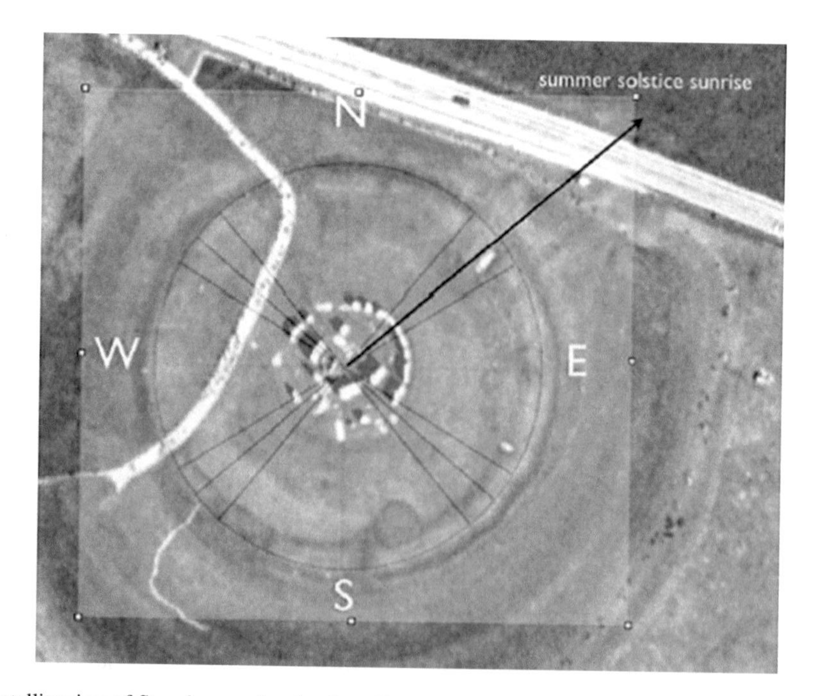


**Fig. 7.1** A satellite view of Stonehenge, showing key alignments with solar solstice sunrise and sunset directions, as well as lunar standstill points. The striking alignment with summer solstice sunrise is visible, as it connects the inner Sarsen Circle with the heel stone at *upper right* (original figure by the author using the NASA World Wind software)

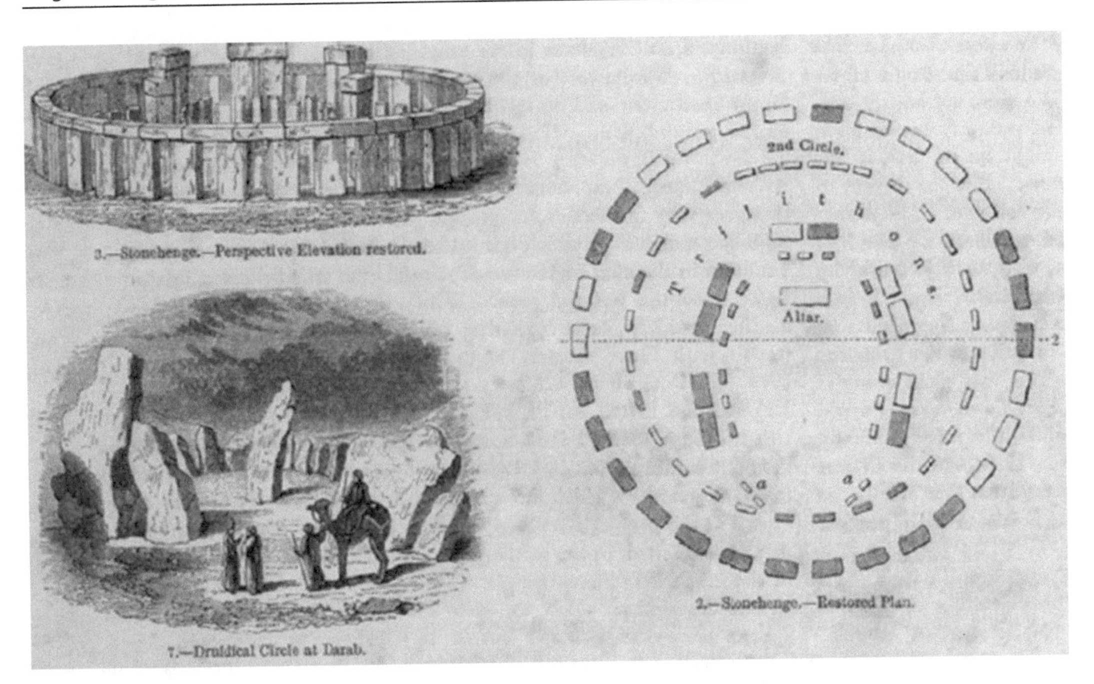


**Fig. 7.2** Depiction of a reconstructed Stonehenge from the work of Stukeley (1740), showing some speculative labels of a "Druidical circle" and "altar" that helped reinforce the early interpretation of Stonehenge as a "Druid altar" (image from the Huntington Library, reprinted with permission)

**Fig. 7.3** Stonehenge summer solstice party shortly after sunrise, June 2004. The monument still attracts crowds each year, who stay up all night at the site and observe the first light at sunrise, celebrating the solstice and enjoying the spectacle of the event (original photograph by the author)

reported four additional alignments related to the Celtic calendar, but was not of the opinion of Stonehenge being an observatory, but rather a ritual center for ancient Britons (Krupp 1983, p. 217) (Figs. 7.1, 7.2, and 7.3).

Perhaps the most extreme of the astronomical interpretations of Stonehenge is found in the work "Stonehenge Decoded" (Hawkins and White 1966). The lead author, Gerard Hawkins, mathematically analyzed the possible alignments between the major stones and computed the probability of these alignments as arising from random chance. Out of 50 possible lines connecting stones and pits,

Hawking found 12 solar alignments and 12 lunar alignments and also proposed that the 56 Aubrey Holes served as a sort of megalithic calculator for counting the years between solar eclipses, which would be expected to recur in a 56-year cycle, consisting of 19, 19, and 18 years. Critics questioned Hawkins calculations, by more accurately accounting for the limited precision of the proposed alignments. Since the alignments appeared to be within only 2°, there is a higher probability of these alignments arising from chance (Hadingham 1984, p. 46).

Further speculation on the astronomical and mathematical significance of Stonehenge was proposed in 1974 by Alexander Thom, who noticed numerical ratios between the separation of stones and the circumference of circles connecting ancient wooden posts known as "Woodhenge." By expressing distances in units of the "megalithic yard" (defined as 2.72 ft) Thom found circumferences of wooden circle had units of 40, 60, 80, 100, 140, and 160 megalithic yards and that triangles connecting the station stones worked out to multiples of 5, 12, and 13, a Pythagorean triple (Ruggles 2005, p. 408).

Regardless of how modern people interpret Stonehenge in its mathematical and astronomical aspects, the site still evokes mystery and wonder, and also still aligns with the Sun and Moon during the solstices. Figure 7.1 shows some of these alignments, using a satellite image of the Stonehenge site, superimposed with lines indicating the locations of solar summer solstice sunrise, sunset, and lunar standstill positions. The definite alignment with summer solstice sunrise is apparent, and the possible coincidence with the standstill positions with large stones is also visible.

New archaeological research, including the first new excavations in 40 years, has provided good evidence of how Stonehenge functioned as a possible sacred site for healing. The excavations show a variety of human bones with a large fraction showing signs of trauma, injury, and disease. The monumental architecture may have had less to do with astronomy and more with curing earthly ills of pilgrims (Odling-Smee 2007). One of the intriguing paradoxes of ancient sites like Stonehenge is that their study has attracted the most modern of technologies—satellite imagery, ground-penetrating radar, and radio-isotopic analysis—to unravel its true purpose. With additional archaeological information, it is clear that while Stonehenge may indeed have been an ancient observatory, it was also part of an extensive "ritual landscape" that included multiple henges and rings of posts in the vicinity of Stonehenge. Examples include a wooden ring of posts 1000 yards Northeast of Stonehenge, and a huge stone wall and settlement known as Durrington Walls, a bit farther in the same direction (Underhill 2011). Careful analysis of the isotopes in skeletons and soil from across the region has enabled archaeologists to trace the origins of some of the people who died near Stonehenge, and some appear to have traveled across Europe to the site, increasing the geographic reach of the site to as far as Switzerland. Clearly Stonehenge has inspired and challenged people for thousands of years, and it seems that only now as our science advances into the twenty-first century we are able to shed new light on the meaning of Stonehenge to ancient people.

## *Callanish*

Standing stone circles around Britain are numerous, but few are as extensive as those of Callanish, a site far north of Scotland in the Island of Lewis, in the Outer Hebrides (Fig. 7.4). These stones form a circle 13 m across and form a long avenue 84 m long which points to the NNE. Despite the suggestion by Sommerville in 1912 that a pair of stones point toward a lunar standstill point, the site offers few astronomical alignments. Sommerville also suggested the avenue that points to the NE could align with the horizon location of Capella in 1800 BC, and the slight tilt from an EW axis of the stones could point toward Altair in 1800 BC. A definitive explanation of the site in astronomical terms, however, is elusive (Hadingham 1975, p. 104).

**Fig. 7.4** The standing stones of Callanish, which include a stone circle and a long avenue facing NNE, have been a source of inspiration and mystery for centuries and possibly are aligned with a lunar standstill point (image from Marta Gutowaska, posted at http://commons.wikimedia.org/wiki/File:Callanish_standing_stones_1.jpg)

## Newgrange

Newgrange is part of a trio of Celtic sites, in a region within 100 km northeast of Dublin, Ireland. The three sites include Newgrange, Knowth, and Dowth, which appear to form a triangle of ancient sacred sites and date from approximately 3000 BC. Newgrange is a huge mound structure nearly 300 ft across, with a 45-foot high wall decorated with quartz stone carved in a variety of geometric patterns. But the most astounding part of Newgrange is inside its long corridor, which leads 62 ft into the center where the inner sanctum is illuminated only once a year by the faint gleams of light of the winter solstice sun.

The alignment of Newgrange with the winter solstice sun is nearly perfect and provides a chilling impression to modern observers, who witness the pattern of light and shadow on the winter solstice today, just as it has happened each year for the past 5000 years (Figs. 7.5 and 7.6). Within the central chamber a vaulted ceiling rises 20 ft above and has a corbelled roof and additional set of three smaller chambers. Inside Newgrange are remains of human and animal bone, burned and unburned, along with jewelry and beads, suggesting that the device was a very elaborate tomb. The scale of the construction is equal to any in the ancient world, with nearly 200,000 tons of stone comprising the mound and 97 enormous kerbstones, 31 of which were decorated with geometric patterns (Scarre 1999, p. 52).

The patterns of spirals and circles on the stones have been interpreted to some to be clues for lunar and solar cycles and some authors have also speculated on alignments between some of the carved stones at the Newgrange site. Regardless of what exactly these patterns mean, the alignment of Newgrange and its winter solstice light show are clear evidence of astronomical skill of the ancient people who constructed this site. Like Stonehenge, Newgrange is part of an extended "ritual landscape" that included dozens of large neolithic monuments, and hundreds of smaller standing stones, balanced stones, or stone tombs known as "dolmens." These ancient sites have attracted scientists and enthusiasts who have been diligently using modern technology for charting their locations and deciphering their functions. The web site known as the Megalithic Portal (http://www.megalithic.co.uk/)

**Fig. 7.5** Newgrange as viewed from the side, with the entrance at the right, which faces the winter solstice sunrise as it has for over 5000 years (image by Jennifer Perry, reprinted with permission)

**Fig. 7.6** The entrance to Newgrange includes a shaft (*at top* decorated with *white stones*) and a larger entryway (*below*), which faces the winter solstice sunrise. In front of the entrance is a large stone decorated with ancient Celtic spiral patterns (photograph by Jennifer Perry, reprinted with permission)

has compiled much of this data, and offers a clearinghouse of GPS data for standing stones and other megaliths throughout the UK and Ireland, with additional data for megaliths from around the world. It is also possible to get the Megalithic Portal app—which allows you to locate over 25,000 ancient sites with the convenience of your cell phone with accurate geographic locations made available via satellite GPS!

## Native American Alignments and Monuments

The Hopi horizon calendar, mentioned earlier, allowed alignments of the Sun with each notch or point on a distant mountain to indicate times for activities such as planting of maize and gathering of berries. These types of observations sometimes also noted the location of moonrise and moonset, with its 18.6-year cycle between extremes on the horizon, suggesting that some Native American groups were aware of the longer cycles of eclipses and planetary motions (Williamson 1984, p. 45). However, Native Americans also created aligned structures and observatories that allowed observations of the Sun and stars and embodied principles within their worldview. We describe some of these structures in the following section.

### *The Anasazi, Hopi, and Pueblo Kiva*

The construction of the kiva gathered together many of the deepest symbols of the Pueblo Native American cosmos, to create a public space that embodies fourness, a connection with the sky and stars, and harmony with the universe (Fig. 7.7). As described in the Acoma Creation Story:

> When they built the kiva, the first put up beams of four different trees. These were the trees that were planted in the underworld for the people to climb up on. In the north, under the foundation they placed yellow turquoise; in the west, blue turquoise; in the south, red, and in the east white turquoise. Prayer sticks are placed at each place so the foundation will be strong and never give way. The walls represent the sky, the beams of the roof (made of wood of the first four trees) represent the Milky Way. The sky looks like a circle, hence the round shape of the kiva.
> (Williamson 1984, p. 138)

**Fig. 7.7** The great house or kiva known as Chetro Ketl, from Chaco Canyon, in New Mexico, USA. The form of this kiva includes the set of niches along the inner wall and four large post-holes for holding up the ceiling, which has not survived (*image source*: http://commons.wikimedia.org/wiki/Image:Chaco_Canyon_Chetro_Ketl_great_kiva_plaza_NPS.jpg)

In the Hopi and Zuni creation story, the people emerge into the present world after being led through four previous earths. Their emergence into the present world is accompanied by a speech from Spider Woman that confirms the central role of the kiva in preserving and embodying the Pueblo cosmology:

> In time you will find the land that is meant for you. But never forget that you came from the lower world for a purpose. When you build your kivas, place a small *sipapuni* (a small hole) there in the floor to remind you where you came from and what you are looking for.
>
> (Courlander 1987, p. 33)

Perhaps the most perfect example of the Native American kiva is the 1000-year-old Casa Rinconada, at the Chaco Canyon site in northwest New Mexico. The Casa Rinconada kiva is perfectly symmetric, "round like the sky," and aligned with the cardinal directions, with each of four posts for the roof arranged in a perfect square. A small opening in the wall connects the interior of the kiva with the summer solstice sunrise direction. A set of 28 niches in the interior wall provides a ceremonial space for offerings, perhaps related to astronomical ceremonies related to the lunar month. The symmetry and geometric perfection of Casa Rinconada is reminiscent of the Greek Parthenon and embodies the principles of Native American cosmology as fully as the Parthenon embodies the principles of Greek mathematical philosophy.

## The Navajo Hogan and Pawnee Lodge

A monument to the universe does not necessarily require impressive tonnage of stones or a huge statue of a departed ruler. The Navajo hogan embodies cosmology on a smaller, family sized scale (Fig. 7.8). The Navajo believed that after their emergence into the present world, Black God and Changing Woman taught the *Dine* (Navajo for "people") the ideal way of life, centered on the activities of First Man and First Woman, and the hogan, their home. The building of a hogan, a simple single family dwelling, incorporated the most lofty symbols of the structure of the universe, with each of four named poles symbolizing the four cardinal directions and the ruling spirit of each

**Fig. 7.8** The Navajo hogan, used by present-day Navajo, embodies the principles of the Navajo cosmology in its design and is aligned so the entrance faces east (*image source*: photo by Wolfgang Staudt from Saarbruecken, Germany, at http://commons.wikimedia.org/wiki/Image:Navajo_Hogan,_Monument_Valley.jpg)

direction–that of the Earth, Mountain Woman, Water Woman, and Corn Woman. These parts of the Hogan and its construction are named and celebrated in the Hogan Song:

> Along below the east, Earth's pole I first lean into position …
> Along below the south, Mountain Woman's pole I next lean into position …
> Along below the west, Water Woman's pole I lean between in position …
> Along below the north, Corn Woman's pole I lean my last in position …
> (Williamson 1984, p. 153)

The Skidi Pawnee followed similar pattern in constructing a family dwelling that mirrored the structure of the universe, with special attention being paid to the relation between the dwelling and four principle stars, which represented the four corners of the sky. The Pawnee lodge was made of earth and was circular like the kiva and contained an opening to the stars in the middle. Access to the lodge was from the side and entrances were designed to face east, in the form of a small tunnel to the inside of the earthen enclosure. Like the kiva and the hogan, four posts supported the lodge and each part of the lodge symbolized a key element of the universe. The floor represented the earth, the roof the sky, and the four posts were associated with the four posts of the sky–the Yellow Star (Capella) , the White star (Sirius), the Big Black Meteoritic Star (Vega), and the Red Star (Antares) (Krupp 1983, p. 237).

## *Solstice Viewing: Shadow and Light in California*

The Chumash tribe of Southern California, along with neighboring tribes in the southwest, used caves and crevices to provide sacred spaces in which to make contact with the stars, the Sun, and the sky world. In some cases rock art provides locations of special significance and particular caves were adorned with sun figures that would be illuminated during the winter solstice sunrise. Three known caves within the Chumash territory that function as solar observatories include the caves of Burrow Flats (Fig. 7.9), Window Cave, and Condor Cave. In all three cases, unique geometry such as a small aperture or roof crevice directs the light of the rising sun into the center of the cave. Within the cave are multiple panels of rock art and sun figures that are cleft by the light as it lights up the day at the solstice. Since the winter solstice is such a key event in the Chumash culture, the shaman priest would make use of these caves to help determine the exact date for the solstice festival. After the Sun was determined to reach its lowest point, the priest would lead the people in a ritual dance involving dancing and chanting around a sun stick. Each year this dance was repeated and happily the Sun reversed its course toward the south without fail during each of the solstice rituals.

**Fig. 7.9** Burrow Flat cave, near Chatsworth, CA, in which a *notch* on the *top of a cave* provides a dagger of light (*visible below*) which crosses through a sun figure on the winter solstice. (From *top right*) View of cave, showing the distinctive notch which provides a "dagger" of light; (*upper right*) beginning of solstice event when light touches center of sun figure; (*lower left*) the "dagger" of light grows and shifts away from the sun figure after sunrise; (*lower right*) closer view of rock art which includes a spiral sun figure and a shamanic stick figure (photos by B. Penprase)

## *Mounds and Stone Circles in Native American Architecture*

One of the major sites in Native American archaeoastronomy is the Big Horn Medicine Wheel, near the northern border between present-day Wyoming and Montana (Fig. 7.10). The site is located at an altitude of nearly 10,000 ft and consists of a very large circle of stones 90 ft in diameter which are connected by a set of 28 spokes, with a central large rock pile or cairn 10 ft in diameter and 6 additional cairns outside the circle. The numerical significance of 28 spokes could relate to the duration of the lunar months, and it is known that other cultures (including the Lakota) used the number 28 in their structures as a sacred number.

**Fig. 7.10** The Bighorn medicine wheel, as viewed from the southwest, showing the set of cairns aligned with stellar rising azimuths and the summer solstice sunrise cairn (*top right*). The line connecting the solstice cairn and the center points toward the summer solstice sunrise (image from http://upload.wikimedia.org/wikipedia/commons/9/91/MedicineWheel.jpg)

The astronomer J. Eddy analyzed the directions of both the 28 "spokes" and lines connecting the seven cairns and found that at least four of the cairns and the major axis of the site all pointed toward astronomically significant horizon azimuths (Eddy 1974). The alignments include the summer solstice sunrise and the rising azimuths of four significant stars visible during summer from this part of Wyoming–Sirius, Aldebaran, Fomalhaut, and Rigel (Fig. 7.11).

While it is compelling to imagine the medicine wheel as an astronomical observatory, some skepticism is in order. A statistical analysis of the alignments on the medicine wheel shows that the alignments are only slightly more significant than those predicted by random chance. If we consider the seven cairns at the site, there are 42 possible pairings. This result is given by the equation $P = n(n-1)$, where $n$ is the number of cairns and $P$ is the number of pointings (Williamson 1984, p. 203). Using the precision of the alignment, one can calculate how many astronomically significant pointings would arise randomly. For example, if you have 42 pairs of stones that define locations on the horizon, each aligned with an accuracy of 1°, 42 out of 360 or about 12% of the horizon is covered by alignments from pairs of cairns.

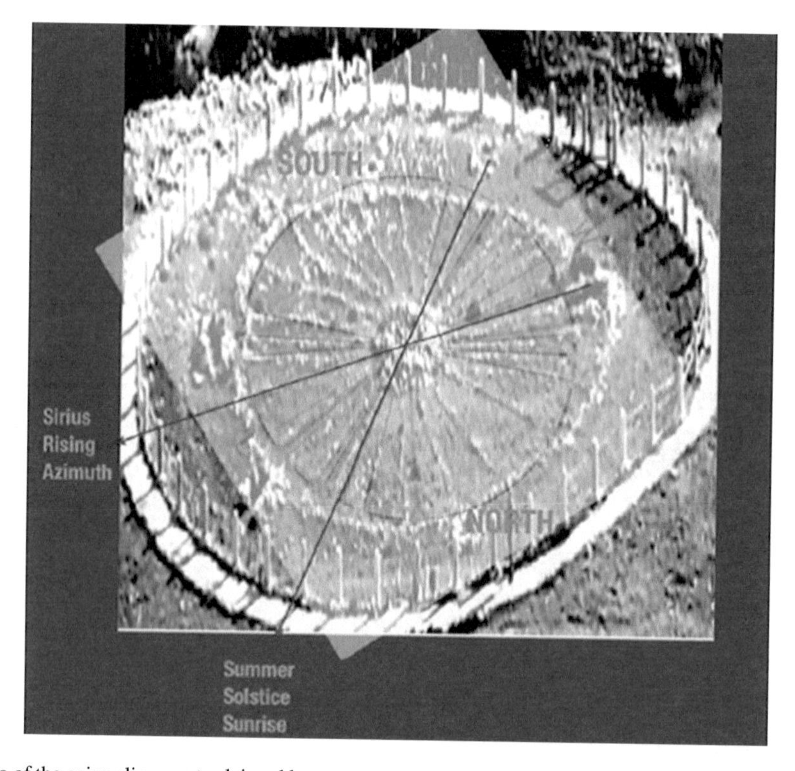


**Fig. 7.11** Two of the cairn alignments claimed by astronomer Jack Eddy in 1974 for the Big Horn medicine wheel (*red lines*), along with other possible astronomically significant directions (*fainter yellow* and *blue lines*). Using the declinations of the nine brightest stars visible from Montana we can determine the locations on the horizon where they rise or set, which are shown with *blue*. Also shown are the solstice rising and setting azimuths for the Moon and Sun (*yellow*). When we calculate stellar and solar solstice horizon positions for rising and setting, about 8.3% of the horizon is covered. It is possible that some of these alignments coincide with pairs of cairns by random chance

Eddy determined that approximately 6 out of the 42 possible directions were astronomically significant or about 14% percent of the pointings. Directions were considered astronomically significant if they coincided with the position of a bright star on the horizon or solstice sunrise or sunset positions. What fraction of the horizon contains a bright star or solar rising or setting location?

Using the assumptions above and an accuracy for an alignment of about 1.5°, one can show that about 8.3% of the horizon is aligned with one of the nine brightest stars or the solar solstice sunrise or sunset positions. This means that the astronomical interpretation of the Bighorn medicine wheel site is only slightly more likely than random coincidence, since a random pointing (within 1.5°) on the horizon will have an 8.3% chance of being "astronomically significant" and Eddy calculated 14% of the sightings were aligned!

One additional piece of evidence, however, suggests that there could be more to the story. Several additional medicine wheels are known in the Rocky Mountain region of the USA, and the provinces of Alberta, and Saskatchewan in Canada. Many of these medicine wheels show the same asymmetric axis aligned with winter solstice sunrise and some of the same stellar alignments. For example, the medicine wheel in Alberta known as Moose Mountain medicine wheel includes a set of seven cairns, with what appears to be the same stellar alignments (Brecher and Feirtag 1979; Krupp 1983). This agreement between several sites removes some of the doubt one may have from considering just the Bighorn medicine wheel, making Eddy's astronomical interpretation more likely.

Regardless of whether the medicine wheels are aligned with stellar rising and setting azimuths, it does seem clear that they provided some celestial purpose, as they do appear in a natural site for observing the sky and are aligned with sunrise at the summer solstice. Perhaps rather than interpreting these as purely astronomical observatories (which could be more a symptom of our Western bias) we should instead recognize their importance as mirroring the larger celestial architecture.

The people who made the medicine wheel are perhaps related to the present-day Arapaho, Cheyenne, or Crow tribes. For these people the sky was central to their way of life and embodied the greatness of the spirit that gave life to the land. As recounted by Joseph Campbell, the metaphor of a "hoop" or circle was powerful for the Sioux and was a central part of the vision of Black Elk, a Oglala Sioux elder:

> In the vision, Black Elk saw that the hoop of his nation was one of many hoops, which is something that we haven't learned at all well yet. He saw the cooperation of all the hoops, all the nations in grand procession. But more than that, the vision was an experience of himself going through the realms of spiritual imagery that were his culture and assimilating their import. It comes as one great statement, which for me is a key statement in the understanding of myth and symbols. He says, "I was myself on the central mountain of the world of the world, the highest place, and I had a vision because I was seeing in the sacred manner of the world." And the sacred central mountain was Harney Peak in South Dakota. And then he says, But the central mountain is everywhere.
> (Campbell and Moyers 1988, p. 111)

Joseph Campbell summarizes Black Elk's conception of God with the quote "God is an intelligible sphere whose center is everywhere and circumference nowhere." Perhaps the medicine wheel is a physical representation of this "intelligible sphere" which allowed the Sioux to become closer both to their conception of God and to the stars above.

## *Cahokia Mounds*

One lost civilization from the present-day American Midwest left behind a number of fascinating aligned structures, which are locked into the cardinal directions and appear to be celestially inspired. The Cahokia site includes over 68 ancient mounds, including one that rises in four terraces to over 100 ft. This largest mound, known as Monks mound, is a massive earthen step pyramid reminiscent of a ziggurat and is the largest prehistoric earthwork in North America. Steps to the mound were built on the south face and the entire structure is carefully aligned to the four cardinal directions. Additional alignments between lines connecting some of the mounds have been found. A "meridian" line between the Monks mound and the smaller mound 72 and mound 66 falls on the north–south line, while additional sightlines between the mounds and wooden posts in the complex pointed toward the equinox sunrise (or due east) (Fig. 7.12). The archaeologist Warren Wittry has found evidence of large wooden posts along the meridian line and on top of Monks mound, which may have helped constrain a precise alignment with the meridian. A large circle of posts 410 ft across in the form of a "woodhenge" has been uncovered on the site and dates AD 1000. Many aspects of the "woodhenge" are similar to the early phase of Stonehenge, which itself was once a wooden circle of posts. Some of the posts appear to mark summer, equinox, and winter sunrise positions, much as the station stones of Stonehenge (Cornelius 1997).

**Fig. 7.12** Monks mound, at the Cahokia site in southern Illinois (*image source*: http://commons.wikimedia.org/wiki/Image:Monks_Mound_in_July.JPG)

## *Chaco Canyon Lunar and Solar Alignments*

Chaco Canyon, in northeastern New Mexico, contains over a dozen major structures, including the buildings of Pueblo Bonita and Casa Rinconada (discussed earlier). The intense construction of the Chaco Canyon site, combined with a lack of evidence of continued occupation by large numbers of permanent inhabitants, lends the site with an aura of mystery.

Many alignments both within and between Chaco Canyon buildings have been proposed, including those of Casa Bonita with an EW line (which is additionally aligned with sunrise and sunset on the equinoxes), and some structures appear to have exterior walls aligned with the extreme positions of moonrise. A list of the alignments of Chaco buildings have been compiled by Sofaer (Morrow and Price 1997), which includes 7 lunar rising alignments along the walls of structures and 21 azimuth alignments between structures along lunar rising azimuths.

Some alignments are a bit speculative and require an observer to stand in a particular location (such as near a panel of rock art) and look toward a distant second building to view the sunrise location. One example is the building known as Wijiji, which is thought to be a site where a viewer in front of a sun painting can see the location of sunrise on either side of the winter solstice, to enable a very accurate measurement of the date of the winter solstice (Williamson 1984, p. 91).

Other Chaco Canyon alignments are quite simple and easy to verify, such as the EW alignment of the south wall of Casa Bonito (see Fig. 7.13) which could have both astronomical importance (facing the equinox sunrise) and practical implications (maximizing solar heating in winter and controlling temperature in summer). The Casa Rinconada kiva, mentioned above, also is clearly aligned with the cardinal directions as its entrance and four posts are perfectly aligned with NSEW directions, and a small window admits the summer solstice sunrise light into one of the interior niches.

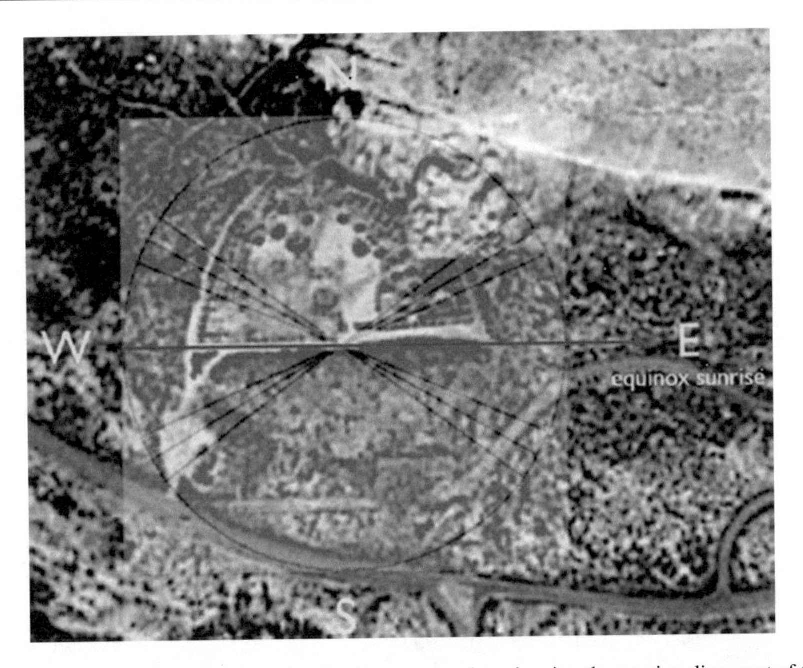


**Fig. 7.13** Satellite image of Chaco Canyon's Pueblo Bonito complex, showing the precise alignment of the south wall with the direction of the equinox sunrise, either for astronomical purposes or for the practical purpose of temperature control using passive solar heating and cooling (original figure by the author, using the NASA World Wind software)

Another amazing feature of Chaco Canyon is a vast network of roads that connects the site with distant settlements. Some of these roads are aligned with cardinal or astronomical directions and appear to be ceremonial roads instead of practical transportation corridors. One road crosses directly over a cliff to preserve its north/south orientation, instead of taking easier routes with more gradual elevation change. Lekson has argued in his work "The Chaco Meridian" that these roads formed a "meridian" which connected two other ancient cities with Chaco, as well as central buildings within Chaco Canyon between the buildings of Pueblo Alto, Casa Rinconada, and Tsin Kletzin (Lekson 1999, p. 83) (Fig. 7.14).

Of course no discussion of Chaco Canyon would be complete without mention of the "Sun Dagger," a unique solar observatory at the top of Fajada Butte (Fig. 7.15). A set of three slabs leaning on a cliff near the top of the butte projects into the dark cliff wall a narrow sliver of light that cuts through a spiral sun figure during the summer solstice. Additional unique patterns of light and shadow during the equinox and winter solstice have been observed, suggesting that the Fajada Butte site functioned as a year-round solar calendar.

**Fig. 7.14** Set of prehistoric roads in the Chaco Canyon region, showing the "Chaco Meridian" which radiates from the central Chaco Canyon site to surrounding kivas or "great houses" (*red circles*). Locations of modern cities are shown (*green circles*), along with state boundaries. Chaco Canyon served as the hub of an ancient civilization and the network of roads (*blue lines*) connected ceremonial kivas along lines aligned with the cardinal and solstice directions (*image source*: http://commons.wikimedia.org/wiki/Image:San_Juan_Basin_Prehistoric_Roads.jpg; satellite imagery from NASA's World Wind software)

**Fig. 7.15** Fajada Butte, in Chaco Canyon, home of the Sun Dagger used by the Anasazi to mark the seasons and solstices (*image source*: http://commons.wikimedia.org/wiki/Image:Chaco_Canyon_Fajada_Butte_summer_stormclouds.jpg)

## Hovenweep Castle

"Hovenweep Castle," a site in Utah built around AD 1200, features a series of masonry buildings that include small windows facing south and west (Fig. 7.16). Some have claimed the windows of the "sun room" within Hovenweep Castle pointed toward the winter solstice and the equinox sunset, while another porthole within a building known as "Unit House" pointed to the summer solstice sunrise. These buildings were built about 1277 AD, a time when the resources in the region in Southeastern Utah were becoming precarious, with evidence for a drought in the region that caused rapidly declining populations. The massive nature of the buildings suggests that they may be useful both for astronomical observations and for defense.

**Fig. 7.16** Masonry buildings of the Hovenweep Castle site in Southeast Utah, viewed just after Summer Solstice Sunrise in 2015. This structure shows small windows above the doorway—these windows built into some of the structures provided views of the summer and winter solstice sunsets (*image by the author*)

Within the "Sun Room" of Hovenweep Castle the rays of light from summer and winter solstice sunset cast beams through carefully placed windows that cross lintels to an interior room. To view the Hovenweep alignments, however, one has to sit in a particular location and view through two portholes, which align with the edge of a small door (Malville 2008, p. 138). Williamson places the odds of the Unit House alignments being a result of coincidence at 1 in 216,000 based on an estimated precision of 1° in the alignments (Williamson 1984, p. 119). However, since many other alignments are possible to draw between other door jams and portholes that do not align, the argument becomes less persuasive. Since the buildings at Hovenweep are over 800 years old, and the people who made them have left no written records, the mystery of the site remains.

One of the most dramatic views at Hovenweep is the summer solstice sunrise near Holly House, where an archaeoastronomy light show still attracts crowds even today—nearly 800 years later. An enormous rock blocks the inside of a small cave in which multiple sun spirals have been inscribed. On the summer solstice sunrise, a dagger of light slices just under the rock ledge and cuts through the sun spirals. The effect lasts for about 20 min, and is only visible on a few mornings each year. Figure 7.17 shows several stages of this event—with an inset image giving a view of the sun spirals.

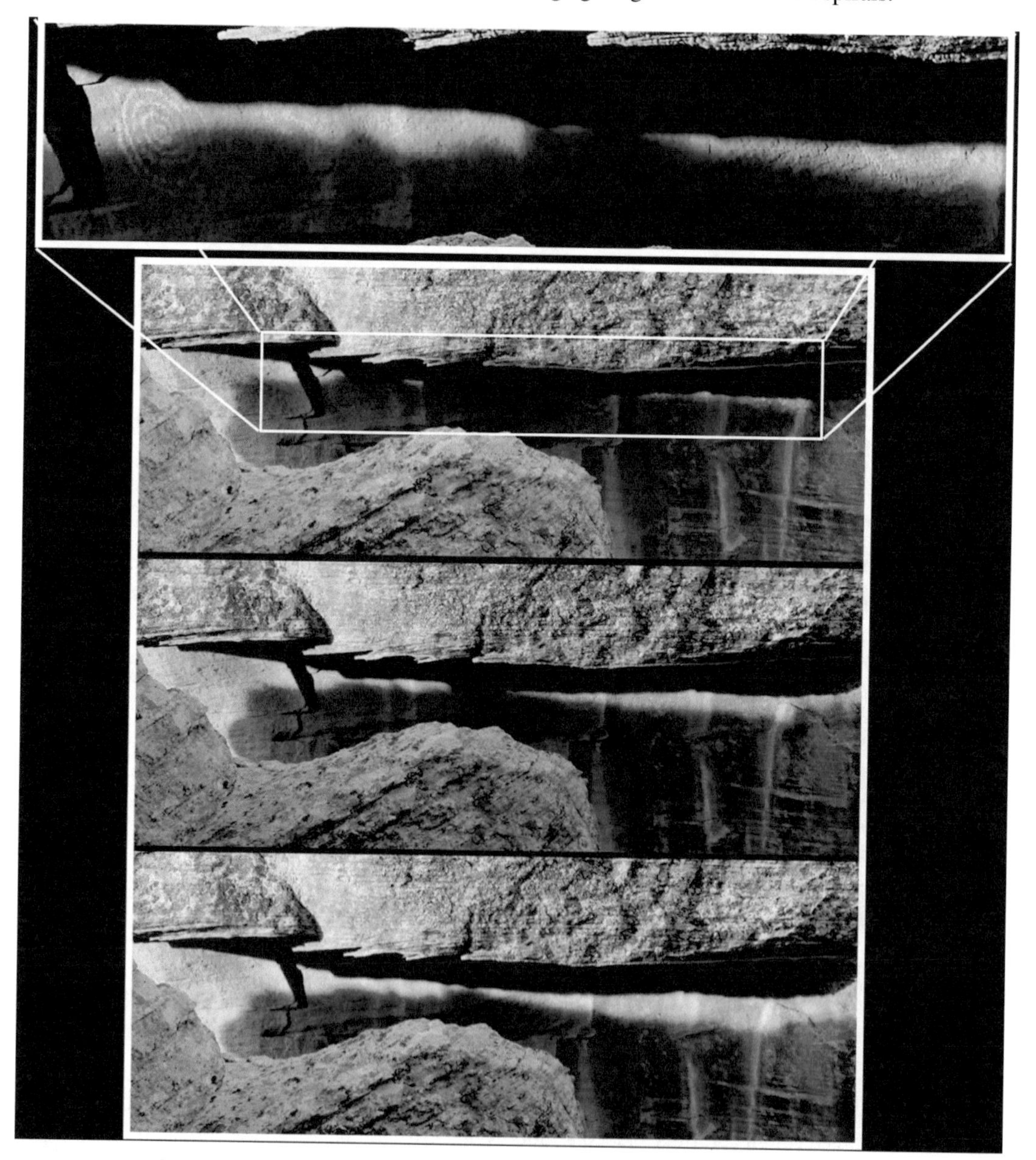

**Fig. 7.17** The Summer Solstice lighting of a rock shelter near Holly House, one of the Hovenweep structures. At the Summer Solstice sunrise, the beam of sunlight is able to rise above a large rock and slice through multiple sun spirals. The inset (*top*) shows the sun spiral, and the three lower panels show different stages of the light show, which makes a dramatic annual spectacle (*image by the author*)

## Celestial Cities from Ancient Civilizations and their Alignments

From ziggurats in Mesopotamia, to the Giza pyramids, to the Mayan step pyramids, all parts of the Earth feature man-made mountains from ancient civilizations. Many of these structures contain within them the blueprint of the cosmos and feature numerical ratios within their dimensions, alignments with stars and the Sun, and unique features that reflect the physical universe of the culture. The intermeshing of the monument with time and space, and with earth and sky, connects each civilization with the larger universe and magnifies the power of the site, which we will explore in the following sections.

### *Ziggurats in Mesopotamia*

The region of present-day Iraq, home to the civilizations of Sumer, Assyria, and Babylonia, is also the home to some of the first pyramids, known as ziggurats. The most famous ziggurat complex is the Ziggurat of Ur, in central Iraq (Figs. 7.18 and 7.19). The Ziggurat of Ur was constructed around 2000 BC and consists of levels with dimensions that follow numerical ratios important to the Sumerians. The proportions of the width and length for the ground level are very close to an exact ratio of 3:2, with dimensions of 62.5 × 43 m, while the second level maintains a ratio of 4:3 with dimensions of 36 × 26 m. Access to the upper levels was from a celestially aligned northeastern face oriented toward the summer solstice sunrise, with a grand staircase facing the first rays of the summer Sun.

**Fig. 7.18** The ziggurat of Ur, as it might have looked when built in 2000 BC (image from http://commons.wikimedia.org/wiki/Image:Ziggurat_of_ur.jpg)

**Fig. 7.19** The Ziggurat of Ur as it appears today, from the direction of the summer solstice sunrise, which would rise on the horizon in the direction of the large staircase to the top level (*image source*: http://commons.wikimedia.org/wiki/File:Ziggarat_of_Ur_001.jpg)

Ziggurats served multiple functions over many centuries. For example, the Ziggurat of Ur was constructed by a Sumerian king named Ur-nammu and restored between 555 and 539 BC by the Babylonian king Nabonidus. The ziggurat seemed to be a temple and funeral complex; tombs at the site were discovered in 1922 by the archaeologist Leaonar Wooley, including the 16 royal graves that contained remains of royal personages and their household servants. With the skeletons were jeweled furniture, chariots, daggers, musical instruments, and even games, suggesting that the purposes of the site served both the living and the departed (Scarre 1999). Other ziggurats in Iraq include those of Ashur, Susa, Elam, and Agade, which are in various states of preservation but show some of the same features as that of Ur. Additional ziggurats exist outside of Mesopotamia, such as the well-preserved Chogha Zanbil complex dating from 1250 BC, and the ziggurat of Sialk, perhaps the oldest ziggurat from 3000 BC. The ziggurats are in present-day Iran and remain as reminders of larger Elamite cities, which existed for thousands of years.

What all the ziggurats have in common is the enormous scale, approaching 100 m along the base, with a height of 30 m or more achieved through three to five layers. The ziggurat usually contained four or more levels and had its base aligned with the cardinal directions or solstice sunrise directions. The height of the ziggurat provided an ideal platform for viewing the stars, above the smoke of the densely crowded city, and also for performing sacrifices and rituals, secure from the crowds below. Often the outside edge of the ziggurat contained subterranean levels where rulers were entombed, and thereby mirrored the structure of the Mesopotamian universe, with multiple layers above earth, and multiple "underworlds" below the surface.

Some of the alignments of the ziggurats with the sky are easily seen from above, and the ruins of the building still point toward the solstice sunrises and sunsets thousands of years later.

## Egyptian Pyramids

Pyramid construction in Egypt began in the Fourth Dynasty (2613–2494 BC) and culminated in the most famous and grandest of the pyramids, the Great Pyramids of Giza. Even today the Giza pyramids are some of the largest man-made objects on the Earth, and building such constructions would tax the limits of even today's technology. The effort required to haul the 2.3 million blocks, ranging in weight from 2.5 to 15 t, to build the 481-foot high structure is mind boggling.

The Pyramids of Giza have numerous astronomical connections, which have been seminal in starting the science of archaeoastronomy. The most obvious includes the excellent alignment of the base of the pyramids with the cardinal directions which is shown in Fig. 7.20. The alignment of the pyramids is extremely accurate and is held within a tolerance of 5.5 arc-min from perfect alignment with an EW direction, with a perfectly square base within a tolerance of only 7.8 in. over 710 ft. The pyramids were also built to provide a physical embodiment of the rays from the Sun, which were believed to emanate from Re the Sun god. Some of the pyramid hieroglyphics tell of the explicit connection between the pyramids and astronomy. In these writings Pharaoh is said to "ascend to the sky among the stars" to "regulate the night," and to take "possession of the sky, its pillars and stars" after death (Krupp 1983, p. 101).

**Fig. 7.20** An aerial view of the Great Pyramids of Giza, with an overlay showing an EW line aligned with the cardinal directions (*red line*), and additional lines showing solstice directions (*gray*). The precise alignment of the bases of the pyramids with cardinal directions is apparent (original figure by the author using NASA's World Wind software)

Studying the internal structure of the pyramids has revealed more subtle astronomical alignments. The American astronomer Virginia Trimble observed that the set of three air shafts within the pyramid that point into the sky at elevations that correspond to the positions in 2700 BC of the stars within Orion's belt, and the circumpolar star Thuban. Since the sky "precesses" over a period of centuries, the pole star Polaris was not located in its present position in the sky. Instead Thuban was the brightest of the circumpolar stars, and therefore a logical target for stellar sighting shaft. These alignments were spectacular and provided a "celestial mooring" for the pyramid that previously had not been known (Fig. 7.21). Trimble pointed out that these shafts had no other known function and that the assignment of their role as "air shafts" did not fully explain why the Egyptian pyramid built a completely two nearly perfectly straight passage over 300 ft long and only 9 × 9 in. across.

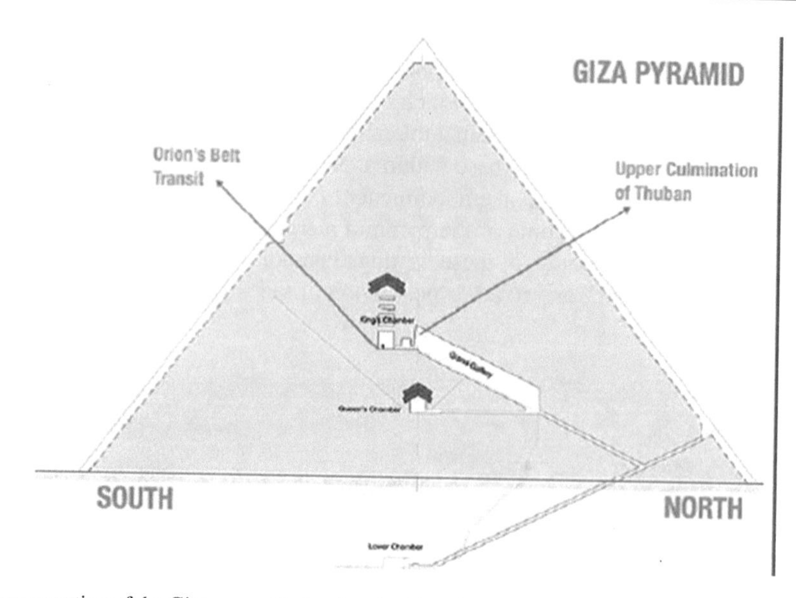


**Fig. 7.21** A cross section of the Giza pyramid showing the directions of "air shafts" that point toward Orion's belt and Thuban (figure by the author using background image by Jeff Dahl at http://commons.wikimedia.org/wiki/File:Great_Pyramid_Diagram.jpg)

The alignment with the stars in Orion was appropriate to the funereal function of the pyramids, since in ancient Egypt, Orion was associated with Osiris, "Lord of Everything" and ruler of the underworld. The pyramid alignment with Thuban is appropriate since the circumpolar region was associated with eternal life and circumpolar stars never set. Thuban was the circumpolar star closest to the north celestial pole in 2700 BC and, therefore, is an appropriate symbol of eternal life (Trimble 1964). Some questions do remain, however, since a third shaft does exist and has not been adequately explained in astronomical terms.

Most of the pyramids at Giza are closely aligned with the cardinal directions. However, recent research has also shown that there is a measurable rotation of some of the pyramids that correlates with their date of construction. The small departure from true EW alignment is typically less than one-third of a degree, and yet can be measured with modern surveying technology. A plot of the deviation from true cardinal directions for 12 of the Giza pyramids against construction date showed a statistically significant linear correlation (Spence 2000).

While it is possible that these small rotations are from random error, the study showed that the offsets correlated very well with the construction dates of the pyramids, which ranged from 2640 BC for the older Djoser pyramid to 1853 BC for the pyramid of Amenemhat III. The excellent correlation led Spence to conclude that the pyramids were actually aligned using stellar sightings and that the shift in the orientation of pyramids with time resulted from the very gradual shift of the stars due to celestial precession! Sightings of various stars in Ursa Major and Draco (where stars like Thuban are found) would cause the pyramids to slowly rotate in their orientation, and since the Egyptian architects were so careful in their work, the pyramids themselves faithfully duplicate this very small shift.

## *The Kaaba in Mecca*

In the Arab world, no site is more sacred than Mecca and at its center is the Kaaba the cubic structure which is the focal point of the pilgrimage or *hajj*, which all devout Muslims are expected to make once in their lifetimes. Embedded in the eastern corner of the Kaaba is the Black Stone, which is possibly the remnant of a meteorite, and it is believed by Muslims to date back to the time of Adam and Eve. The site of the Kaaba itself is celestially aligned, since the structure housing the black stone is offset from the cardinal directions so that one wall is parallel to the line connecting the winter solstice sunset and the summer solstice sunrise. The other side of the building faces toward the horizon where the bright star Canopus rises. The new Abraj Al-Bait Towers, or Makkah Royal Clock Tower Hotel, also add additional astronomical aspects to the site. The Towers rise 610 m over the South side of the site, and include viewing decks and an observatory for moon sightings along with the world's largest clock face.

Astronomical alignments are found into other structures throughout the Islamic world–early mosques in Iraq featured prayer walls that face the winter solstice sunset direction, while many mosques in Egypt were built with their prayer walls facing the direction of the summer solstice sunrise. By aligning their buildings in this way, the architect can incorporate both an alignment toward Mecca (the sacred direction known as the *qibla*) and the same alignment of the sky as the Kaaba itself (Ruggles 2005) (Fig. 7.22).

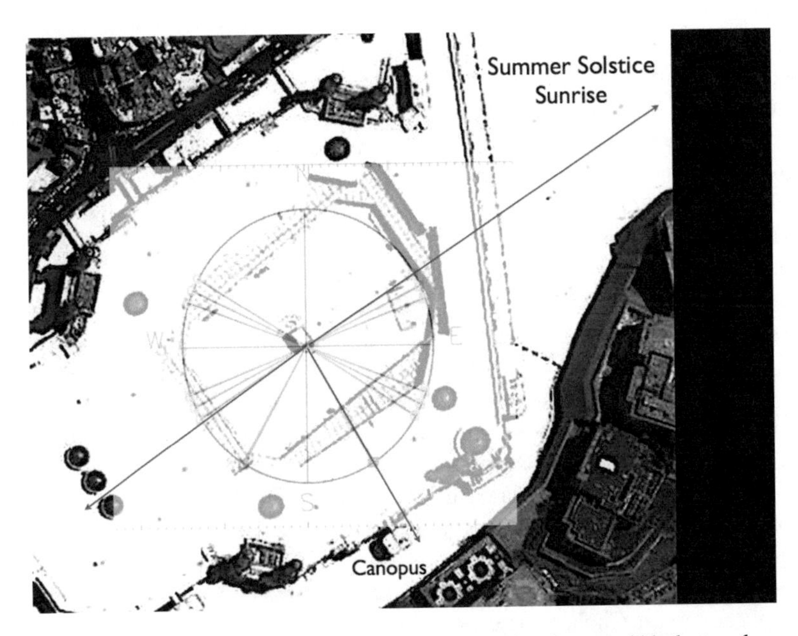


**Fig. 7.22** A satellite view of the central square in Mecca building housing the sacred black stone known as the Kaaba. The structure is aligned with two celestial directions–the summer solstice sunrise and winter solstice sunset, and one side faces the horizon in the direction of the bright star Canopus (original figure by the author, using NASA's World Wind software)

Across the world, ancient mosques were built to face toward mecca in what is called the *qibla* direction. In many cases the orientation of the mosque set the orientation of the entire city, and so entire street grids are oriented in the *qibla* direction. Interesting exceptions include some ancient mosques in Islamic Spain and Syria which faced South, the Egyptian Mosque of Amr in Fustat (near Cairo) which faces the winter sunrise, and early mosques in Iraq which faced the winter sunset. In some cases ancient Tunisian cities inherited the orientations favored by Roman constructions and the mosques are aligned in angles that appear at right angles to the winter solstice sunrise, while ancient Spanish mosques were converted Christian churches which favored cardinal direction orientations (Bonine 1995). While the majority of mosques face the correct *qibla* direction, in many cases coincidental orientations arise from a mix of prior cultural influences—themselves influenced by astronomy.

## Alignments of Ancient Indian Temples

Across India are literally thousands of temples, built during a variety of ancient empires. These Indian temples, like most sacred structures, have celestial symbolism within the architecture and iconography. The temples of India typically feature a main entrance that points East or an angle offset from East, with an axis that connects the entrance to the most important shrines and which provides a symmetry axis upon which the other buildings are aligned. An analysis of the orientation of these temples shows that most Indian temples appear to be carefully aligned to face East, or an angle slightly north of East that most commonly points toward the Summer Solstice sunrise.

Malville (2015) did a careful analysis of the orientations of temples and shrines surrounding the ancient capital city of Vijayanagara, in Karnataka, India near the Hampi ruins. Vijayanagara was the center of a giant empire that ruled much of southern India until it was destroyed in 1565 AD. At its peak, it contained perhaps 500,000 people and includes shrines and temples that were built over nearly 1000 years. Malville was able to map the locations and the orientation angles of over 150 temples, shrines, and palace buildings in the region of Vijayanagara, using theodolite readings and GPS data. He found that palace buildings were aligned along an East–West direction, and constructed along a very precisely aligned North–South axis that connected the most important imperial buildings. The larger temples were preferentially aligned to face East, but many of the buildings were tilted slightly north of East, with a mean angle of 17° away from East–West, with most of the smaller shrines pointed 15–17° north of East. At the latitude of the site (15.3 N), this orientation is inconsistent with a solstitial direction, and is not placed well for any of the major stellar azimuths. Melville concluded that the 15–17° shift is best explained by the azimuth of sunrise on days where the sun makes a zenith crossing, which at this latitude is 16°.

During field work in Tamil Nadu, a group of Yale-NUS College students, Yale-NUS College anthropology professor Bernard Bate, and the author took measurements of temple complexes built during the ancient Chola empire which ruled over a swath of Southern India between 600 and 800 AD. These massive temples include some of the largest and most famous in all of India—Brihadisvara Temple, Gangaikonda Cholapuram, and Airavatesvara Temple among them. They were built both as a place of worship and a sign of the increasing power and status of each of the Chola emperors. The largest of these temples, the Brihadisvara temple in Thanjavur, was built by the third emperor of a Chola dynasty, Rajarnja I between 985 and 1014 AD and towers 59.82 m above the main axis (Pichard 1995). Like the Egyptian pyramids, such massive constructions embodied the highest levels of technology and construction of their era and were carefully laid out in celestial directions. Satellite and on-site measurements revealed most of the largest Chola temples were very precisely laid out along an East–West axis. A notable exception is the largest and one of the last Chola temples—the Brihadisvara Temple, which is tilted 18° north of east and faces toward a direction that is not aligned well with summer solstice sunrise, or zenith crossing sunrise. An examination of the stellar associations with this azimuth revealed that the stars of the Pleiades rise 18° north of east, during the period of construction and occupation of the site in 985–1014 AD. The alignment of Brihadisvara toward the rising azimuth of the Pleiades or *krittika* nakshatra makes more sense when one realizes that these stars are first visible during the first month of the traditional Tamil calendar and also are associated with the god Murugan, the eldest son of Shiva.

To provide further data on the alignment of temples in Tamil Nadu, we investigated the alignments of 31 temples and smaller shrines using satellite imagery, following some of the analysis of Malville (2015). Of the 31 Tamil temples and shrines, the largest number were aligned East–West (11), and several were aligned toward summer solstice sunrise (5). Of the remaining temples, a few (3) were aligned toward the zenith sun rising azimuth (which at the mean latitude of our temple sample appears toward 12° North of the East–West line). Like other sites from ancient India, our sample of Tamil temples and shrines encode the sky in their orientations, and are designed along axes that face winter solstice sunrise, the zenith sunrise azimuth, with the largest of the temples, Brihadisvara, facing the rising of the *krittika* or Pleiades stars (Fig. 7.23).

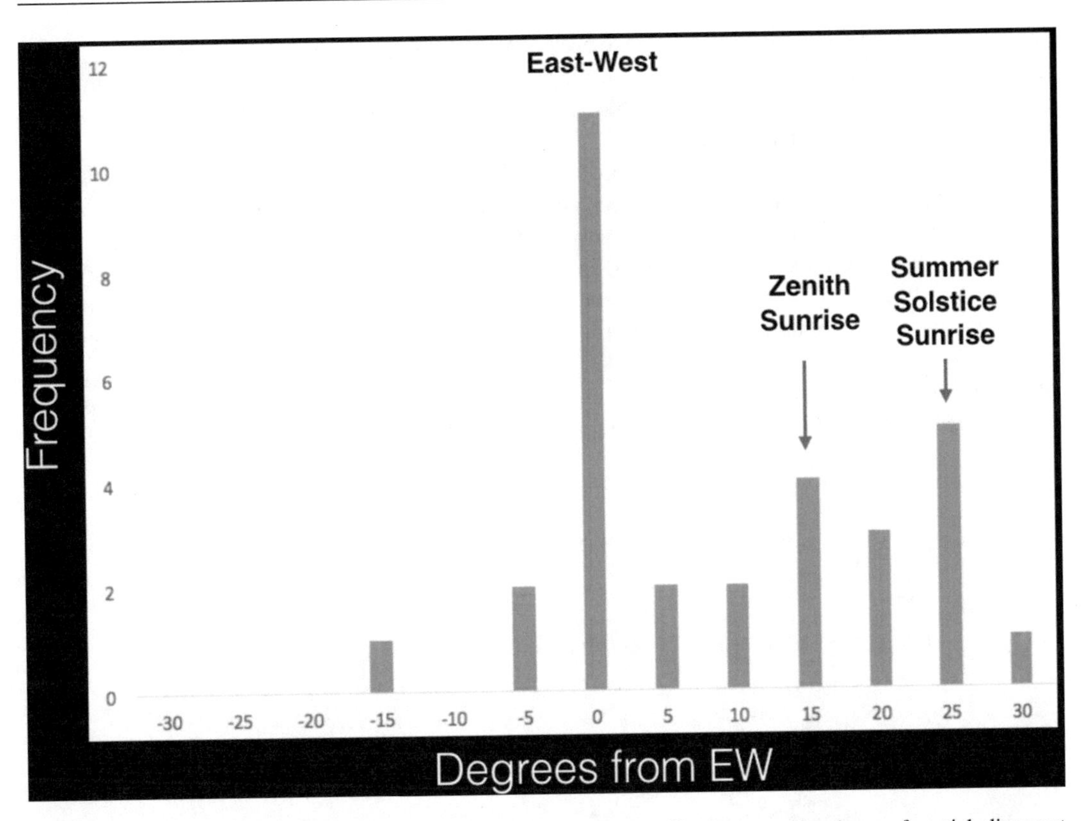


**Fig. 7.23** Analysis of 31 temple orientations in Tamil Nadu using satellite imagery, showing preferential alignment along an East–West axis—with the main axis and entrance facing the equinox sunrise. Other preferred orientations include entrances facing summer solstice sunrise, and the azimuth for sunrise on the dates of a zenith crossing

## *Teotihuacan*

Across the world, in a valley north of modern Mexico City, lies the ruins of the ancient city of Teotihuacan, which features an even larger pyramid (in terms of mass and area) than those in Egypt. The city-state of Teotihuacan appears to have arisen in about 600 BC and rose to its peak between AD 200 and 750. Later observers have given suggestive names to the two main pyramids of Teotihuacan, calling them the "Temple of the Sun" and the "Temple of the Moon." Unlike the Egyptian pyramids, no writings survive to describe the functions of the pyramids of Teotihuacan and its long avenue that connects the city to thousands of abandoned shops and dwellings. The buildings and stones do survive, however, along with the horizon and stars.

Present-day descendants of the Teotihuacan civilization (which predates the Aztec and Mayan cultures) speak of these temples as "the place where time began" and still recognize shrines deep under the centers of the two main pyramids. The Teotihuacan pyramids seem to connect underworld, earth, and sky with a cosmological model in stone that mirrors the universe of this ancient culture. Teotihuacan shares with the Mesopotamian ziggurat and the Egyptian pyramids the universal human urge to commemorate the larger cosmos of its people.

The pyramids of Teotihuacan are a stirring testament to the physical and spiritual power of these ancient people and contain over 12,648,008 square feet of brick and rubble in its bulk, making it larger in bulk than the biggest of the Giza pyramids, but not as tall, since its peak is only 75 m. The pyramids rise to the sky much as a ziggurat, in stepped fashion, with five terraces and 700-foot long sides. The main part of the Sun pyramid appears to have been constructed between AD 150 and 225 over a "sacred cave" and over several earlier structures. The Maya civilization also had the practice constructing multiple pyramids on top of the same site, perhaps from a shared cultural heritage. In the core of the pyramid is a set of four chambers that reach 300 ft eastward and 20 ft below the main pyramid (Scarre 1999). The smaller pyramid of the Moon, constructed 50 years later, rises at the north end of the avenue and this smaller pyramid appears to have been constructed 50 years later than the pyramid of the Sun.

The structure of Teotihuacan's ruins suggests that it may have been one of the largest cities on the Earth at its peak in AD 500, with over 200,000 inhabitants, similar in size to the ancient city of Rome (Fig. 7.24). By AD 750 it appears to have been abandoned. In this ancient city are 600 pyramids, 500 workshop areas, a huge marketplace, and over 2000 apartment complexes, all arrayed along its main avenue, known today as the "Avenida de la Muerte" or "Avenue of the Dead." The *Avenue de la Muerte* is aligned 15.5° east off of a north–south axis and features the larger pyramid of the Sun in the middle and the pyramid of the Moon on the north side.

**Fig. 7.24** View toward the Temple of the Sun from the Temple of the Moon; the vast scale of the main avenue known as "Via del Muerte" or Avenue of the Dead is apparent (*image source*: http://commons.wikimedia.org/wiki/Image:Avenue_of_the_Dead_at_Teotihuacan.jpg)

The Avenue del Muerte of Teotihuacan clearly provided a ceremonial boulevard for the ancient city dwellers, but unlike most ancient cities, it appears to not be aligned with either the cardinal directions or important solstice positions on the horizon.

Some have argued that the misalignment of Teotihuacan's main avenue is intentional and allows the EW axis to align not with the cardinal directions, but with the horizon position of the brightest star in the sky, Sirius, which at these latitudes and during AD 150 rises 16.5 south of east, or with the Pleiades, which appears to have set about 16.5° north of west. One additional compelling argument for the Pleiades alignment is that during AD 150 the heliacal rising of the Pleiades appears close to the solstice.

One unique aspect of Teotihuacan is the presence of carved patterns known as the "pecked crosses" which appear near the avenue of the dead and in many other locations in the surrounding countryside. Aveni and others have cataloged over 70 of the pecked crosses in the region near the city and lines connecting the crosses appear to form astronomically interesting sightlines along cardinal directions, as well as with summer solstice sunrise over nearby peaks, and connect sightlines between the city and the peaks of Cerro El Chapin, Cerro Colorado, and Picacho Peak (Krupp 1983). The most interesting of the "pecked cross" alignments is found between the Avenue of the Dead and a site to the NW. If one constructs a line connecting the pecked cross as a perpendicular to the Avenue it points directly toward the setting azimuth of the Pleiades. The massive Temple of the Sun also faces this direction and the entrance to its subterranean inner sanctum aligns with the Pleiades setting point.

## *Mayan Celestial Cities*

Within the Mayan cities to the south of Teotihuacan one finds a number of smaller and steeper pyramids compared to Teotihuacan. The layout of most Mayan cities defies a simple explanation in terms of alignments and lacks the same structure and unity as the city of Teotihuacan. Mayan ceremonial ball courts and large step pyramids appear to be jumbled in random directions relative to the sky and to each other. In some cases, however, the orientations of key buildings do appear to offer views toward astronomically significant directions.

Statistical studies of the building alignment across the Mayan territory detect distinct clustering of alignments about an orientation skewed approximately 12° from the cardinal directions at many sites, with a secondary peak at about 25° from cardinal directions (Aveni and Hartung 1986). A histogram of the angle of building alignments across Mesoamerica, shown in Fig. 7.25, shows the two distinct peaks at 12 and 25°, indicated by dashed lines.

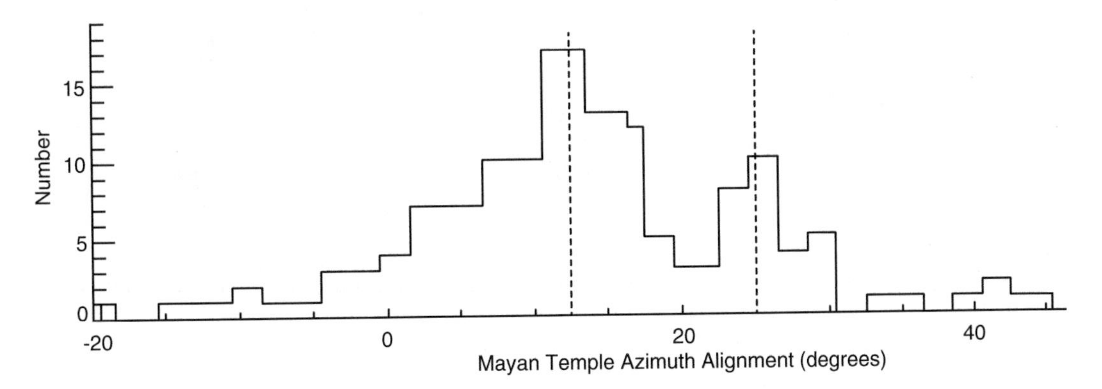


**Fig. 7.25** A histogram of building alignments across Mesoamerica, for a wide cross section of sites, showing two distinct peaks in alignments at approximately 12–14° from cardinal directions (i.e., 12° E of N or 12° north or south of E) and 25° from cardinal directions (figure by the author using data from Aveni and Hartung 1986)

The ancient Mayan city of Copan shows three groupings of building orientations, with the Plaza and ball court inclined about 8° from cardinal directions, a stairway close to cardinal directions, and a group of buildings with 7–9° alignments (Aveni and Hartung 1986).

As in the case of Teotihuacan, it is possible to find astronomical explanations for such alignments. The Mayan astronomy and calendar placed a heavy emphasis on the planet Venus, which was associated with the serpent god Quetzalcoatl. The two peaks in the alignment distribution seem to coincide with key horizon locations of Venus and additional symbols on the buildings make this astronomical interpretation more likely.

## Venus Alignments

The dazzling spectacle of Venus rising on the horizon inspired the Maya to rotate some of their key buildings to face the "standstill" locations of the planet. One example is the Palace of the Governor, at the Mayan city of Uxmal, Mexico (Fig. 7.26). In this temple, the central plaza contains a stairway and a doorway and stela which define a sightline 28° south of east and points toward the location of the 6 km distant Nohpat site. This direction precisely matches the azimuth direction of the "Venus standstill" rising azimuth as viewed in AD 750. The astronomical interpretation is reinforced by the fact that the building facade contains over 350 Venus hieroglyphic symbols and carvings of direction gods with numbers consistent with the location and days of disappearance of Venus before re-emergence at the standstill location.

**Fig. 7.26** The Palace of the Governor, at Uxmal, which faces 28° south of east and aligns its view toward the extreme "standstill" position of Venus on the horizon (*image source*: http://commons.wikimedia.org/wiki/File:Uxmal-Palace-of-the-Governor.jpg)

Another striking example of Mayan Venus-based astronomical alignments is found at the city of Chichen Itza, in the front face of the building known as the Caracol (Fig. 7.27). The Caracol's monumental staircase is aligned with the northernmost setting point of Venus. A round building on the top level, known as the Caracol or "Snail," includes windows that point toward both northern and southern extreme setting points of Venus. Additional windows of the Caracol also seem to point in the directions of winter solstice sunset and summer solstice sunrise (Aveni and Hartung 1986).

**Fig. 7.27** The Caracol of Chichen Itza, featuring the "observatory"-type structure which has windows pointing toward winter and summer solstice sunsets, as well as the northernmost setting point of the planet Venus. The entire front face of the building is oriented toward the northernmost Venus setting azimuth (*image source*: http://commons.wikimedia.org/wiki/Image:Caracol_in_chichén_itzá.jpg)

## Solar Alignments

The Mayan city of Chichen Itza features the famous step pyramid known as the El Castillo, perhaps the most famous of the Mayan step pyramids (Fig. 7.28). Within the architecture of Castillo is one of the most dramatic Mayan building alignments. The front steps of the Castillo face the summer solstice sunrise and provide a fantastic show of light and shadow during the equinox, when the shadows of the steps slither down the front of the pyramid, suggestive to many of the serpent god Quetzalcoatl.

**Fig. 7.28** View of the Mayan step pyramid of Chichen Itza, the famous "El Castillo" which is aligned with the summer solstice sunrise and features a "serpent-like" shadow on its steps during the equinox sunrise (*image source*: http://commons.wikimedia.org/wiki/Image:El_Castillo.jpg)

While one of the peaks in building alignments (25–28° from cardinal directions) seems to correlate with key Venus positions, the other peak (12–14°) seems related to solar alignment. This alignment is not a usual sunrise or sunset direction, but instead is related to the date when the sun passes overhead. At the near-equatorial latitudes of the Mayan cities (15–21°), the Sun passes directly overhead twice each year, in times separated by about a lunar month from the summer solstice. The 14° skew of many Mayan buildings matches the azimuth of the sunrise just before the first zenith passage each year (Sharer and Traxler 2006). While such an interpretation is a bit speculative, it would be consistent with the high degree of astronomical knowledge of the Maya.

## The Aztec City of Tenochtitlan and Its Temple Alignment

The Aztec capital city of Tenochtitlan was a vibrant center of commerce and ceremony that included huge plazas, floating buildings, and a network of interconnected canals centered on Lake Texcoco, which was navigated by hundreds of canoes and barges. Upon encountering Tenochtitlan, the Spanish *conquistadores* were filled with awe, since the Aztec capitol rivaled or surpassed any of the cities in Europe they had seen. In the words of the conquistador Bernal Diaz:

> As we got among the houses and saw what a large town it was, larger than any we had yet seen, we were struck with admiration. It looked like a garden with luxuriant vegetation, and the streets were so full of men and women who had come to see us, that we gave thanks to God at having discovered such a country.
>
> (Díaz del Castillo and Cohen 1963, p. 164)

The Aztec culture is famous for its emphasis on sacrifice, and the role of the Templo Mayor as a center of sacrifice ritual is well documented. Many of the sacrifice rituals derive from a belief in the need to provide fresh blood as a fuel for the current Sun. In the Aztec creation the current universe or the Sun was made possible by the sacrifice of a god and the Aztec recreated this sacrifice frequently in their rituals, to help sustain and renew the Sun. One of the main festivals for the Aztec occurred in the middle of spring and was described by a Spanish chronicle of 1532 to take place "when the sun stood in the middle of the Temple of Huitzilopochtli, which was at the equinox."

The plan of Tenochtitlan radiated from the Templo Mayor, which defined the axis around which the city was built and was at the center of four quadrants roughly aligned with the cardinal directions. Like Teotihuacan, the main temple and city seems to be skewed slightly (approximately 7.5°) from either the cardinal directions or the expected solstice directions. A study of the exposed foundations of the Templo Mayor allowed archaeoastronomers to measure the 7.5° shift in the alignment, causing archaeoastronomer Anthony Aveni to ask whether this shift was intentional and had astronomical motivation.

Aveni put these cultural factors together with some new calculations of solar elevations and realized that for the Sun to be in the "middle of the Temple" it would have to rise above the main bulk of the step pyramid to appear between the twin towers. At the equinox, an audience viewing the Sun in this position from the base of the pyramid would be facing 7.5° south of east, precisely the offset of the temple from the cardinal directions. Apparently the Aztec temple and the rest of the capital city of Tenochtitlan was rotated slightly to enable the first rays of the equinox sun to shine onto the central plaza. According to one Spanish chronicler, the Templo Mayor "was a little out of the straight, Montezuma wished to pull it down and set it right." Alignments of cities of the scale of this Aztec capital usually are not accidental, Aveni argued, and the slight rotation of the city helped "set right" the city (Krupp 1983, p. 266).

## The Forbidden City of China

Perhaps the most impressive celestially aligned city is the imperial center of China, sometimes known as the "Forbidden City" (Fig. 7.29). Like the cities of Tenochtitlan and Cuzco, the Chinese capital of Beijing was built around a central temple aligned with the central axis of space, which was defined by the sky. The Forbidden City was built during the Ming Dynasty and is precisely aligned along a north/south axis, with the residence for the emperor in the northernmost part.

As mentioned earlier, the center of Chinese star maps was the circumpolar zone, known as the "Forbidden Purple Palace," which contained the primary reference stars used to define the *hsiu* or lunar mansions of the Chinese astronomers and astrologers. Likewise the northern district of the capitol was reserved for the emperor, and his palace and lodgings were known as the "Purple Forbidden City" or *Zi jin cheng* in Chinese (Krupp 1983), which we shorten to the "Forbidden City."

The role of the emperor was to reconcile earth and sky and to rule with justice to preserve the "mandate of heaven." A Sung dynasty (AD 1193) document, later inscribed in stone, gives instructions to new emperors of China:

> All manifestations of the Spirit emanate from Heaven.. This evolves into the sun and moon, divides into the five planets, arranges in order as the twenty-eight mansions, and meets to form the directors and the circumpolar stars. All of these being involved in the immutable reason, are also in harmony with the rational principle in Man …
> (Kelley and Milone 2005, p. 331)

The emperor's role was to be the hub, and center of the harmony of heaven. The circumpolar stars revolved around the central pole star and the activities of the Royal Palace centered and revolved around the northern part of the Forbidden City.

Each of the entrances or gates within the city was given names that reinforce the emperor's "mandate of heaven." The famous *Tian an men* square itself is named for the *Tian an* or Heavenly Peace gate (*men*), which was thought to arise from concordance between the emperor and the sky. The path into the central Throne Room goes past the *Qian* ("Straight Toward the Sun") gate, which forms a line

**Fig. 7.29** The central square of the Forbidden City, in modern Beijing (*image source*: http://commons.wikimedia.org/wiki/Image:Forbidden_City1.JPG)

with the "Gate of Heavenly Peace" (*Tian an men*), the "Meridian" gate, and the "Gate of Supreme Harmony." All of these gates aligned with the meridian and led to the farthest north point in the palace, where the emperor would live.

The set of five gates also mirrors the Chinese belief in five elements in the universe (wood, water, metal, fire, earth), five colors (blue, black, white, red, yellow), five directions (east, north, west, south, center), five virtues, five sacred animals, five planets, and many other instances of five. The meeting point between the emperor and the public was in a central pavilion, known as the Hall of Supreme Harmony, which was slightly south of his residence, just as the circumpolar sky would meet the Earth just south of the pole star (Krupp 1983; Kelley and Milone 2005).

As the Sun moved through the various quadrants of the sky during the seasons of the year, the emperor performed astronomically timed rituals in a counter clockwise rotation about the capital at the beginning of each season. The emperor presided over ceremonies for spring equinox, summer solstice, fall equinox, and winter solstice that were suffused with the appropriate color and directional symbolism for each occasion (Krupp 1983, p. 193).

New year rituals were performed in the northern part of the Imperial City, Beijing's temple of Heaven, itself designed with cosmic symbolism of 4 central posts and 24 outer pillars. In spring, the emperor would perform a ceremonial digging of the first furrow and sacrifice to assure good harvests. Near the winter solstice, the emperor would visit the Round Mound (*Huan qui tan*), located near the Temple of Heaven. During the ceremony he would approach the mound from the south and enter the top of the three terraces to light a ceremonial fire. At this time the emperor would also make sacrificial offerings of jade, silk, and other valuables to the sky, as well as solemnly recount the highlights of the passing year. With these ceremonies, the emperor physically brings the affairs of the Earth in accord with the larger celestial harmony.

## Angkor Wat

The ancient Khmer empire of present-day Cambodia built its capital city of Angkor in the ninth century and its central temple complex (one of hundreds) is known as Angkor Wat (Figs. 7.30, 7.31, 7.32). The temple was built by the emperor Suryavarman II between 1113 and 1149 AD. The temple was constructed according to the principles of the Hindu temple architecture, with five main towers and an extensive set of galleries carved with scenes from the Hindu epic Mahabharata, along with statues of various Hindu gods and heroic sculptures of the emperor Suryavarman II. Angkor Wat sprawls within a moat-enclosed region 1.3 × 1.5 km in size, with a 190 m wide moat. In the center rises the main temple, contained within a terrace of 332 × 258 m, rising to a series of five towers housed on a 75-meter square, with the tallest spire reaching 65-m in height (Fletcher et al. 2015; Freeman and Jacques 2003).

**Fig. 7.30** Angkor Wat from the air, from a photograph taken by Charles J. Sharp. Angkor Wat embodied the cosmology of the ancient Hindu civilization that built it with several dimensions of the compound in accord with timescales of Hindu cosmology (*image source*: http://commons.wikimedia.org/wiki/File:Angkor-Wat-from-the-air.JPG)

**Fig. 7.31** The ancient Khmer city of Angkor Wat includes long avenues that connect the temple complex with a surrounding moat. The dimensions of Angkor Wat are astounding—the main complex has dimension of 1.3 by 1.5 km, with a moat of 190 m width. The scale of the construction enabled an entire city to live within the moat and to also be close to ceremonial observations of the sky and other rituals. Viewing porticos within the many buildings of Angkor Wat (right) allow vistas of sunrise behind the main temple at winter solstice. Slight asymmetries in the layout of Angkor Wat enable the equinox sunrise to be viewed after clearing the main towers. (Photos by the author) (Fig. 7.32).

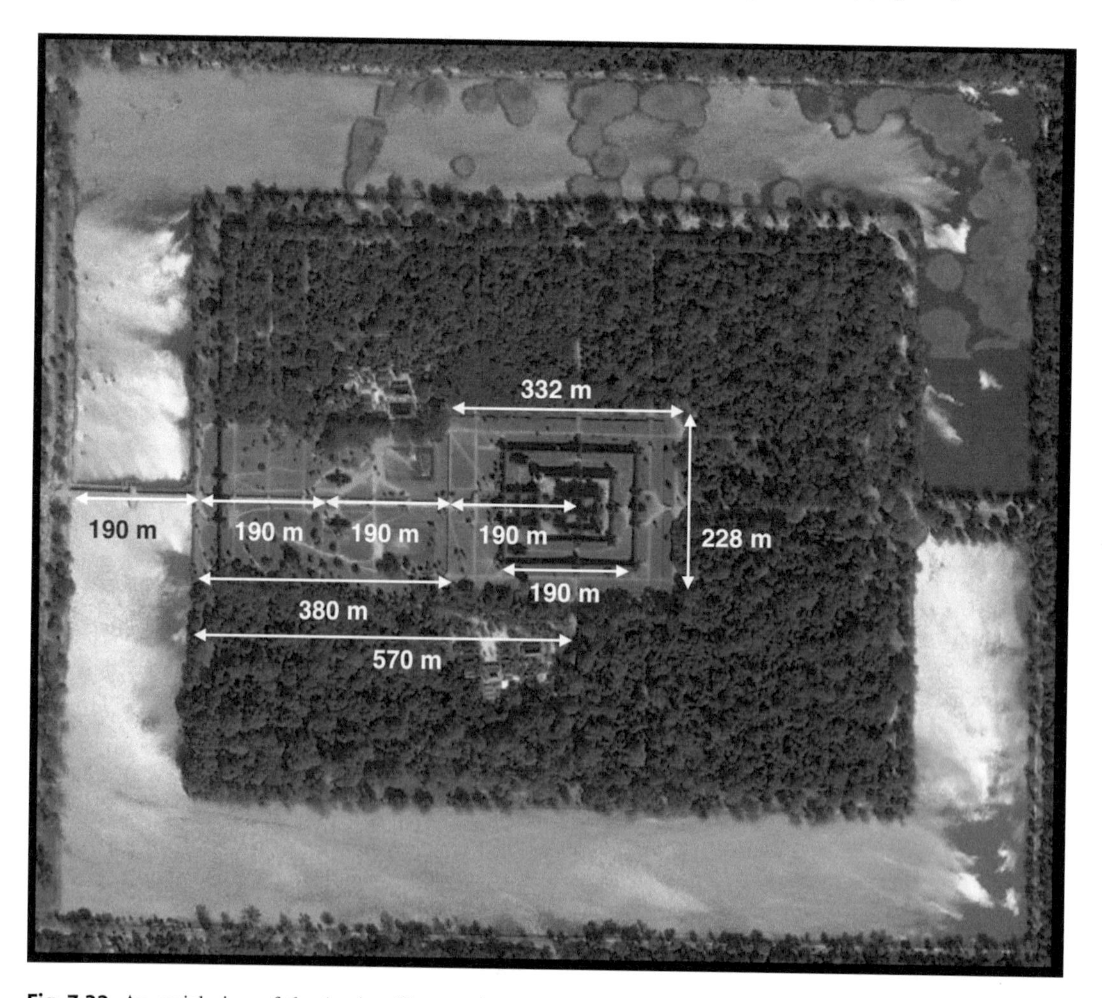


**Fig. 7.32** An aerial view of the Angkor Wat temple complex, showing the recurrent dimension of 190-m which is repeated in integer multiples within the design of the structure. The moat itself is 190-ms wide, while the distance between the inner edge of the moat and the inner enclosure wall is twice that distance or 380-m, and the distance to the center of the tallest spire is three times this distance or 570-m. The design encodes the 1:2:3 ratio which is also the ratio of duration of time within the Hindu *kalpas* or ages of the universe

Inside the moat are four concentric rectangular enclosures. Helicopter-borne lidar observations can peer through the canopy and vegetation to reveal evidence that the outermost enclosure housed hundreds of families, with houses arranged on a neat grid pattern and small ponds and canals connecting the households. The city sprawled well beyond the Angkor Wat moat, and at its peak near 1200 AD was comparable in size to modern day Phnom Penh. New lidar maps outside of the moat have revealed hidden networks of hydraulic works, hundreds of small ponds and large "serpentine" maze-like formations of unknown function (Fletcher et al. 2015; Evans 2016). These new observations have discovered additional roads and canals connecting cities and houses well beyond the Angkor Wat enclosures, and in some cases have discovered entirely new cities. The conclusion from this modern research is that Angkor Wat housed a much larger population than had previously been estimated, and that the various temple complexes surrounding Angkor Wat were part of an extended urban geography that at its peak near 1200 AD may have housed 750,000 people, with grids of streets, and extensive hydraulic works.

Like most celestial cities, Angkor Wat contains many astronomically inspired symbols and alignments. The temples near Angkor Wat were dedicated to the god Vishnu, and therefore faced West to allow for viewing of sunset from the main temple. Angkor Wat includes an unequal number of columns on the North and South sides of the temple—with 20 pillars in the North and 18 in the South and a slight shift from the main E-W axis, perhaps to enable sunrise viewing from the West entrance (Kelley and Milone 2005, p. 301). The astronomer Robert Stencel reported many numerological coincidences between the dimensions of the structure and key numbers in the Hindu timekeeping system by drawing lines within the complex connecting major structures (Stencel 1976). When the dimensions were expressed in units of 0.435 m (which was the assumed size of the Cambodian cubit or *hat*) Stencel found that the temple included dimensions corresponding to the number of months in the year (12 *hat* for the distance between steps) and the number of lunar mansions in Hindu astronomy (27 *hat* for the lengths of nine chambers in the central tower).

More tantalizing coincidences were found outside the temple, where dimensions appeared to be proportional to the lengths of the four *yugas*, the long cycle of time within Hindu cosmology. The moat's width and distances between the entrance and the outer wall and central tower appear to coincide with numerical ratios of 439, 867, and 1296 *hat*, which closely match the proportions of the lengths of the first three *yugas* of Hindu cosmology (432,000, 864,000, and 1,296,000 years or ratios of 1:2:3). Whether these numbers were by design or are an interesting numerical coincidence is hard to know for sure, but if they were intentional, Angkor Wat would provide a unique connection between celestial architecture and cosmic timekeeping (Kelley and Milone 2005, p. 302). The use of numerical ratios within temple dimensions was also common in the Chola and Pallava Hindu dynasties, who through trade and conquest from India brought both Hinduism and temple design to the Khmer empire. Khmer cities surrounding Angkor Wat are also aligned with an East–West axis, with a few notable exception. Lidar images of nearby temple complexes (Fig. 7.33) show orientations within 1–2 degrees of East–West, with the exception of Preah Khan of Kampong Svay, which is tilted 28° away from East, which is too far north to face Summer Solstice sunrise. Despite centuries of study, and the use of the latest modern technologies, the temples near Angkor Wat are still shrouded in mystery.

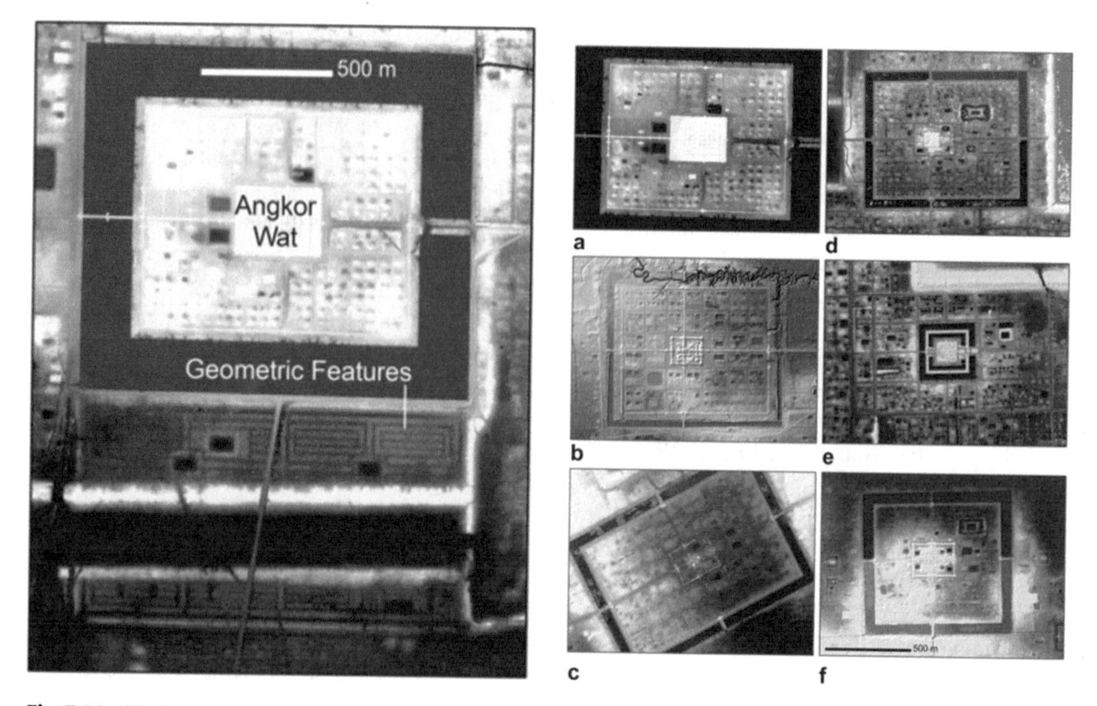


**Fig. 7.33** (*Top*) Using helicopter-borne lidar, which allows archeologists to peer through the jungle canopy and other vegetation, maps reveal the intricate networks of roads, ponds, and mounts that fill the interior of Angkor Wat's outermost enclosure, giving evidence of habitation by hundreds of households. A mysterious geometric "serpentine" feature to the South of Angkor Wat was also discovered from lidar observations. (*Bottom*) Panels showing the six largest temples near Angkor Wat with lidar mapping. These images reveal grids of ponds and roads within moat-enclosed central square and a general alignment along East–West axes. The notable exception in panel (c) is Preah Khan of Kampong Svay, with a North–East facing entrance 28° from East–West. This direction is not obviously explained by astronomical considerations and is too far north for a Summer Solstice Sunrise direction at this latitude. (Images from Evans 2016; reproduced according to terms of a Creative Commons license)

## Temples and the Embodiment of Cosmic Order

Often it is difficult to decode the alignments and cosmological significance of structures from ancient civilizations, such as Teotihuacan or Angkor Wat, since in these cases there are no written records to verify our interpretation. However, we do have many examples of temples within the European architectural tradition in which the design includes aspects of the well-documented cosmology of the European, Roman, or Greek cultures. Three examples are presented here: the Pantheon or Imperial Rome, the Parthenon of Ancient Athens, and Cologne Cathedral in Germany. In all three examples the buildings exhibit geometric symbolism, numerical dimensions, and celestial alignment that make them as embodiments of cosmic order.

### *The Pantheon of Rome*

The Roman Emperor Agrippa built the Pantheon in 27 BC. Inscriptions on the temple indicate that the temple was built to glorify the *gens Iulia*, or royal family, and was dedicated to Mars and Venus, who were thought to be related to the royal family. The emperor Hadrian restored the building in AD 126 several decades after it was damaged severely by a fire. The building is aligned to the north and includes the largest masonry dome in the world, which contains proportions that preserve an exact equality between the diameter of the circular base (142 ft) and the height of the dome, which rises 142 ft above the ground (Figs. 7.34 and 7.35).

**Fig. 7.34** The Pantheon of Rome, which embodies the Roman cosmology in its architecture and creates a celestial atmosphere from the Oculus, which admits natural light into the dome (*image source*: http://upload.wikimedia.org/wikipedia/commons/f/f5/PantheonRoma.jpg)

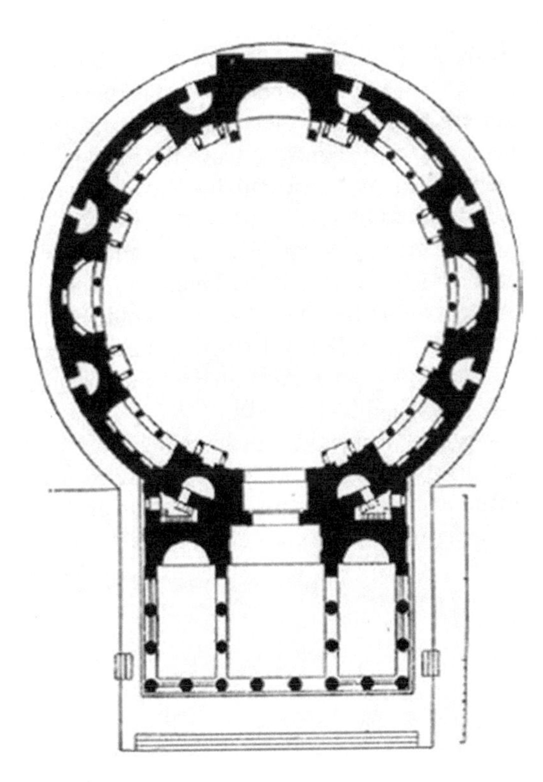

**Fig. 7.35** Plan view of the Pantheon (*image source*: http://commons.wikimedia.org/wiki/Image:Dehio_1_Pantheon_Floor_plan.jpg)

Tapering the thickness of the material to provide thinner walls toward the top reduced the enormous weight of the dome. The dome ranges in thickness from 21 ft at the base to only about 4 ft thick at the top, and the material at the top was mixed with lightweight pumice stone to further reduce the weight. At the top of the dome is a circular opening known as the Oculus that fills the building with a beam of sunlight.

Celestial symbolism within the building includes the dome's resemblance to the spherical universe favored in Greece and Rome, with an interior space that is reminiscent of the sky. Large square patterns in the roof were once gilded and perhaps also decorated with stars. Within the building are also seven niches, which are conjectured to have housed seven planetary deities, further strengthening the celestial symbolism. Cassius Dio wrote in his "History of Rome" (ca. AD 200) that the building was called the pantheon because "it received among the images which decorated it the statues of many gods, including Mars and Venus; but my own opinion of the name is that, because of its vaulted roof, it resembles the heavens" (MacDonald 2002).

Since Roman times, the Pantheon has been converted into a Christian church and was in continuous use throughout the medieval period. Boniface IV dedicated the building as a church of S. Maria ad Martyres in AD 609, and exquisite sarcophagi, statues, and paintings have been installed, including the sarcophagi of Leo X and the painter Rafael. The niches now include paintings by Camassei, a chapel commemorating a sixteenth century group of artists and musicians, a bust of Raphael, a statue known as the Madonna del Sasso, and other paintings. Despite the additional Christian and Renaissance art, the Pantheon retains its original astronomically inspired and celestial atmosphere.

## *The Parthenon of Athens*

The Parthenon is perhaps the best known structure of classical Greece and was built between 447 and 438 BCE. The temple was dedicated to the Greek goddess Athena in her two aspects as Athena Polios ("of the city") and Athena Parthenos ("young maiden"). The Parthenon is a remarkable embodiment of the Greek aesthetic values within pre-Socratic philosophy and includes many geometric ratios and curved surfaces that reflect a sense of order consistent with the Greek geometrical cosmology. Curved surfaces include columns that are tapered slightly to make the illusion of being narrow-waisted, and sides which rise slightly in the middle to make the building appear more dramatic, and to use perspective to enhance the appearance of the building.

The Parthenon building is supported by 46 large columns and occupies a rectangular base 101 ft wide by 228 ft long. A 40-foot high statue of Athena occupied the inner cella of the temple. The height of the building (accentuated by curved surfaces) retained the "golden ratio" of 9:4. Several groupings of the three dimensions, width (W), height (H), and length (L) can be expressed in terms of the ratio 9:4, using width = 30.88 m, height = 13.72 m, and length = 69.5 m. The ratios that include the "golden ratio" include $W/H = 9/4$, $L/W = 9/4$, and $L/H = (9/4)^2$.

Additional numerical ratios can be found within the building, since if one divides the length of the Parthenon into three segments there is a ratio of 3:4 between the dimensions of each segment and the building's width. While the Parthenon is not aligned with the Sun or stars, it does embody the Greek mathematical sensibility that underlined Greek cosmology and astronomy (Figs. 7.36 and 7.37).

**Fig. 7.36** The Parthenon, one of the most celebrated celestial structures of the ancient world, embodies the Greek fascination with numbers and geometry (*image source*: http://upload.wikimedia.org/wikipedia/commons/d/dd/2004_02_29_Athènes.JPG)

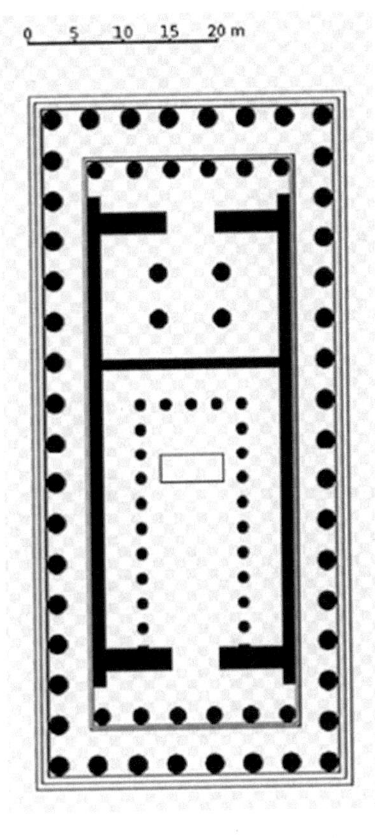


**Fig. 7.37** Plan view of the Parthenon (*image source*: http://upload.wikimedia.org/wikipedia/commons/2/21/Parthenon-top-view.svg)

## Cologne Cathedral

The Cologne cathedral is one of the finest examples of a medieval cathedral that embodies the cosmology of the medieval Christian church. Like many cathedrals, the Cologne cathedral was the work of centuries and dozens of generations of stonemasons, laborers, and artisans. Work began in 1248 and was not completed until 1880, a period of over 630 years! The loft of the cathedral is greater than any medieval church, featuring the highest Gothic vault in the world, and a ratio of height to width of 3.6:1. When it was completed it was the tallest structure in the world, and includes two towers that are 157 m high. Like most cathedrals of the time, it contains a floor plan in the shape of a cross with an entrance aligned to the east.

The practice of orienting church and cathedral entrances to the east is traditional in Europe and also forms the basis of the word "orient" which refers to the eastern-facing entrance. Like most medieval cathedrals, the structure itself provides instruction in the Bible in the form of carved reliefs and stained glass images that portray the major events in the Bible. The floor plan is in the form of a cross, with the altar at the "head" of the cross, which underlines the role of the cathedral as an embodiment of the cosmic order of the Christian theology (Figs. 7.38 and 7.39).

**Fig. 7.38** View of the Cologne cathedral at night (*image source*: http://commons.wikimedia.org/wiki/Image:Koelner_Dom_bei_Nacht_1_RB.JPG)

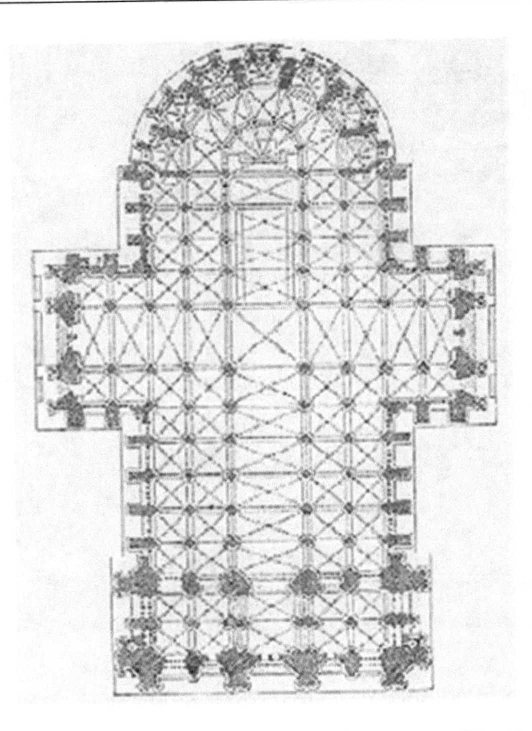

**Fig. 7.39** Plan view of the Cologne cathedral (*image source*: http://commons.wikimedia.org/wiki/Image:DomKoelnGrundriss.jpg)

## *The Chola Temples of Tamil Nadu, India*

The cathedrals of the ancient Indian Chola empire are found in Tamil Nadu, India, and they are magnificent embodiments of the symmetries and mathematical ideas of this expansive Indian empire. The two latest Chola temples, Brihadisvara Temple and Gangaikonda Cholapuram, were constructed between 980 and 1030 AD, and represent the culmination of architectural principles began within the earlier Pallava dynasty between 600 and 800 AD. Brihadisvara Temple and Gangaikonda Cholapuram were built within 30 years of each other, with the larger Brihadisvara Temple constructed by the Chola emperor Rajarnja I between 985 and 1014 AD, and the Gangaikonda Cholapuram constructed by his son Rajendra after his father's death in 1020–1040 AD. The name "Gangaikonda Cholapuram, means 'the town of the Chola who took the Ganges'" and refers to Rajendra's conquests (Pichard 1995). Both temples embody important and beautiful symmetries. The length of Brihadisvara is twice its width, and the interior courtyard is divided into two square spaces, with the tallest tower placed in the center of one of these squares, and the large *nandi* or bull statue placed in the other. Straight lines also can be drawn connecting the center of the main tower where the Shiva altar is located to the two towers that flank the entrance and the exterior of the temple, and their heights gracefully rise to provide a linear connection across the main axis (see Fig. 7.40). Additional symmetries are built into the height

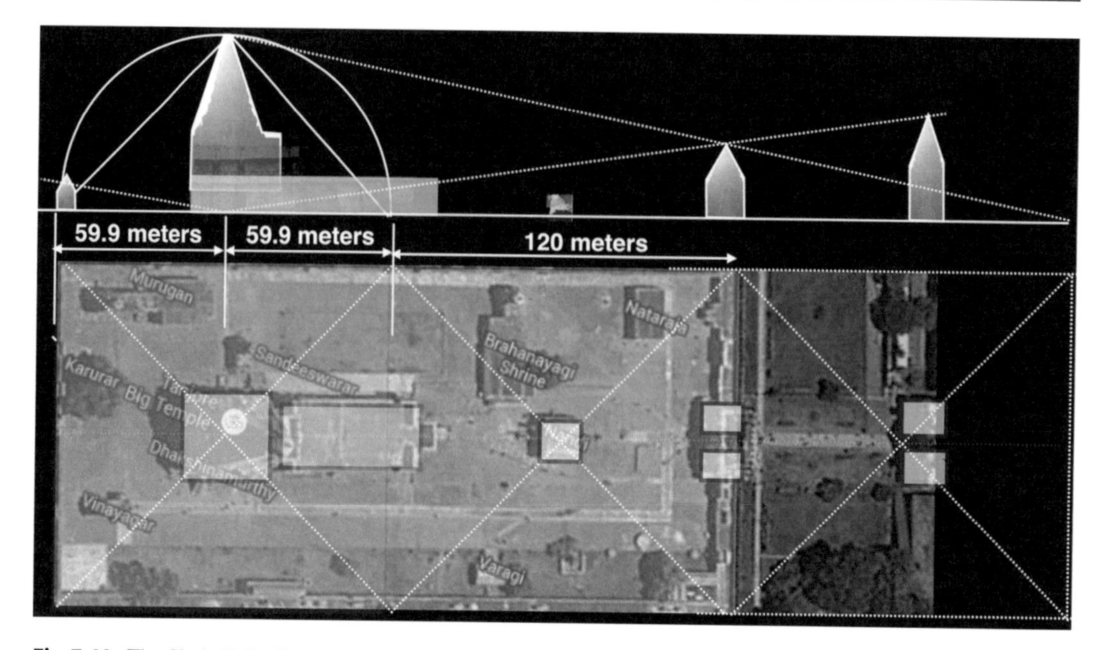


**Fig. 7.40** The Chola Brihadisvara Temple, shown from above and from the side in plan view, and with key symmetries indicated. The layout of the temple includes several squares of equal size, with the main compound enclosed by a wall with a 2:1 aspect ratio. A circle of radius 59.8 m can circumscribe both the square units of the temple layout and the side view of the main Shiva temple, giving it symmetry both along the main axis and with its vertical dimension. In addition, lines of sight from the base of the tower complete straight lines that connect to the tops of the towers at the wall entrance and the easternmost gate (Figure adapted from Pichard 1995)

of the main tower, which matches the radius of an inscribed circle that fits within the square of the main temple, and also which connects towers on the Western end of the temple yard with the entrance to the Shiva temple. A similar concert of geometry is visible within the Gangaikonda Cholapuram temple (Fig. 7.41). Gangaikonda Cholapuram is built with a rectangular geometry, in which inscribed circles intersect with the axis of the main tower, and like Brihadisvara, the main tower has a height of 51-m which is equal to the radius of the inscribed circles that fit within the temple yard. Additional inscribed circles connect from corners of the temple compound to altars within the Shiva temple.

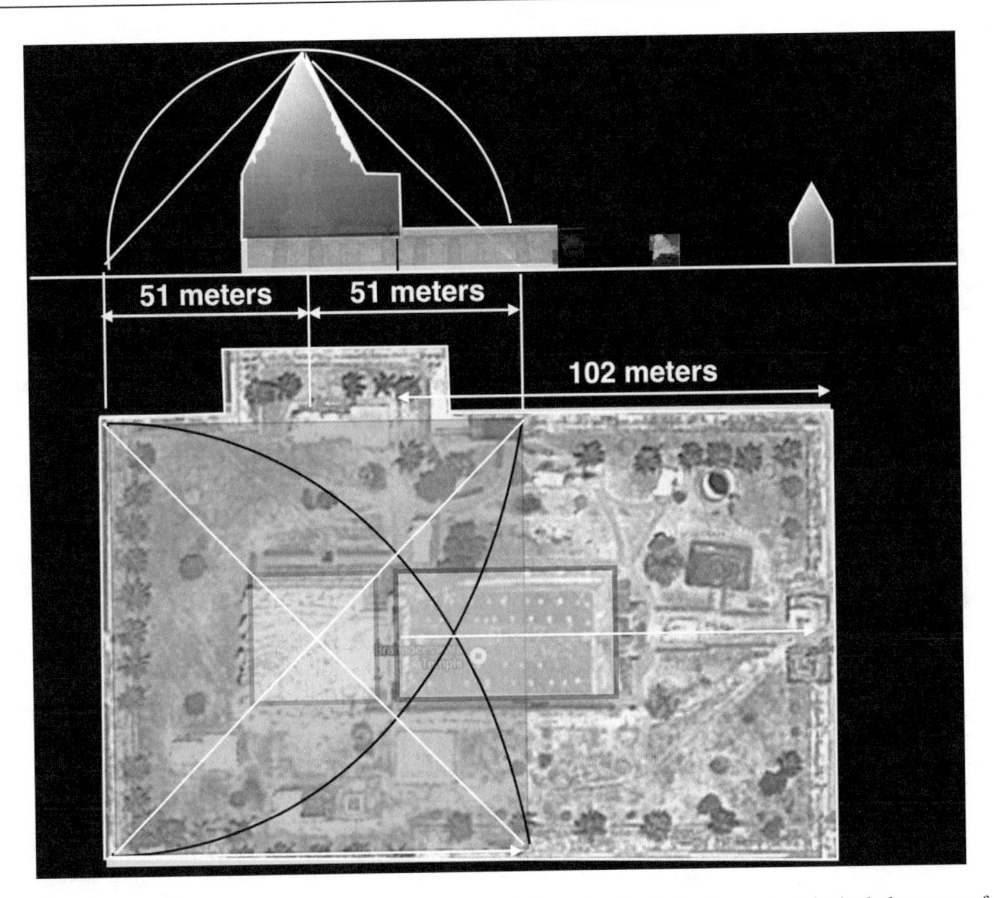


**Fig. 7.41** The Chola Gangaikonda Cholapuram temple, built just after the Brihadisvara temple, includes some of the same design principles of its predecessor, with the unit size scaled down slightly to 51 m. The beautiful symmetries include a unit square that can be circumscribed by a circle with a 51 m radius both horizontally and vertically and a slightly foreshortened courtyard which includes the same dimensions from the west wall of the central temple to the easternmost gate (Figure adapted from Pichard 1995)

## The Limits of Alignment

Astronomical calculations can shed light on cultures that otherwise left little or no written records and highlight the universality of astronomy among civilizations. In other cases, an overemphasis on astronomical or mathematical interpretation risks imposing an analytical and astronomical mindsets on ancient cultures that may be more a reflection on the modern authors than the ancient culture. In this section, we will examine a few of the more speculative claims of authors who posit that ancient people were able to construct sets of temples or pyramids within patterns that create elaborate maps mirroring constellations or to bring numerical ratios to the landscape.

Carl Sagan's excellent advice, "extraordinary claims require extraordinary evidence," is good guidance, as a scientific examination of such claims requires skepticism. Three examples are presented– the numerical ratios and alignments of lines connecting megalithic sites in Great Britain, the connections between Egyptian pyramids known in popular books as the "Orion Mystery," and the Nazca lines in Peru.

## Linear Geometry in Britain

The astronomical underpinnings within the megalithic sites of Stonehenge, Callinash, and Newgrange have been discussed above. Some authors have considered alignments and dimensions *between* megalithic sites, using sightlines which amount to thousands of miles, with separations that show interesting numerical ratios of a supposed ancient unit known as the megalithic yard. Alexander Thom in 1973 observed numerical ratios of separations of megaliths within stone circles, which included several triangles with sides that have numerical ratios of 1:3:10. In some cases these inscribed triangles intersected important stones, while in other cases it was hard to see what the lines are connecting. Much larger triangles have been proposed to connect megalithic sites, including one huge triangle that connects Stonehenge, Lundy Island (off the west coast of England), and the quarry site for the Blue Stones of Stonehenge in Wales. A triangle drawn on a map showing these three sites forms a large triangle of lengths 5, 12, and 13—a Pythagorean triple, according to the Welsh archaeoastronomer Robin Heath (Heath 2001, p. 39).

Heath also has developed a clever construction in which one can use this same triangle to derive the number of lunations in a solar year, which he claims is built into megalithic sites. Within the 5:12:13 triangle it is possible to divide the short side into a smaller triangle with sides of 3:12:12.368. The hypotenuse of this triangle has a length (12.368) that corresponds to the number of lunar months in a solar year.

If the ancient Britons had this mathematical skill, it is proposed that the separations of these widely separated sites embody the numerical ratios of the lunar calendar and geometry. However, this hypothesis brings many unanswered questions—how do we know the Britons cared about such obscure mathematical details? Even if they were able to determine the ratios why would they bother to build stone monuments to encode the distances? And how would ancient people be able to survey and lay out such lengthy sightlines? Could there be an ulterior motive in suggesting that early Britons were so mathematically advanced? Until such questions can be answered, we have to resist such speculative theories.

## Star Maps from the Pyramids: The Orion Mystery

The popular imagination has been fired by many speculative theories about the origins of pyramids and other ancient monuments. One theory which has many of the trappings of a scientific theory with little of the actual scientific rigor is that of the "Orion Mystery" popularized by the author Robert Bauval in the book of the same name (Bauval 1995). The theory posits that the pattern of the pyramids on the land mirrored that of the stars within the Orion constellation. The location of the Nile River relative to the "Orion" figure of pyramids is said to be similar to the Milky Way relative to the Orion constellations. Additional pyramids such as the Bent Pyramid and Red Pyramid, which do not match stars within Orion, are attributed to the constellation of Taurus. One look at the placement of these pyramids on the plains of Giza compared to a map of the Orion constellation shows that the match is not convincing (see Fig. 7.42). It is hard to imagine that Egyptians over centuries would conspire to construct an ill-matching constellation figure from millions of giant stone blocks! Nevertheless, the theory has gained a lot of attention and seems to be widely discussed in the popular literature.

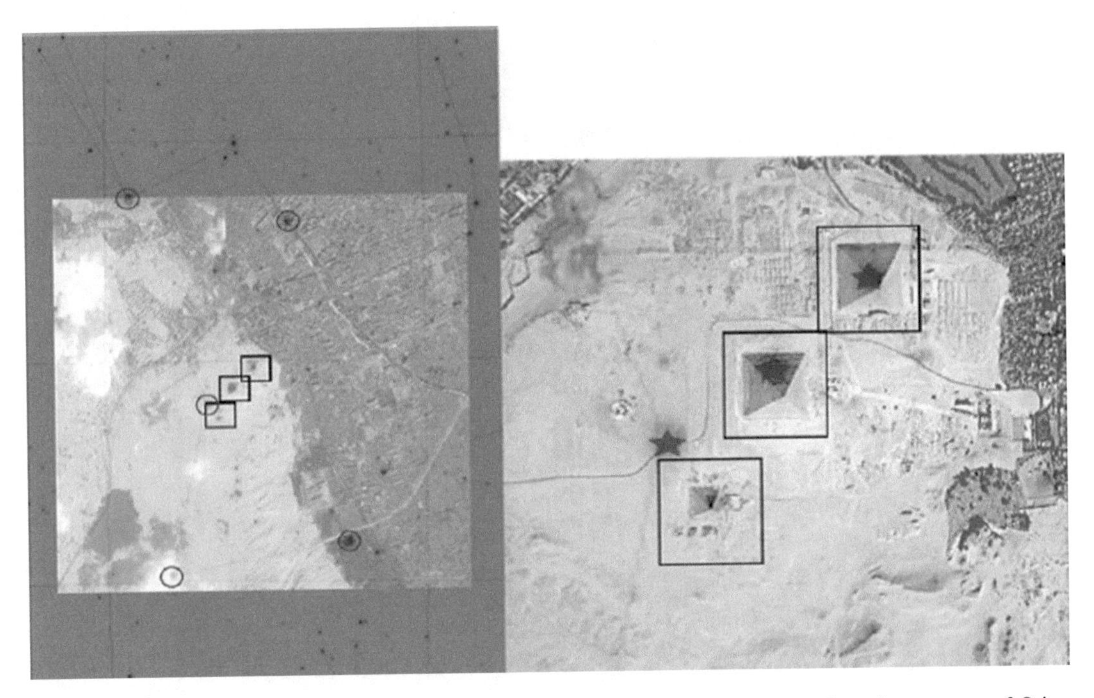

**Fig. 7.42** One of the most mysterious aspects of the "Orion Mystery" is that an overlay of a star map of Orion causes only two of the seven principal stars to appear at the locations of the Giza Pyramids. The *black boxes* show the locations of the three main Giza pyramids, while *red circles* show locations of the superimposed Orion stars from an accurate star map. Only two of the seven stars in Orion line up toward pyramids! The *right image* is a zoom of the central region, showing that the alignment of the belt stars (*red stars*) against pyramids (*boxes*) is not very convincing (original figure by the author, using satellite imagery from NASA's World Wind software, and star map from the Voyager IV program)

## The Nazca Lines

The Nazca Valley of southern Peru features a vast expanse of arid land that was decorated with a number of stone outlines in the shapes of birds, insects, monkeys, human-like figures, and spirals and zigzag patterns (Fig. 7.43). The figures are subtle and barely visible when standing in the valley, as they are formed from moving rocks sideways to form lines and patterns hundreds of meters across. The effort in coordinating the movement of the rocks must have required a large number of people and some advanced planning, but no written record exists to describe the meaning or intent of the figures.

**Fig. 7.43** One of the Nazca figures, as viewed from the air (*image source*: http://upload.wikimedia.org/wikipedia/commons/c/c1/Nazca-lineas-perro-c01.jpg)

Naturally an enigmatic site such as Nazca has inspired a number of fanciful theories, such as the use of the figures for ancient airborne craft, even perhaps from extraterrestrial visitors who might have used them as landing strips! However, a more sober examination of the figures suggests that they could have served a number of more mundane purposes. Aveni identifies five possible explanations for the Nazca figures–as markers of calendrical events, a "teaching tool" to train builders of more complex structures, a marker of sources of water, paths for ritual walking, and as drawings intended as offerings to gods or "sky people" (who presumably would have been able to see them) (Aveni 1990).

One attempt to identify astronomical alignments of Nazca was attempted by Maria Reiche, who suggested that various figures pointed toward rising points of stars such as Orion and other stars. Gerard Hawkins claimed further alignments, including the Pleiades rising azimuth, but acknowledged that the alignments at the site were consistent with those arising from random chance. More recent analyses have considered all the figures at the site and concluded that only a few of the directions had a chance of astronomical significance, and that the few figures that did align were the less dramatic lines, leaving no astronomical explanation for the figures of animals and other creatures (Ruggles 2005). There is, however, a correlation between the orientation of the figures and lines which were either parallel or perpendicular to directions of water flow, lending support to the theory that these figures were connected with rituals related to water or rain (Kelley and Milone 2005, p. 456).

Perhaps the above examples suggest that a certain degree of skepticism about astronomical alignments is warranted, despite the demonstrated tendency of humans from around the world to align temples, cities, and houses with astronomical directions. As one final cautionary example, the ancient Norse site of Lindholm Hoje, shown in Fig. 7.44, includes literally hundreds of ancient standing stones, dozens of which most certainly align with astronomically important directions (stellar rising positions, solstice sunrise, etc.). However, the site is known to have nothing whatever to do with astronomy, as the stones were intended to trace the shapes of ancient Norse warships, and not to point toward the stars!

**Fig. 7.44** The site of Lindholm Hoje, in Denmark, which provides a cautionary tale for would-be archaeoastronomers (*Image source*: http://commons.wikimedia.org/wiki/Image:Lindholm_Høje,_den_29_april_2008,_billede_3.jpg)

## References


Aveni, A., and H. Hartung. 1986. Maya city planning and the calendar. *Transactions of the American Philosophical Society* 76(7): 1–87.

Aveni, A.F. 1990. *The lines of Nazca*. Philadelphia, PA: American Philosophical Society.

Bauval, R. 1995. *The Orion Mystery: Unlocking the secrets of the Pyramids*. New York, NY: Crown Trade Paperbacks.

Bonine, M.E. 1995. Romans, astronomy and the qibla: Urban form and orientation of islamic cities of Tunisia. In *African cultural astronomy – Current archaeoastronomy and ethnoastronomy research in Africa*, ed. J. Holbrook et al. New York: Springer.

Brecher, K., and M. Feirtag. 1979. *Astronomy of the ancients*. Cambridge, MA: MIT Press.

Campbell, J., and B.D. Moyers. 1988. *The power of myth*. New York: Doubleday.

Castleden, R. 1993. *The making of Stonehenge*. London; New York: Routledge.

Cornelius, G. 1997. *The starlore handbook: An essential guide to the night sky*. San Francisco, CA: Chronicle Books.

Courlander, H. 1987. *The fourth world of the Hopis: The epic story of the Hopi Indians as preserved in their legends and traditions*. Albuquerque, NM: University of New Mexico Press.

Díaz del Castillo, B., and J.M. Cohen. 1963. *The conquest of New Spain*. Baltimore, MD: Penguin Books.

Eddy, J. 1974. Astronomical alignment of the Big Horn Medicine Wheel. *Science* 194: 1035–1043.

Evans, D. 2016. Airborne laser scanning as a method for exploring long-term socio-ecological dynamics in Cambodia. *Journal of Archaeological Science* 74: 164–175.

Fletcher, R., D. Evans, C. Pottier, and C. Rachna. 2015. Angkor Wat: an introduction. *Antiquity* 89: 1388–1401.

Freeman, M., and C. Jacques. 2003. *Ancient Angkor*. Bangkok: River Books.

Hadingham, E. 1975. *Circles and standing stones*. London: Heinemann.

———. 1984. *Early man and the cosmos*. New York: Walker.

Hawkins, G.S., and J.B. White. 1966. *Stonehenge decoded*. London: Souvenier.

Heath, R. 2001. *Stonehenge—Ancient temple of Britain*. New York, NY: Walker and Company.

Kelley, D.H., and E.F. Milone. 2005. *Exploring ancient skies: An encyclopedic survey of archaeoastronomy*. New York: Springer.

Krupp, E.C. 1983. *Echoes of the ancient skies: The astronomy of lost civilizations*. New York: Harper & Row.

Lekson, S.H. 1999. *The Chaco meridian: Centers of political power in the ancient Southwest*. Walnut Creek, CA: AltaMira Press.

MacDonald, W.L. 2002. *The Pantheon: Design, meaning, and progeny*. Cambridge, MA: Harvard University Press.
Malville, J.M. 2008. *A guide to prehistoric astronomy in the southwest*. Boulder: Johnson Books.
Malville, J.M. 2015. Astronomy of Indian cities, temples, and pilgrimage centers. In *Handbook of archaeoastronomy and ethnoastronomy*, ed. C.L.N. Ruggles. New York: Springer.
Morrow, B.H., and V.B. Price. 1997. *Anasazi architecture and American design*. Albuquerque, NM: University of New Mexico Press.
Odling-Smee, L. 2007. Dig links Stonehenge to circle of life. *Nature* 445(7128): 574–574.
Pichard, P. 1995. *Tanjavur Brhadisvara – An architectural study*. New Delhi: Indira Gandhi National Centre for the Arts.
Ruggles, C.L.N. 2005. *Ancient astronomy: An encyclopedia of cosmologies and myth*. ABC-CLIO: Santa Barbara, CA.
Scarre, C. 1999. *The seventy wonders of the ancient world: The great monuments and how they were built*. London: Thames & Hudson.
Sharer, R.J., and L.P. Traxler. 2006. *The ancient Maya*. Stanford, CA: Stanford University Press.
Spence, K. 2000. Ancient Egyptian chronology and the astronomical orientation of the pyramids. *Nature* 408: 320–324.
Stencel, R. 1976. Astronomy and cosmology at Angkor Wat. *Science* 193(4250): 281–287.
Stukeley, W. 1740. *Stonehenge, a temple Restor'd to the British druids*. London: W. Innys and R. Manby.
Trimble, V. 1964. Astronomical investigations concerning the so-called air-shafts of Cheops' Pyramid. *Mitteilungen des Instituts fur Orientforschung Akademie der Wissenschaften zu Berlin* 10: 183–187.
Underhill, W. 2011. Putting Stonehenge in its place. *Scientific American* 304(3): 48.
Williamson, R.A. 1984. *Living the sky: The cosmos of the American Indian*. Boston: Houghton Mifflin.
Wolfe, L.M. 1979. *John of the mountains: the unpublished journals of John Muir*. Madison: University of Wisconsin Press.

# Chapter 8

# The Archaeoastronomy of Modern Civilization

*To one who has been long in city pent,*
*'Tis very sweet to look into the fair*
*And open face of heaven, — to breathe a prayer*
*Full in the smile of the blue firmament.*

John Keats, *Sonnet XIV*

*Towered cities please us then,*
*And the busy hum of men.*

John Milton, L'Allegro (1631)

The desire to align buildings to recreate the cosmic order in the works of humans is not confined to the past. The desire is universal across the Earth and can be seen within our current societies. It is possible to analyze our present-day civilization much as we have considered earlier civilizations. We find that our civilization, like others before us, responds strongly to the power of the stars, with structures and cities aligned with the Sun and stars. The surprising survival of astrology, and other astronomical symbols within our twenty-first century civilization also, makes archaeoastronomy (ancient astronomy) something we can study in the present day. In this chapter, we examine the ways in which our modern civilization practices ancient astronomy even in the twenty-first century.

## The Archaeoastronomy of Modern Cities

### *The Astronomy of Skyscrapers*

The most obvious artifact of our civilization (which future archaeoastronomers would perhaps try to analyze) is the skyscraper. The cities of our modern civilization seem disconnected from the sky—and many of us have trouble observing the skies in cities both from our rushed pace of life and the over-abundance of artificial lighting. Yet a closer examination shows that modern buildings (like many ancient structures) contain symbolic elements and celestial alignments that connect them to the sky or to the belief systems of our culture. The instinct of humans to build structures to bring them closer to the sky seems universal—across cultures and time—from ziggurats to skyscrapers.

Astronomical number symbolism, international rivalry, the laws of physics and commerce all intersect in the designs of skyscrapers. Each building is a compromise of all of these forces. Of the top ten tallest skyscrapers at the time of this edition (2016), six show strong elements of numerical or astronomical


B.E. Penprase, *The Power of Stars*, DOI 10.1007/978-3-319-52597-6_8

symbolism, and nearly all are aligned with important cardinal or astronomical directions. The tallest *planned* building, the Jeddah tower, and the current tallest completed building, the Burj Khalifa both rise to magnificent heights. Topping out at over 3200 feet, or just over a kilometer, the Jeddah tower has been designed to defy gravity—and leaves little room for astronomical or numerical elements. Likewise for the Burj Khalifa, which slices the sky like a needle and rises 2700 feet above Dubai. The Shanghai Tower, the current number two building, is similarly optimized for physics, and spirals gracefully upwards above Shanghai with a transparent skin that helps distribute weight and heat efficiently.

The number three tallest building, the Makkah Royal Clock Tower, has very strong astronomical design elements, however. The tower rises 1970 feet above Mecca at the south side of the *kaaba*. This building includes a giant clock face (the largest in the world) and includes two moon viewing galleries. At number four, the new World Trade Center rises 1776 feet above Manhattan with the year of the US Declaration of Independence built into its height. The number five highest skyscraper, the Taipei 101 Tower, contains 101 floors and rises 508 meters above the ground. This building design is inspired by bamboo in its organic shape and is sectioned into eight parts, based on the belief within China that eight is a lucky number. Like an ancient ziggurat, the tower rises to the sky in sections, and is aligned with the cardinal directions (www.emparis.com) (Fig. 8.1).

**Fig. 8.1** *Left:* The Taipei 101 tower, which includes eight sections based on numerological beliefs within China and also is celestially aligned with cardinal directions. *Right:* Three of the largest buildings in the world reaching into the night sky. From *left* to *right* are the Shanghai Financial Center (currently number 6), with the full moon visible near the "moon gate" at *top*; the Jin Mao tower (*center*; currently number 16) and the Shanghai tower, the second tallest building in the world (photos by the author)

The sixth tallest building, the Shanghai Financial Center in Mainland China, was designed with extensive astronomical symbolism. The Shanghai tower rises to 492 m in 101 floors. The original design included a circular "moon gate"—a dramatic round opening in the top floors intended to invoke an image of the Moon. However, this design raised objections within China, as it reminded many of the Japanese flag (itself an image of the rising sun), and a simpler trapezoidal opening was substituted. The large opening in the building faces to the northeast and defines an axis that is close to the lunar standstill point. A future archaeoastronomer might mistakenly conclude that this building was part of an ancient astronomical observatory!

The eighth and ninth tallest buildings are the two Petronas towers, in Kuala Lumpur, Malaysia. The two towers rise in 88 floors to a height of 452 m, and are stylized to be reminiscent of Islamic art. The cross section of the building itself is astronomically inspired, as it has the form of an eight-point Islamic star. The orientation of the base of the building and the line connecting the peaks of the two towers define two non-cardinal axes that could challenge a future archaeoastronomer.

The eleventh tallest building in the world is the Sears Tower (renamed the Willis tower), which rises to 442 m in 110 floors. The shape of the Sears Tower hardly evokes astronomical symbolism and appears more boxlike (perhaps invoking the boxes of merchandise to be delivered from the Sears catalog!). However as the network of streets for the entire city of Chicago is firmly rooted into a cardinally aligned grid, the Sears Tower shares Chicago's astronomical alignments and faces the equinox sunrise and sunset. The sixteenth highest tower, China's Jin Mao tower, is loaded with numerical symbolism. The 88 floors are divided into 16 segments, each of which is 1/8th shorter than the segment before it, and the entire structure is supported by eight exterior "composite super columns" and eight exterior steel columns. The Jin Mao tower could be interpreted as a temple to the number 8, the Chinese symbol of good fortune (World.com 2009).

Clearly the universal human urge to build celestial structures, including pyramids, never goes out of style.

## *Alignments of Modern Cities*

Celestial alignment of cities is not confined to the ancient Inca or the Chaco Canyon civilization. Many of the largest cities of the world, Mumbai, Tokyo, Istanbul, and Sao Paulo, are confined closely to geographic contours of a coastal bay, island, or available dry land. More often than not, these practical considerations and established contours of a city dominate its alignment with the cardinal directions. In such cases, cities from space appear almost as large organisms, with tentacles of expressways, and veins of roadways that crisscross in a tangled mix. City blocks appear much as cells within a tissue in a microscope (Fig. 8.2).

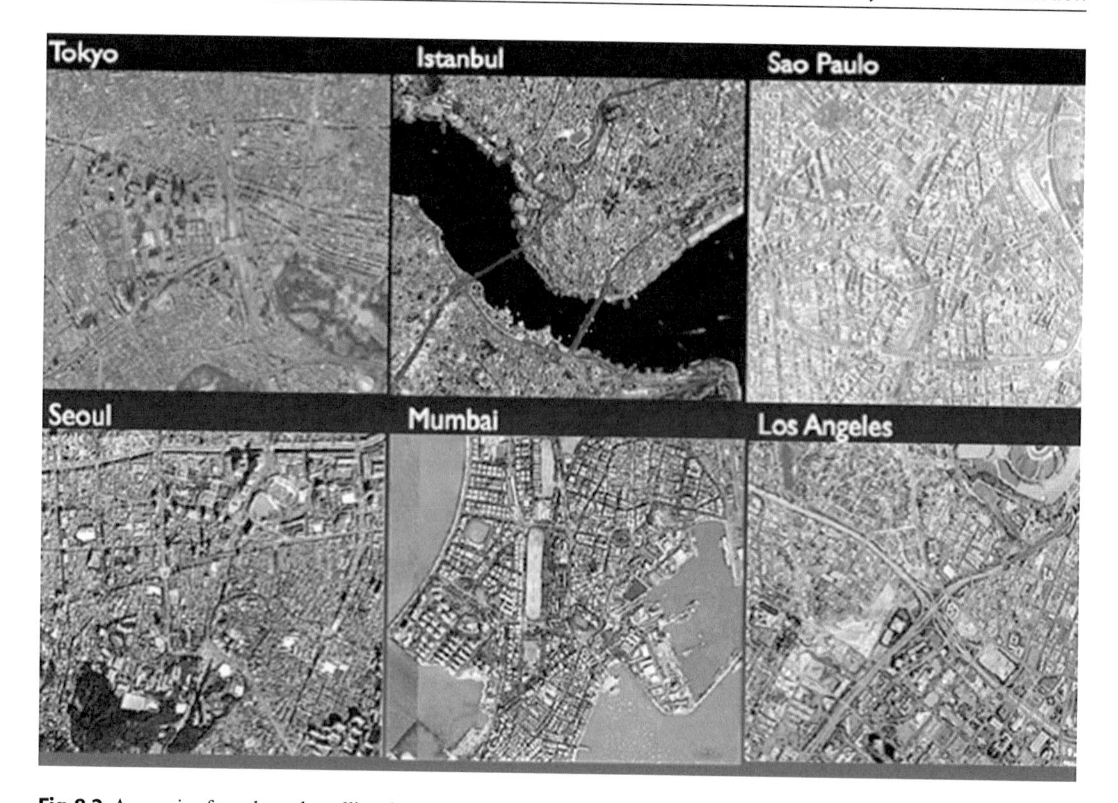


**Fig. 8.2** A mosaic of condensed satellite views of the largest cities of the world. Each image is taken from a simulated height of 3 km, using the NASA World Wind satellite imaging program to give a sense of the "texture" and orientation of modern cities on a consistent coordinate grid and scale. Most city sections appear to have grown organically from the ground, following paths of ancient lanes, banks of rivers, or coastlines (original figure by the author using NASA's WorldWinds software)

In some few cases, however, order crystallizes in the city. Major boulevards can set the axis of a city into order, as is the case for Paris or Buenos Aires, or an ancient city core can set the alignment, as is the case for Beijing (aligned to the ancient "Celestial Forbidden City." In other cases, alignments of modern cities come about as the result of a disaster, such as in the case of Chicago (rebuilt after the fire of 1872) or San Francisco (rebuilt after the earthquake of 1902) (Fig. 8.3).

**Fig. 8.3** A montage of satellite views of the largest cities of the world that show alignment. These modern cities in some cases show striking alignments, for a variety of reasons, ranging from rebuilding after a disaster (Chicago, San Francisco) and intentional astronomical alignment such as Beijing (original figure by the author, using NASA's World Wind software)

While in most cases the celestial alignments of modern cities are not planned, these alignments nevertheless manage to tie our modern cities to the sky, enabling equinox sunrise and sunset to traverse our major east—west boulevards and to give a glimpse of the North Star between the rows of tall buildings along the astronomically aligned avenues. Most "gridded" cities align their streets with cardinal directions, but New York City shows an interesting skew of approximately 28.9°, which is close to the winter solstice sunrise azimuth (31.7°).

Figure 8.4 shows this alignment from above, with the directions of solstice sunrise and sunsets, along with lunar standstill points. During the winter solstice, the alignments of the streets in Manhattan enable the sunrise to rise perfectly between the tall buildings, due to a 28.9° offset from an east/west axis. This alignment, sometimes called "Manhattanhenge" is coincidental, but nevertheless provides a fascinating archaeoastronomy case study in a modern city!

Another interesting (coincidental) alignment can be found within the city of Paris, where the line connecting the Arc de Triomphe and the Eiffel tower is also close to the winter solstice sunrise azimuth. Interestingly, one of the best-aligned cities of modern times is that of Las Vegas, shown in Fig. 8.3. Las Vegas clearly would present an interesting challenge to any future archaeoastronomer. In addition to a very precise North-South alignment of its central avenue ("the Strip") a hypothetical future archaeoastronomer would have to explain the purposes of the large monuments of Las Vegas, their curious placement along the major ceremonial avenue, the presence of a pyramid, towers, and artificial waterways of unknown purpose, and the interesting triangular geometry of many of the casinos—their exact function apparently somehow connected to an "ancient" numerological ritual known as "gambling!"

**Fig. 8.4** The slightly askew Manhattan street grid incidentally creates an archaeoastronomical alignment sometimes known as "Manhattanhenge." During the winter solstice sunrise and summer solstice sunset, the sunrise can be viewed along any of Manhattan's many streets, which are inclined 28.9° off from the cardinal directions. The *left panel* shows a satellite view of Manhattan with the solstice azimuth shown, while the *right panel* shows a street level view of a summer sunset in Manhattan (*image source*: NASA World Wind software; http://en.wikipedia.org/wiki/Image:Manhattanhenge2.jpg)

## Celestial Structures of Today's Civilization

While many of the alignments of large-scale modern cities are a combination of history, accident, and intention, some structures of modern civilization were built for the express purpose of commemorating or observing the sky. Even today, modern people are recreating the equivalent of Stonehenge, Chaco Canyon, and Teotihuacan in their own original ways. We offer a sampling of some of the modern celestial structures, some which will last for original of years and others that are more reflective of the ephemera of present-day culture.

### *James Turrell: Skyspace and the Rodin Crater Project*

James Turrell is a modern artist, living his dream of creating structures that invoke the power of the open sky. Turrell's focus is light, and the perception of color and time that comes from observing the intersection of light and sky. "There is no object in my work," according to Turrell, who adds, "There never was. There is no image within it." (Kosky 2013). The goal is to induce in a viewer "another kind of seeing" where "Light is not so much something that reveals as it is itself the revelation" (Heldmore 2013; Kosky 2013). The process of giving in to the abstract and formless colored light in a Turrell installation has been compared to the experience ancient people felt when they viewed the sky—and felt the "Power of Stars" above them. As eloquently described by Kosky,

> Offering neither object to grasp nor image to see, Turrell's art separates what modern forms of lighting connect—seeing from controlling, touching from mastering—and introduces us into a domain where our gaze makes the transition to wonder and admiration. In Turrell's work, then, aesthetic vision can be understood as the heir to contemplation. This makes his art a useful site for recovering something operative in contemplative traditions but repressed from mainstream interpretations of modernity—the leisure and freedom of human being and its correlative, the wonder and beauty of the world.
>
> (Kosky 2013)

With over 82 installations across the world—including both public spaces and installations in private residences, Turrell has explored these interfaces between observer and light, earth and sky. Many of the installations allow Turrell to frame the vastness of the open sky with a large ceiling bathed in colored light that bring many interesting optical effects within the eye of the viewer, while the viewer contemplates the vastness of space and time. One recent installation known as Skyspace was recently installed in 2007 at his alma mater, Pomona College (Turrell graduated from Pomona College in 1965 with a B.A. in psychology and a minor in Math). The Skyspace consists of a roof washed in pastel colors generated from a large bank of computer-controlled LEDs. A large rectangular opening allows the sky to be viewed in the colored frame. At sunset the LEDs cycle through a slow progression of colors and complement the changing colors of the night sky. A small rectangular reflecting pool in the center of the installation complements the show above (Fig. 8.5).

Perhaps the most impressive project that Turrell is undertaking is the remodeling of an extinct volcano, known as the Roden Crater (Fig. 8.6). Many ancient civilizations (with thousands of workers) would not have embarked on as ambitious a program, but James Turrell on his own is building a space within the crater which provides multiple viewing areas for visitors in which the night sky is framed by tunnels and windows carved into the crater's hard volcanic rock. The viewing is known by Turrell as "celestial vaulting" and is inspired by visits to ancient astronomically significant sites in which beams of light illuminate central chambers during solstice periods (such as Newgrange or the Chumash Burrow flat site). Turrell's crater also provides a dramatic viewing site for observing the sky (such as an ancient ziggurat or Mayan step pyramid)

**Fig. 8.5** The Skyspace, an installation by James Turrell at Pomona College, which features a ceiling bathed in colored lights that frames the night sky (copyright James Turrell, photo by Florian Holzherr, 2005, reprinted with permission)

**Fig. 8.6** A selection of images of the Roden Crater project, by James Turrell. At *left top*, one of the chambers of the crater, known as the East Portal, illuminated by a beam of light from a long shaft cut into the rock. *Upper right* is a view of the "Crater's Eye" which creates a dramatic view of the night sky and an amazing range of lighting within the structure. At *lower left* is one of the galleries in the structure, and *lower right* is a series of views of the crater from the side (copyright James Turrell, photo by Florian Holzherr, 2005, reprinted with permission)

**Fig. 8.7** *Top:* Roden Crater from the air, showing the top of the volcano where the artist James Turrell has created his "Crater's Eye" for viewing the sky from within the crater. The caldera works to provide unusual lighting for the viewer and presents an impressive array of light and shadow within the structure. *Bottom:* View from within the crater showing the "Crater's Eye Exterior Plaza" which admits light into the network of passageways and galleries within the crater (copyright James Turrell, photo by Florian Holzherr, 2005, reprinted with permission)

Within the Roden crater is a chamber in which an 874-foot long tunnel feeds light into the room and provides a beam of light that changes color and shape with the course of the day and during different seasons. A second room provides a huge viewing area aligned with the southern lunar standstill (Fig. 8.7).

The entire room within the Roden Crater acts as a giant camera obscura, in which the beam of moonlight creates an 8 ft diameter image on a gigantic marble monolith 13 ft wide and 15.5 ft tall. This marble slab is thought to be the largest quarried slab in North America, which harkens also to Stonehenge in its scale! The structure is intended to last for ages and to continue to create its dazzling array of light and shadow for thousands of years, just as in the ancient megalithic structures described earlier.

## Carhenge

Another modern archaeoastronomy construction that harkens to Stonehenge is the structure known as "Carhenge" which features an automotive tribute to Stonehenge, created to commemorate the June 1987 summer solstice by Jim Reindeers (Fig. 8.8). Carhenge features 38 vehicles arranged in a 96 foot diameter circle, to create several intact triptychs and a heel stone (a 1962 Cadillac). The neighbors of the site, in Alliance Nebraska, are not known to use the site for solstice ceremonies, but a visitor center has been constructed for passing (automotive) Druids interested in a quick ritual or photograph! Another North American automotive monument vaguely reminiscent of Stonehenge is known as Cadillac Ranch, in Amarillo, Texas. This monument features a linear arrangement of 1960s era Cadillacs buried nose down in the Texas soil and was created in 1974 by Stanley Marsh IV (Kirby et al. 2009).

**Fig. 8.8** Carhenge, one of several North American monuments combining modern artifacts and ancient archaeoastronomy (*image source*: http://en.wikipedia.org/wiki/File:A_Yool_Carhenge1_02Sep03_exif.jpg)

## Modern Stonehenges

Across North America and beyond, modern humans have recreated Stonehenge with varying degrees of accuracy. Samuel Hill, a pacifist and renowned engineer of bridges and roads in the Pacific Northwest, built a large replica of Stonehenge in 1918 at Maryhill, Washington (Fig. 8.9). Hill's Stonehenge was built to protest the fallen soldiers in World War I from nearby Klikitat County and was intended to make a statement about the larger human sacrifice of World War I. The project took 12 years to complete and was finished just before Mr. Hill's death in 1930 (Kirby et al. 2009).

Additional North American Stonehenges include a half-scale replica at the University of Missouri, Rolla, which was made from solid granite, a Stonehenge II in Kerrville Texas, a 60% sized replica of Stonehenge made by a retired oilman Al Shepperd in his pasture from plaster-covered steel, and a "Foamhenge" made by artist Mark Cline of Natural Bridge Virginia from realistic looking Styrofoam blocks. The collection of monuments in North America may lack some of the mass, durability, and tradition of their counterpart in England, but they do include some of the same inspiration and spirit and reflect our modern need to make monuments of lasting meaning.

**Fig. 8.9** The North American Stonehenge at Maryhill, Washington, built in 1918 as a tribute to soldiers lost in World War I (*image source*: http://en.wikipedia.org/wiki/File:Maryhill-WA-Stonehenge.jpg)

## Modern Sundials

One of the most common celestial structures throughout the ages is the sundial. Since the beginning of civilization, humans have created sundials and modern people have added their own contributions to the art form. Modern sundials often feature original materials, computer-aided design, and abstract and geometric styling. Sometimes modern sundials are building sized or only vaguely appear as sundials. Below is a sampling of a few of the more inspired and modern sundials (Fig. 8.10).

**Fig. 8.10** A few modern sundials: (clockwise, from *top left*) Reclining sundial, in Recifes, Brazil; sundial designed in 1965 by Henry Moore, now in front of the Adler planetarium, Chicago; sundial from London with a nautical theme, water column sundial commissioned to Pomona College artist Sheila Pinkel for Santa Ana College (*image sources*: http://commons.wikimedia.org/wiki/File:Mocoroh_analemmatos01.jpg, http://commons.wikimedia.org/wiki/File:2003-08-14_Adler_Planetarium_sundial.jpg; photograph of London sundial by the author, and Pinkel sundial courtesy of Sheila Pinkel)

## The Survival of Ancient Beliefs in Modern Times

Astrology is the wooly ancient forebearer of astronomy, much as alchemy is the ancestor of modern chemistry. Yet unlike alchemy, astrology still persists and is practiced by enthusiasts around the world. Daily astrology columns are read in newspapers, celebrity astrologers are well paid and in demand, and astrology has even played a role in the decisions of recent presidents. It is possible to have an astronomical horoscope (based on technology from 700 BC) automatically downloaded to one's cell phone or "twittered" to one's handheld computer (technology from AD 2010). What can account for the persistence of astrology and how has astrology developed over the centuries?

What most of us consider to be astrology in the USA or Europe is actually derived from the ancient astrological practices of Egypt and Babylon and modified by Greek and medieval scientists. Below is a brief description of how astrology manifests itself in our modern civilization, in the different cultures of the world.

## A Brief History of Western Astrology

### Babylonian Astrology

The first records of astrology coincide with earliest known writings of the Mesopotamian cultures. Some of the earlier writings include astronomical texts known as the Venus tablets of Ammizaduga, which date from 1500 BC, and which list dates and times over a 21-year period for Venus to rise and set, and corresponding omens and weather patterns on the Earth. Other early astrological texts include the “Three Star” tablets that date back to 1000 BC to record lists of three stars that heliacally rise in each of 12 months. Some of the most detailed early writings in astrology come from 650 BC, from the Assyrian king Ashubanipal. These texts describe how astrologer-priests, or *ummanu*, would document many of the conditions of the natural world, and the corresponding omens associated with them, as well as advice to the king based on their reading of the skies.

In the early days of astrology, the stars were just one part of the process of learning from the face of nature the hidden will of the gods. The motions of birds, animal entrails, and weather patterns were also used with planets and stars to read the future. As in many cultures (such as the Chinese and Mayan) Babylonians believed that the rule of the emperor was thought to rest upon the good will of the gods, who would indicate their displeasure in signs in the night sky. Each of the major gods was assigned places in the sky in the form of planets—Marduk, the ruler of the sky, was associated with Jupiter; Nergal, god of the underworld, was associated with Mars; Nabu, god of writing, was associated with Mercury; Ishtar, goddess of love and fertility, was associated with the planet Venus.

The early practice of astrology was largely for royal divination and was combined with other observations of nature to provide long lists of omens to help guide the ruler (Fig. 8.11). Astrologers would specialize in both the reading of omens and prescribing spells and rituals for dispelling ill portents. Especially bad was the occurrence of an eclipse, which required the king to install a temporary “dummy king” to rule on the day, while the actual king would hide to prevent his deposition. One of

**Fig. 8.11** Rulers of Mesopotamia used astrology and astronomical symbolism to consolidate their power. The Stela of Shamshi-Adad V from Nimrud (northern Iraq) originates from about 820 BC and shows the king near the symbols of the main gods of Assyria, all of which had astronomical symbolism. Near his hand is the symbol of the gods Ishtar, Adad, Sin, Shamash, and Ashur (from *bottom to top*) (photograph taken at British Museum by the author)

the early Mesopotamian forecasts from the seventh century BC reinforces the idea of an eclipse being dangerous for kings:

> When on the first of the month of Nisan the rising sun appears red like a torch, white clouds rise from it, and the wind blows from the east, then there will be a solar eclipse on the 28th or 29th day of the month, the king will die that very month, and his son will ascend to the throne.
>
> (Cramer 1954, p. 5)

One of the first complete texts of astrology (or astronomy) known is the set of tablets dating from the seventh century BC known as the Mul Apin. These tablets include a catalog of over 70 stars, with dates of heliacal risings, and the first use of the "zodiac" which is the region of stars in the "path of the Moon." A group of 18 star groupings are included in the Mul Apin, which form the basis of an early zodiac, each grouping matching a band approximately 20° across. A later table from 410 BC includes a text that refers only to reduced set of the 12 familiar "zodiacal constellation" most familiar to Greek, Roman, and later European astrology (Whitfield 2001, p. 21).

In the final years of the Babylonian culture, from 400 BC to 235 BC, astrology had evolved from a tool of the royal court to a more personal art of fortune-telling. The first extant horoscope dates from 410 BC, based on using modern computers to recreate the unique combination of planetary positions. By the Seleucid period, of 235 BC, astrology became "democratized" and was widely applied outside the royal court. One example of a personal astrological forecast includes a report of the positions of the Sun, the Moon, Jupiter, and other planets on the zodiac, and the corresponding meaning of the configuration as seen in 235 BC:

> Year 77 (of the Seleucid era), the fourth day, in the last part of the night, Aristokrates was born. That day: Moon in Leo, Sun 12° 30° of Gemini, Jupiter in 18 Sagittarius. The place of Jupiter means his life will be regular, he will become rich, he will grow old, his days will be numerous.
>
> (Whitfield 2001, p. 24)

### Greek and Roman Astrology

By the time of the conquest of Mesopotamia by Persian and later Greek forces, the role of astrology had diminished in Mesopotamia, but took root in Greece and Rome, where it found many prominent advocates, including some of the leading philosophers of antiquity. The connections between the East and Greece allowed the astronomical discoveries and records of the Babylonians to be adopted along with many aspects of the mystical religious traditions of Zoroastrianism and the practices of astrology.

Plato himself, in his work *Timaeus*, argued that the creator or *demiurge* had formed the human soul from the same material from which he made the stars. Plato describes the task the demiurge performed as he created the human divine souls that establish a one-to-one correspondence between souls on-earth and stars, and establishes the destinies of individuals:

> He (the demiurge) spake, and once more into the cup in which he had previously mingled the soul of the universe he poured the remains of the elements, and mingled them in much the same manner; they were not, however, pure as before, but diluted to the second and third degree. And having made it he divided the whole mixture into souls equal in number to the stars, and assigned each soul to a star; and having there placed them as in a chariot, he showed them the nature of the universe, and declared to them the laws of destiny, according to which their first birth would be one and the same for all.
>
> (Archer-Hind 1888, p. 28)

In Plato's philosophy, the entire world was a single living spirit or "world soul" and this mystical outlook, which favored pre-destiny and an intimate connection with the stars, was very compatible with the Eastern practices of astrology.

Aristotle also indicated that astronomical events could directly affect matters on the Earth, giving further support to the theory of astrology among the Greeks. His idea of "coming-into-being" and "destruction" which is described in the work "On Generation and Corruption" attributes these two major effects to the Sun's motion along the ecliptic, giving an authoritative basis for astrology for centuries to come.

By 400 BC, as a result of the adoption of Mesopotamian astrological techniques, the Greek planets had been renamed from their earlier naturalistic names that describe their appearance (Phainon, Phaeton, Pyroeis, Phosphoros, and Stilbon for Saturn, Jupiter, Mars, Venus, and Mercury, respectively) to names of gods that paralleled those of the Babylonian astrologers (Star of Cronos, Zeus, Ares, Aphrodite, and Hermes for Saturn, Jupiter, Mars, Venus, and Mercury, respectively).

By the first century BC, astrology had taken root across the Mediterranean world, all the way from Greece to Antioch and from Alexandria to Rome. The Astronomicon, a compilation of the techniques of astrology written Marcus Manilius and Vettius Velens in the first century AD, includes a complete description of the terminology and techniques of Greek astrology.

The basic tool of Greek and Roman astrology is an individual's horoscope. The horoscope was a chart of the positions of planets at birth, which was compared with the locations of zodiacal constellations, their relations to four cardinal points. The horoscope also included a ring of locations in the sky known as the "mundane houses," which were believed to correspond to aspects of an individual's life (which included enemies, friends, honors, travel, death, marriage). The chart was drawn relative to a horizontal line, which represented the horizon, with an eastern point known as the "ascendant" and a western point known as a "descendant." The addition of a transit point ("mid-heaven") and a nadir point ("Immum Coeli") divided the local horizon and sky into four quadrants, in which planets could be charted (Fig. 8.12). For the date of birth the ring of zodiacal constellations would appear at a fixed angle relative to the chart, with certain constellations at each of the four points, and various planets in each of the constellations and houses, depending on the timings of their orbits.

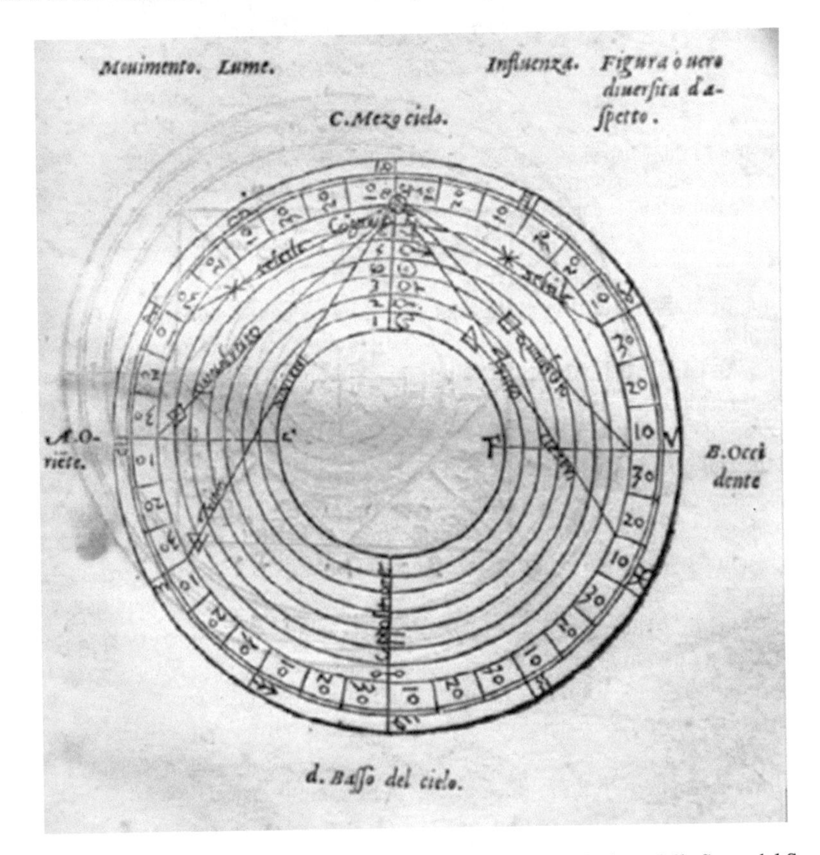


**Fig. 8.12** Diagram from a European work on astrology, "Annotationi sopra la lettione della Spera del Sacro Bosco," by Mauro of Florence (1571), which shows the main elements of the astrological chart developed by the Greek and Roman astrologers. The ascendant is shown (*horizontal line*) relative to houses (*outer ring*) and the planets are marked on concentric circles (image courtesy of the Claremont Colleges Special collections, Wagner collection)

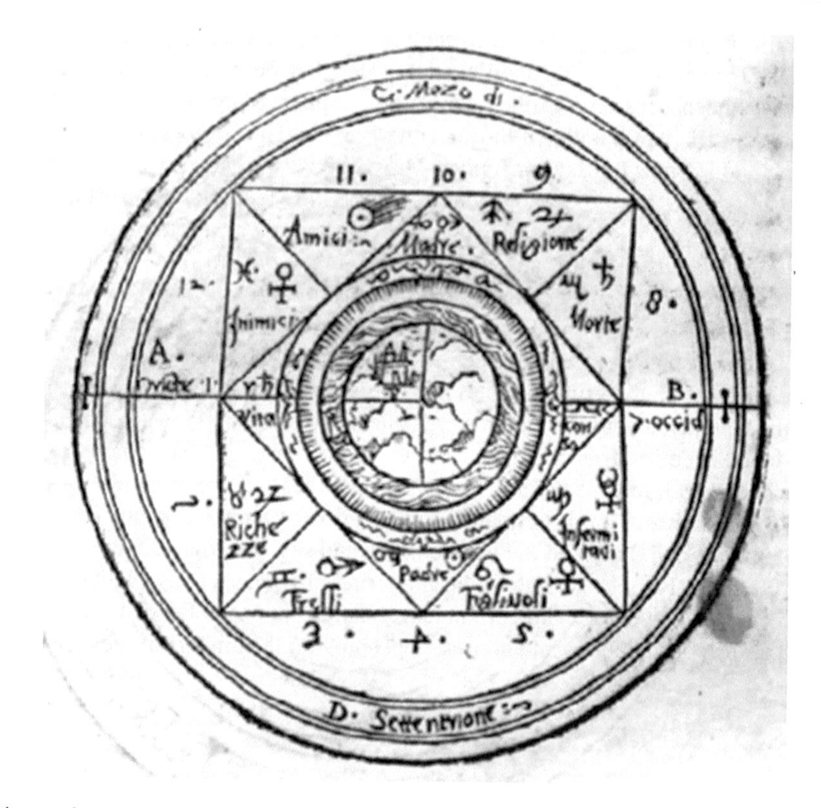


**Fig. 8.13** Diagram from a European work on astrology, "Annotationi sopra la lettione della Spera del Sacro Bosco," by Mauro of Florence (1571), which shows divisions of a horoscope into key elements of a person's life, with various planets ruling each of these divisions (image courtesy of the Claremont Colleges Special collections, Wagner collection)

A complete list of houses, starting at the ascendant, includes the categories "Enemies," "Friends," "Honors," "Travel," "Death," and "Marriage." Then below the horizon from west to east included the houses of "Health", "Children," "Parents," "Siblings," "Wealth," and "Life" (Whitfield 2001, p. 44). There is no apparent logic or explanation of the assignment of these houses, but they probably were adapted from some of the many ancient Babylonian and Greek astrological omens. The most important part of the diagram is the eastern point, which was known as the *horoskopos* or "hour observer."

The relative positions of planets, stars, the Sun and the Moon determined more esoteric quantities such as the "*exaltation*," "*decan*," and "*lots*" of a horoscope. The *exaltation* referred to the coincidence of a house and a constellation within the chart, while the *decans* were a set of stars rising helically at each date of the year (just as in ancient Egypt), and *lots* were mathematical constructions of positions on the chart thought to have great symbolic meaning. Examples of *lots* include the "Lot of Fortune," which was a point at an angle above the ascendant equal to the angle between the Moon and Sun, and the "prorogator," a location on the chart thought to predict the length of the subject's life. The final piece of information in a horoscope was the connection between the planets (which could form shapes such as triangles or the trine, squares, sextiles, or oppositions) and the locations of these planets relative to the horizon and dividing lines in the " mundane houses (Fig. 8.13)."

Additional intricacies of astrology described by Manilus include the *paranatellonta* and *dodecatemoria*. The *paranatellonata* were additional celestial objects thought to have additional powers, such as the Hyades, which if it appears near the ascendant is thought to indicate a particular omen, such as "those born at this time do not enjoy peace and quiet, but they seek crowds of people and the bustle of affairs" (Whitfield 2001, p. 50). The *dodecatemoria* were twelve small segments of about 2.5° where the chart was especially significant (the name comes from "dividing into 12 parts"). Interestingly, astrology in the time of Manilus did not depend on the Sun's position when a person was born, as is common in twentieth century popular astrology, but instead on the constellation of the ascendant. While ancient astrology offered complex machinery for calculating a person's fortune, it was much less specific about the exact mechanism that explained the operation of astrology. The basic underlying principle came down to lists of omens (much as in the case of Babylonian times) and aesthetic judgments about the placement of objects within a chart. Triangular shapes within a chart were judged to be harmonious, while shapes between other objects, such as squares and hexagons, were considered less auspicious. Each planet was thought to radiate a "sphere of virtue" of about 10° that produced effects when it coincided with other objects or key parts of the sky (Whitfield 2001, p. 55)

With the work of Ptolemy (AD 130–170), astrology was placed on something of a scientific footing, at least by the understanding of his time. His two key works, *Tetrabiblios* and *Almagest*, explored links between astrology and the known elements of the universe. Ptolemy perfected the system of prediction of planetary positions, using the most accurate available planetary data, and a complete arsenal of epicycles, deferents, and eccentrics to describe precisely the locations of planets as needed for astrological charts. The astronomical contents of the Almagest provided a contribution of lasting value for both astrologers and astronomers, and survived for over 1000 years as the leading text in astronomy.

Ptolemy's cosmology, much like those of the early pre-Socratics, mixed earth, air, fire, and water in different proportions on the Earth to produce changes. However, Ptolemy extended his system of elements to the heavenly bodies and correlates each planet with one or two of the properties of the elements on the Earth. Heat from the Sun, moisture from the Moon, coolness from Saturn, heat and moistness from Jupiter, and dryness and heat from Mars were all thought to change properties on the Earth depending on the positions of the planets. Additional connections were made between astronomical objects and parts of the body, providing a connection between astrology and medicine.

While it may be hard to imagine today, astrologers in Ptolemy's time imagined Saturn commanding the spleen, the right ear, and the bones, while Jupiter ruled over the lungs and arteries (Fig. 8.14). Astrological forecasting at the time of birth was thought to predict the health of a subject, partly based on the medical effects of planets on organs in the baby. Astrological charts at the time of sickness could be used to prescribe particular herbal remedies to counter the effects of "malevolent" planets. Additional connections between heavenly bodies and human development are put forth in the Tetrabiblios, resulting in the concept of the "seven ages of man"—each part of life being ruled by a different celestial object. The ages of a person are ruled in turn by the moon (0–4 years), then Mercury (5–14 years), then Venus (14–22), then the Sun (22–41), then Mars (41–65), then Jupiter and Saturn for the final years of life (Whitfield 2001, p. 65).

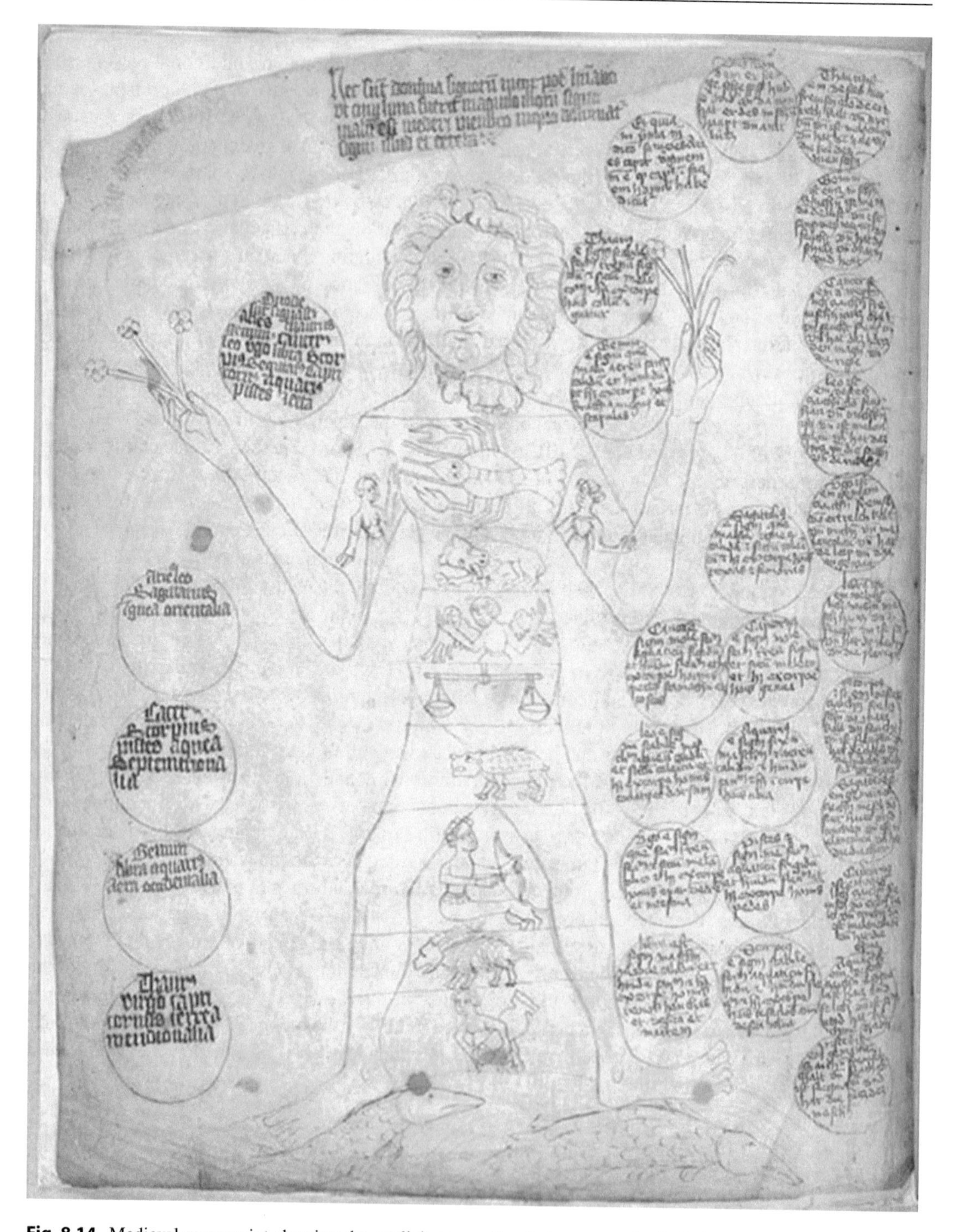

**Fig. 8.14** Medieval manuscript showing the explicit connection between astrological signs and body parts. Within medieval astrology was a planet-based medicine system, whereby each astrological sign "held sway" over different organs (from the "Encyclopedic manuscript containing allegorical and medical drawings,", south Germany: ca 1410; image part of Library of Congress "Heavens" exhibit at http://www.loc.gov/exhibits/world/images/s128.jpg)

It is also important to remember that astrology was not only the province of Babylonian, Greek, or Roman culture. Nearly all of the ancient cultures of the world had some form of astronomy, and a strong motivation for many of these cultures was to tell the future from the stars and planets. In the Aztec and Mayan cultures, astrology was integrated with the timekeeping and calendrical rituals, as large numbers of omens were associated with each of the named days in the 260-day cycle, with additional associations with astronomical events such as comets, eclipses, and meteor showers. The Inca used stars to help perform weather forecasts, based on the visibility of Pleiades stars or dark cloud constellations.

Among the present-day civilizations of modern times, additional forms of astrology outside of the Babylonian astrology are also recognized. Two of the most advanced and widely practiced systems are from India and China. Like the astrology of Europe, the Indian and Chinese systems developed from early scientific examinations of the sky and were the precursors of more modern astronomical investigations. And like the astrology of Europe, Indian and Chinese astrology is still practiced and used to set important dates such as weddings or to time the best date for large business dealings. As such the Indian astrology and the Chinese astrology are also part of the archaeoastronomy of modern civilization.

## Indian Astrology

India can claim a much more ancient astronomical and astrological tradition than Mesopotamia, which is rooted in the Vedic teachings that are dated to 1500 BC and perhaps earlier. The Vedic text known as Atharva Veda includes many different omens and purification rituals that are based on astronomy. Additional astronomical and cosmological ideas are put forth in the Rig Veda. The early Vedic tradition, like Mesopotamian and Greek and Roman tradition, included observations of clouds, haloes around the Sun and Moon, comets and meteors in the list of omens (Whitfield 2001, p. 32).

Many of the principles of Indian astrology were compiled in an ancient text known as the Yavanajataka, which was written by Sphujidhavaja in AD 270. This work includes some translation of Greek astrology literature into Sanskrit, but also mentions many of the uniquely Indian contributions to astrology. Two major differences between Indian and Mesopotamian astrology are the importance placed on stellar associations with the Moon, known as *naksatras*, and the locations of the two lunar "nodes," which point toward particular stellar regions. The naksatras, discussed earlier, are a group of 27 constellations that serve as landmarks of the sky, based on the 27.3-day lunar sidereal month, against which the positions of the Moon or planets or lunar nodes are charted.

The "nodes" of the Moon are locations where the orbit of the Moon crosses the ecliptic plane. Since the Moon's orbit is inclined by approximately 5° above the ecliptic plane, the intersection of the Moon's orbit and the ecliptic defines a line, which points in a discernable direction (Fig. 8.15). This "line of nodes" was known to Babylonian and Greek astronomers and was the basis of prediction of eclipses, which are known to occur only when the full or new moon phase coincides with the line of nodes. The line of nodes points toward the sun approximately every six months, as the earth (and moon) orbit around the sun. However, since the "line of nodes" precesses 18.84° each year, the timing of possible eclipses during an "eclipse season" is 9 days short of six months later or 173.3 days. During each of these "eclipse seasons" a full or new moon brings the possibility of either a lunar or solar eclipse, respectively. A complete cycle of the line of nodes across the sky defines the "Draconic Year" which is about 18.14 years.

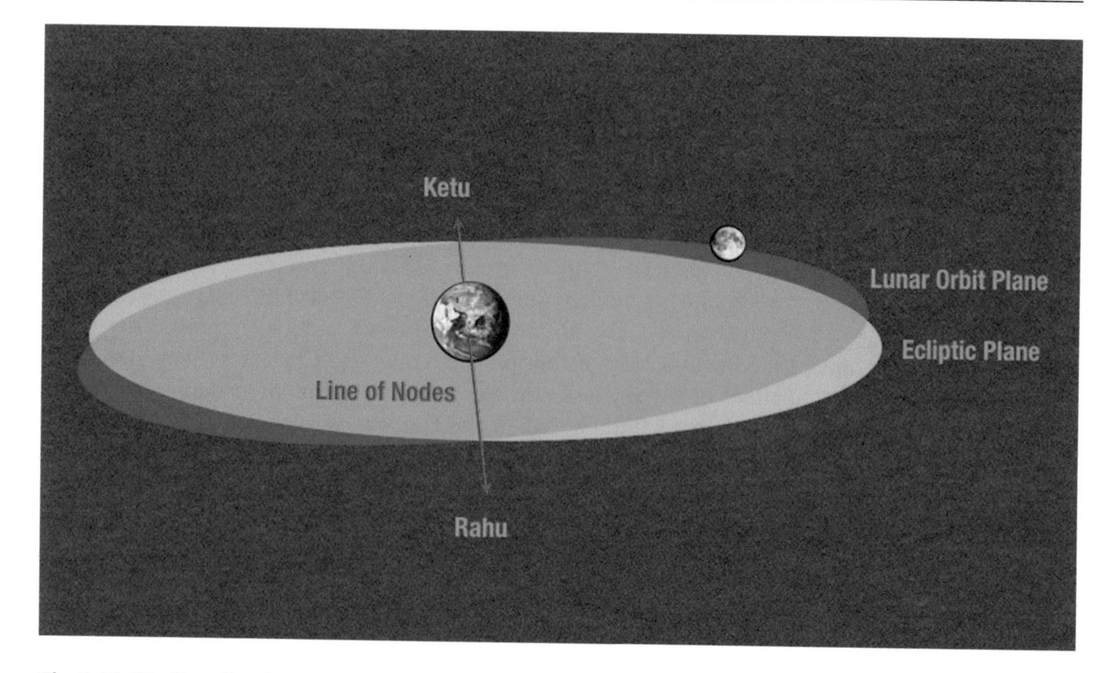


**Fig. 8.15** The line of nodes, as defined by the intersection of the slightly inclined (5°) lunar orbital plane and the ecliptic plane. When the Sun is in the direction of the line of nodes, which happens twice a year, an "eclipse season" results. In Hindu astrology the directions pointed to by the line of nodes are given names "Rahu" and "Ketu" and are important locations in their system (figure by the author)

The Indian or Vedic astrology makes use of the two node points, which are known as *Rahu* and *Ketu* for the ascendant and descendant nodes, respectively (Ojha 1996). Like the medieval "lot," the node points are variable locations on the zodiac defined mathematically and not identified with locations of planets or stars.

Planets, known collectively in Hindu astrology as *grahas* (for "grasp"), are charted relative to the node points and naksatras to comprise an astrological chart. Some effort is made to provide accurate astronomical information to help build an accurate chart, with attention to the ascendant, which in Sanskrit is known as the Lagnam, which is used to define a set of "houses" relative to this point, much as in Babylonian astrology.

Many additional time units developed by Indian astronomers are employed to provide additional complexity and detail in developing a chart. The date of birth is set within a 60-year Jupiter cycle, a 12 lunar month calendar, the phase of the Moon (known as the "Tithi"), and the time of birth expressed in the units of "Ghatis" and "Palas" (corresponding to units of time of 24 min and 24 s, respectively; Ojha 1996, p. 27). The position of the Moon and Sun defines an additional location known as the "Yoga" which is the result of adding the longitudes of both bodies. All of this information is placed on a chart, traditionally a grid much in appearance like a "tic-tac-toe" grid, with planet positions and times indicated (Fig. 8.16).

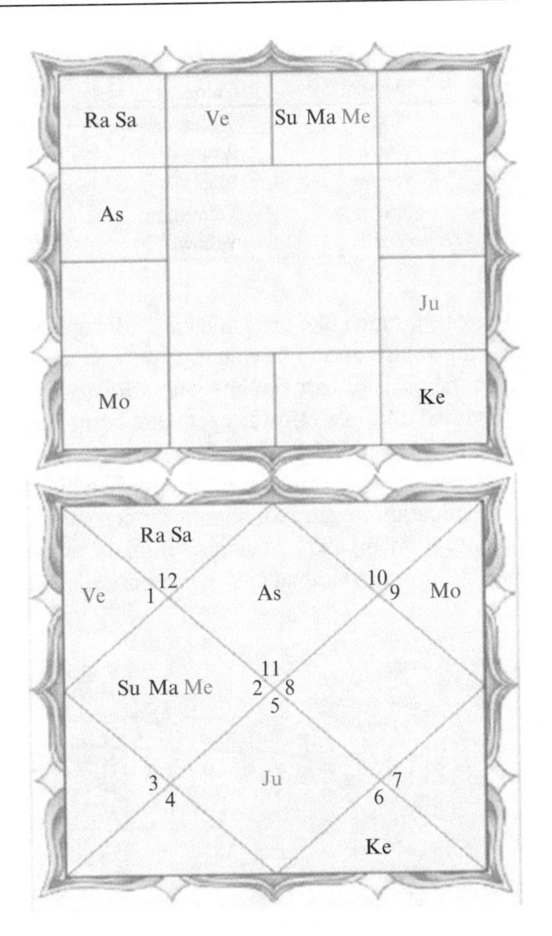


**Fig. 8.16** Southern (*left*) and northern (*right*) Indian astrological charts. The diagrams are made to show the locations of planets within the different zodiac signs in fixed locations for the southern notation (*left*), while the northern notation is a schematic of the sky at the moment of birth, with various planets marked according to their positions on the horizon (*image source*: wikimedia.org)

Once a chart is constructed, the Indian astrologer can then read into the various combinations of the above parameters particular omens based on long lists compiled over centuries. Like the Mesopotamian lists of omens, the particular origin of these lists is not known, and there is not much information on the particular mechanism by which the stars, node points, and other details affect human life on the Earth. Nevertheless, most Indian couples contemplating marriage will consult an astrologer to be sure their birth dates, and wedding dates provide auspicious omens. Likewise, a baby born in India will often have a chart drawn that will offer many predictions about the child's future.

## Chinese Astrology

Chinese astrology mixes elements from traditional Chinese folklore and philosophy with some of the calendrical elements of Chinese astronomy. Within the Chinese tradition is an emphasis on "five-ness" that divided the world into five elements and associated these elements with five colors, animals, and the "five luminaries" or planets. A list of these associations is shown in Table 8.1 and forms the basis of an astrological system.

**Table 8.1** Chinese planet associations with elements and colors

| Planet | Element | Color | Animal |
|---|---|---|---|
| Venus | Metal | White | Tiger |
| Jupiter | Wood | Azure | Dragon |
| Mercury | Water | Black | Tortoise |
| Mars | Fire | Vermillion | Bird |
| Saturn | Earth | Yellow | Dragon |

The Chinese astrological forecast, much like the Indian and Mesopotamian forecast, depended on the moment of birth to define the future of an individual. Chinese astrology tried to assess an individual's birth date and time on the basis of associations with *yin/yang* and one of the five elements. The birth year located an individual within a 60-year cycle that combined the 12-year Jupiter-based *Sheti* cycle (with the familiar zodiac animals) with associations with each of the five elements. Subdivision of the year by lunar month and even hour of birth gave a higher resolution to the forecast and were the basis of the determination of the individual's "inner animal" and "secret animal. The detailed combination of yin/yang, element, and associated "animals" was put together by the astrologer and used to predict properties of the individual and their future life (Eberhard 1986; Lau and Lau 2007) (Figs. 8.17 and 8.18).

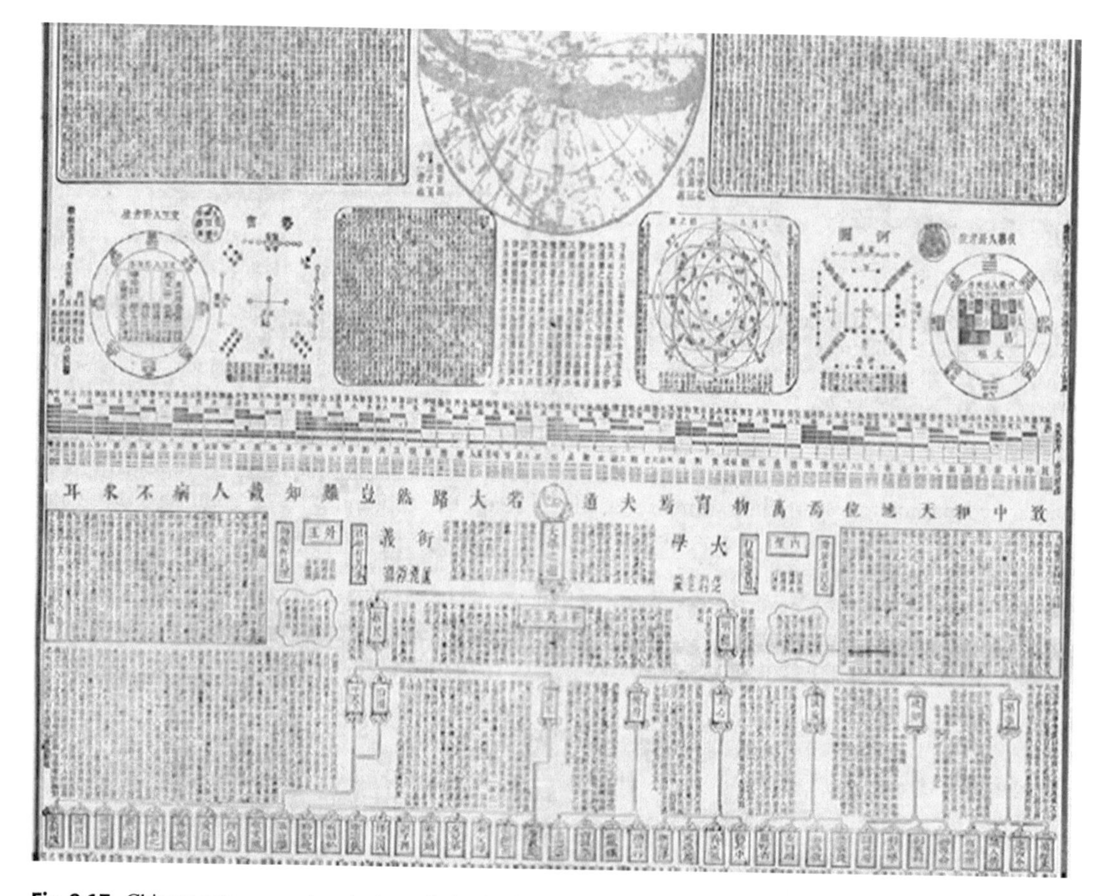

**Fig. 8.17** Chinese astronomy played a key role in shaping the various forms of philosophy and in turn provided inputs for a form of astrology. that incorporated elements of natural philosophy and mysticism. This figure shows a compendium of Chinese knowledge, including astronomy (red star map at *top*), and the "eight phenomena" which describe motions of the Moon and planets, as well as basic interactions of elements on the Earth, and a summary of Taoist and Confucian thought (Library of Congress, Map Division, from Shaoxing, 1722, "San Cai iy guan tu" by Lu Wefan)

**Fig. 8.18** Tibetan Buddhism included many aspects of early Chinese thought in its divination, including the 12 traditional animals of Chinese astrology (shown in the *central wheel*), Taoist hexagrams (shown in the petals of the *central circle*), directional animals, and color symbolism (from *Srid pa ho* (Divination Chart), Tibet, late twentieth century; image from Library of Congress, "World Treasures" exhibit)

## European Astrology from Medieval to Modern Times

After Ptolemy, astrology adapted to the rise of Christianity and also was transformed by Islamic astronomy. The basic machinery of Ptolemy was supplemented with the works of Islamic astronomers such as al-Kindi (801–866), who commissioned translations into Arabic of the Greek works of Plato, Aristotle, and Ptolemy, and Abu Mashar (787–886), who provided additional computations of periodicities of conjunctions of Jupiter and Saturn that he believed provided a cyclical theory for explaining history.

The Islamic astronomers also perfected the use of the astrolabe, which was able to provide accurate measurements of time during evening hours, to predict the positions of constellations at any latitude, and to provide measurements of planetary positions. The astrolabe worked much as a modern planisphere and provided a beautifully carved top layer or *rete* that rotated against a background grid of coordinates known as a climate plate. The top layer included points to indicate positions of bright stars, which could then be located on the climate plate to show their positions at various times during the night. Conversely, a sighting of a stellar position could be dialed into the astrolabe and used to measure the hour of night from the climate plate. A convenient sighting device, known as the *alidade*, was attached to the two disks and could be pointed at bright stars, and a ruled outer edge would provide readings of their positions (Fig. 8.19).

Islamic astronomers were experts in the sighting of stars to enable accurate determination of one's position on the Earth, for both navigation and prayer to Mecca. The aim of Islamic astrology and astronomy was less concerned with forecasting the future of a ruler or individual and more concerned with interpreting larger trends in world history. However, by the thirteenth century, Islamic astrology was in decline and in conflict with prevailing religious beliefs.

**Fig. 8.19** A display of Persian and Arabic astrolabes from the fifteenth to the eighteenth centuries, from the Cambridge Museum of Science and Technology. Such devices were used by Islamic navigators to sight stars and to find the time and directions anywhere on the Earth from the stars (photo by the author)

Conflicts between astrology and Christian theology also threatened the demise of astrology in the European Medieval period, but philosophers such as Origen and Boethius helped revive astrology in the twelfth century. One key event was the conjunction of planets in 1186 which was the subject of study from "all the world's prognosticators, Spanish and Sicilian, Greek and Latin," according to the medieval English writer Roger of Hovedon. Astrology seemed at odds with Christian doctrines, but clever arguments by medieval authors were constructed to justify astrology as complementary to Christianity. One interesting argument, put forward by the philosopher Bernard, is that the stars had to have a function, since God created them, and this function was to foretell the future. In the words of Bernard:

> I would have you behold the sky, inscribed with a multiform variety of images, which like a book with open pages containing the future in cryptic letters, I have revealed to the eyes of the more learned.
> (Whitfield 2001, p. 98)

Another key event for medieval astrology was the arrival of the plague, which offered many opportunities for *ex post facto* astrological interpretations. In 1348 the medical faculty of the University of Paris explained how a triple conjunction of the planets Mars, Jupiter, and Saturn 3 years earlier induced the plague (on March 28, 1345), based on the results of Jupiter's "warm, humid vapors," which were "set fire" by Mars (Fig. 8.20). In the words of Paris Medical Faculty, the conjunction caused "pernicious corruption of the surrounding air, as well as other signs of mortality, famine and other catastrophes" (Whitfield 2001, p. 113). While these medieval astrologers could not predict the arrival of the plague, they could certainly explain events after the fact!

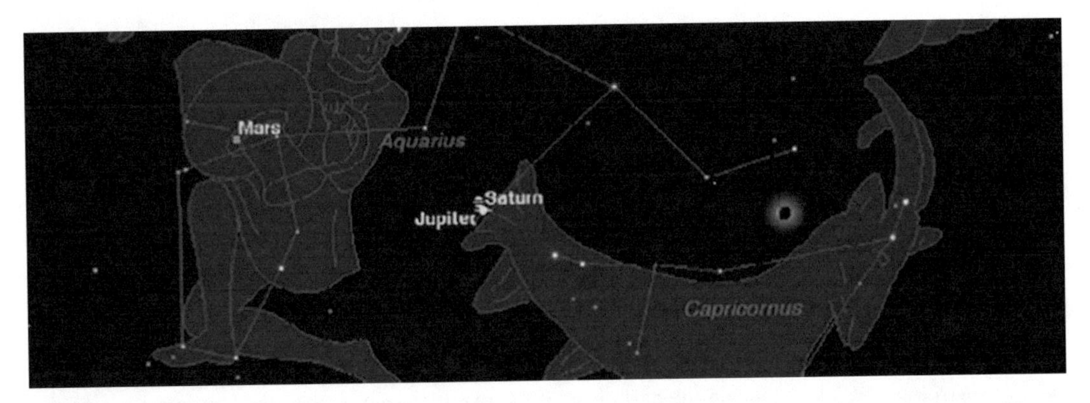


**Fig. 8.20** The conjunction of March 28, 1345, which in the view of some medieval astrologers brought the plague to Europe when Jupiter's vapors were "set fire" by Mars (original figure by the author using the Voyager IV computer program)

Astrology also gives us the root of our word "influenza" or the "flu" based on the ancient belief that disease can be induced from the influence of celestial bodies. Both astrological medicine and alchemy received widespread interest and support in this period, as medieval scientists struggled to explain the turmoil of their time as the results of planetary effects. Each of the planets had an association not only with the early Greek elements but also with metals and with humors of the body.

The widespread belief in astrology influenced many of the major culture figures in medieval and Renaissance Europe. Chaucer was thought to include astrologically based allegories in some of his Canterbury tales, while Shakespeare made numerous astrological references in his works:

> It is the stars,
> The stars above us that govern our conditions,
> Else one self mate and mate could not beget
> Such different issues.
> Shakespeare, King Lear,
> Act IV, Scene III

Even leading minds such as Tycho Brahe and Johannes Kepler were both practicing astrologers during part of their careers. However, Kepler moved away from astrology as he began to discover what he believed were the more important effects of musical harmony and the cosmic geometry of celestial bodies (Whitfield 2001, p. 181).

Just as Ptolemy predicted a decline within the "ages of man," the age of astrology was destined for a period of decline. By 1674, the leading astronomers of the time, such as Flamsteed, were actively attacking astrology as obsolete. An unpublished preface to "Hecker's Tables" condemned the "vanity of astrology" as well as its inaccuracy, inconsistency, and lack of predictive power. Isaac Newton, while fascinated with alchemy, was indifferent to astrology and saw no use for the practice, perhaps since his new physics had much greater predictive power than any horoscope. Astrology's influence in Europe faded steadily during the seventeenth and eighteenth centrles and appeared headed for extinction.

Instead of being a practice for royalty, aristocrats, or leading scientists, astrology by the 1780s was most widely found in popular magazines for poorly educated masses. Magazines such as "*Astrologer's Magazine*" (began in 1783) and "*The Conjurer's Magazine*" kept astrology in print during the period, as did selected publications such as Sibley's (1790) work "*Complete Illustration of the Science of Astrology*." However, the intellectual respectability of astrology never recovered its luster, and astrology from the eighteenth century onward had to survive outside of the mainstream of scientific and intellectual discourse.

## Why Do People Believe Astrology in the Twenty-First Century?

The decline of astrology seemed assured with the rise of science. And yet somehow astrology seems to have survived to this day and by some measures is enjoying something of a revival. What are the predictive powers of astrology? And why do people still believe in this arcane science, after so many thousands of years? Clearly science has moved well beyond theories of elements composed of moisture, earth, fire, and air—we now have a periodic table of the elements and know precisely how atoms are constructed. Medicine too no longer attaches astrological omens to each body part, assuming malfunctions of a spleen are from Saturn's influence—we now have diagnostic equipment, surgery, and pharmaceuticals that completely replaced the medieval medicine based on astrology. As so many aspects of our technology have advanced, why has the belief in astrology survived and not been replaced by the clearly superior predictive and practical power of medicine, physics, and astronomy?

To answer this question, perhaps we should begin with a description of some attempts to determine whether astrology "works." During the early part of the twentieth century, several attempts to analyze astrology on a scientific basis were made. An English journal of astrology was founded in 1895 known as "*Monthly Astrology*" which contained several prognostications on the coming century and articles on the philosophy of astrology. The journal confidently predicted that no European war would begin up until June 1914, when the editor was suddenly convicted for "fortune-telling." By August 1914, World War I had begun and the editor, just released from his imprisonment, was able to revise his predictions to show that war indeed was what the stars were showing.

One German journal of astrology, known as *Kosmobiologie*, was founded in 1928 and attempted to provide scientific studies to validate astrology. A Swiss astrologer, Karl Ernest Krafft, attempted to show the validity of astrology by claiming that statistical trends existed in the horoscopes of thousands of people who were in the same professions and their birth signs. Despite the fact that Nazis persecuted astronomers early in their regime, the Nazis later employed several astrologers (including Krafft). Krafft himself was not successful in convincing the Nazis and eventually was imprisoned and died in a concentration camp. Further statistical work by the French statistician Michel Gauquelin re-analyzed Kraffts work and found that the methodology was invalid and no convincing correlation existed between sun signs and professions (although some correlations were seen between professions and the "ascendant" signs) (Whitfield 2001, p. 201). Despite the mixed result of scientific analyses of astrology (and the unfortunate results for some of its practitioners in the early twentieth century), many people are still convinced of the validity of astrology.

University and college students studying astronomy often are unclear about the difference between astrology and astronomy and come with varying degrees of belief in astrology. For this reason, the Astronomical Society of the Pacific (ASP) has provided its members with an "Astrology Defense Kit" useful for new instructors of astronomy. In this kit are several arguments and questions to ask believers in astrology, along with some class activities. Below we summarize some of the arguments put forth by ASP executive director Andrew Fraknoi in the form of "embarrassing questions" about astrology:

> Why do horoscopes use the moment of birth and not conception? If astrologers can foretell the future, why aren't they richer? If the stars exert such powerful forces, how can the thin layer of the womb shield a child from them? How can horoscopes be accurate without knowledge of the outer planets Neptune and Uranus? Why do planets have a larger force on a baby than objects closer? (i.e. the gravitational pull of a doctor delivering a baby is six times larger than that of Mars). How is it that astrological force does not depend on distance, and if this is so, why is there no influence from stars and galaxies?
>
> (Fraknoi and Schatz 2000)

Another line of argument against astrology might focus on the diversity of outcomes of the large number of people who are born on the same day—certainly if astrology is valid, any twins born should share the same fate, which is clearly not the case for countless twins. Likewise, the astrological "sun signs" of modern twentieth century astrology are based on dates which are obsolete due to celestial precession. If you were born within the dates of most astrology columns, your sign is actually off by one position due to the precession of the equinoxes. For example, the dates of September 2–20 for "Virgo" actually are the dates for which the Sun is in the constellation Libra in the twenty-first century. Some students are upset to learn that all of the astrology columns they were reading for most of their lives were for the wrong sign!

The other strike against astrology, which prevents it from becoming a science, is that it has no mechanisms for determining validity and rejecting material when it is shown to be untrue. For this reason, multiple schools of astrology exist both within North America and Europe and in Asia. How can all of these types of astrology be valid, when they often provide contradictory predictions? Some of these other astrologies, such as the Indian (Vedic) astrology and Chinese astrology, had important early roles in developing the science of astronomy and yet also survive to give independent and often different systems for predicting the future. Can these all be right? Or should astrology, like alchemy, and ancient medicine based on herbs, leeches, and humors, but put aside in favor of modern science? In the words of the late Carl Sagan:

> Think of how many people rely on these prophecies, however vague, however unfulfilled, to support or prop up their beliefs. Yet has there ever been a religion with the prophetic accuracy and reliability of science?
>
> (Sagan 1996)

## References

Archer-Hind, R.D. 1888. *The Timaeus of Plato*. London: McMillan.

Cramer, F.H. 1954. *Astrology in Roman law and politics*. Philadelphia, PA: American Philosophical Society.

Eberhard, W. 1986. *A dictionary of Chinese symbols: hidden symbols in Chinese life and thought*. New York: Routledge & Kegan Paul.

Fraknoi, A., and D. Schatz. 2000. *More universe at your fingertips: an astronomy activity and resource notebook*. San Francisco, CA: Project Astro, Astronomical Society of the Pacific.

Heldmore, E. 2013. Seeing the light. *New York Times, Travel*, May 9.

Kirby, D., Smith, K., and M. Wilkins. 2009. Roadside America. From http://www.roadsideamerica.com/.

Kosky, J.L. 2013. Contemplative recovery—the artwork of James Turrell. *CrossCurrents* 63: 44–61.

Lau, T., and K. Lau. 2007. *The handbook of Chinese horoscopes*. New York: HarperResource.

Ojha, G.K. 1996. *Predictive astrology of the Hindus*. D.B. Taraporevala: Bombay.

Sagan, C. 1996. *The demon-haunted world: science as a candle in the dark*. New York: Random House.

Whitfield, P. 2001. *Astrology: a history*. New York: Abrams.

# Chapter 9

# The Development of Modern Cosmology

*There are great men who are great men among small men, but there are also great men who are great among great men ... Napoleon and other great men of his type, they were makers of empires, but there is an order of men who get beyond that. They are not makers of empires, but they are makers of universes, and when they have made those universes, their hands are unstained by the blood of any human being on earth...*
George Bernard Shaw, in a speech of 28 October 1930 (Danielson 2000, p. 392).

*The larger the island of knowledge, the greater the shoreline of wonder*
(Ralph W. Sockman)

The development of modern cosmology was the result of the work of dedicated scientists over several centuries, who gave enormous sacrifices to bring us the perspective we have today. Our generation is fortunate to have answers to most of the questions about the physical nature of the universe. As a species, we have seen atoms, stars, galaxies, and clusters of galaxies, and know how to determine their motions, temperatures, gravitational fields, and how to predict their future physical state using the laws of physics. We have also peered with giant telescopes to the edge of the universe, and this exploration has led us to even more questions—what is the universe made of? Where did the Big Bang come from? Individuals who have made these questions their life's work are in the words of Bernard Shaw, "makers of universes." The process that makes it possible is the scientific method. In this chapter, we will examine what it is about the scientific method that makes it possible to describe the history of our universe based on physical laws valid in all locations at all times. We will also see how key individuals such as Copernicus, Brahe, Kepler, Galileo, and Newton applied this technique to make the first physical cosmology, and how observers like Herschel, Huggins, Hale, and Hubble applied the latest tools of science and expanded our vision to reach to the edge of the universe.

## Scientific Cosmology

Open up most books on modern scientific cosmology from the twenty-first century, and the title of the book will be simply "Cosmology" or an "Introduction to Cosmology." Have you ever wondered why we now have just "Cosmology" and not "North American Cosmology" or "French Cosmology" or "Australian Cosmology?" As the lack of qualifier might imply, modern scientists believe that the tools and technologies at their fingertips have put them in contact with answers that are independent of time and place.

We no longer talk about Chinese or American or French cosmology, just cosmology. We no longer talk about twenty-first century cosmology, since we are living in the twenty-first century. But is our knowledge of the universe independent of our culture and our place in history? Or is it an outgrowth




B.E. Penprase, *The Power of Stars*, DOI 10.1007/978-3-319-52597-6_9

of our belief system? How do we know we have found "the cosmology" and not just "a cosmology" for our present level of understanding and belief?

A modern scientist enjoys many tools (including the scientific method itself) for selecting theories that conform to experimental realities. Since these tools (telescopes, atomic clocks, particle accelerators) can be duplicated on any part of the Earth or any place in the universe, and since the laws of physics are the same everywhere, our results should be independent location on the Earth. This explains the lack of "French" and "Russian" cosmologies—science has created a universal understanding that transcends geography and culture.

It should be pointed out that any scientific theory is subject to continuous refinement and that our modern physics has produced exquisite precision in measuring and predicting the motions and state of matter and in exploring distant reaches of space and measuring the ages of stars, the masses of galaxies, the motions of matter around black holes, and countless other wonders. Four basic principles are needed for one to proceed with a scientific cosmology, that are listed below:

**Four principles of the scientific method:**

1. We can learn the laws of physics by applying experiment and mathematics to derive reproducible results.
2. Experiments enable development of new theories that can be tested by further experiment.
3. Any theory which does not provide predictions that match experimental results and which is not reproducible is discarded.
4. The laws of physics are universal and apply in all places and times.

The last of these is sometimes known as the "cosmological principle" and is the bedrock principle for modern scientific cosmology. We have to assume that remote galaxies and stars have the same physical components (atoms, protons, and electrons) and obey the same physical laws (gravity, electricity and magnetism, strong and weak force) as the matter here on Earth.

Scientists also employ a version of "Ockham's razor," named after a fourteenth century logician named William of Ockham and sometimes also known as the *lex parsimoniae* or "law of parsimony." Ockham's razor allows scientists to select between theories when definitive proof or disproof is not an option. The razor allows one to "cut out" all those which are not the simplest, and keep only the most succinct, and simple of the possible theories.

Science, while dealing in facts, also has a strong aesthetic sensibility that selects theories that are beautiful. Many have dismissed the scientific method as "reductionist" and have missed the great beauty in the scientific approach to the universe. Does a scientific approach to the universe decrease our appreciation of nature? One excellent answer comes from Richard Feynman, one of the leading physicists of the twentieth century (Fig. 9.1).

> Poets say science takes away from the beauty of the stars — mere globs of gas atoms. Nothing is "mere". I too can see the stars on a desert night, and feel them. But do I see less or more? The vastness of the heavens stretches my imagination — stuck on this carousel my little eye can catch one-million-year-old light. A vast pattern — of which I am a part… What is the pattern or the meaning or the why? It does not do harm to the mystery to know a little more about it. For far more marvelous is the truth than any artists of the past imagined it. Why do the poets of the present not speak of it? What men are poets who can speak of Jupiter if he were a man, but if he is an immense spinning sphere of methane and ammonia must be silent?
>
> (Feynman 1970)

**Fig. 9.1** Richard Feynman, one of the builders of twentieth century physics sharing his joy of physics with Caltech students (*left*) and Albert Einstein (*right*) observing at Mount Wilson with Edwin Hubble and Walter Adams, where the expansion of the universe was discovered. Both men were profound and original thinkers that helped to shape our modern conceptions of matter, space, and time (courtesy of the archives, California Institute of Technology)

Scientific research is also informed by the moral values of the scientist as one is required to remain honest in the gathering and assessment of data. This effort can be not only very challenging but also very rewarding. In the words of Albert Einstein:

> The ideals that have lighted my way, and time after time have given me new courage to face life cheerfully, have been Kindness, Beauty, and Truth. Without the sense of kinship with men of like mind, without the occupation with the objective world, the eternally unattainable in the field of art and scientific endeavors, life would have seemed empty to me. The trite objects of human efforts - possessions, outward success, luxury — have always seemed to me contemptible.
>
> (Einstein 1933)

Science offers verifiable proof that something is true. This certainty comes from the ladder of proven facts and known assumptions, much as a mountain climber is able to scale an extremely high peak given well-tested steps, rope segments, and belaying points. Like a mountain climber, a scientist is able to gain a view "from the summit" that is exhilarating. Unlike a legal trial (which also offers examination of evidence and derives a verdict), a scientific investigation does not rely on eyewitnesses, memory, or conjecture, but instead relies on reproducible and objective measurements. It is important also to note that science research is almost as much about asking questions than it is about answers, and more about venturing out into what is not (yet) known—than a curation of facts.

The dynamic and exploratory nature of science is driven by a form of "knowledgeable ignorance" which in the words of Stuart Firestein (2012), "gets us into the lab early and keeps us up late" and gives us a taste for the "widening circle of ignorance" which "frames the unknown." The quest for the unknown also leads us to the awareness that science is never finished—and that for every answer science uncovers, several more questions are asked. In the words of R.W. Stockman—"The larger the island of knowledge, the longer the shoreline of wonder."

Despite the many successes of science and technology, too many modern citizens trying to learn about the universe (including some of the practicing astrophysicists!), this is an unsettling time. Most non-scientists have not seen the experimental data of physics and astronomy but instead assume on faith that the scientists working with the data have a clear, unambiguous, and exact understanding of the composition of the universe. Indeed, many of the conclusions of modern cosmology do result from exacting measurements of galaxy motions and cosmic background radiation. However, the interpretation of these results is less exact—while it is widely agreed that the fractional composition by mass of the universe is known, the exact nature of the mass that comprises most of the universe is currently a matter of speculation.

To the best of our knowledge, the universe is composed mostly of dark matter and dark energy (95% or more), along with only a sprinkling (5% or less) of ordinary matter like protons and electrons (Fig. 9.2). Our current scientific understanding cannot explain what dark matter and dark energy are, only that they exist in proportions of about 25% and 70% of the mass of the universe, respectively! It is also humbling to know that all the stars, galaxies, and visible matter in the universe account for a trifling 0.5% based on current observations.

This is an unsettling situation indeed and may cause some to wonder—is our modern scientific method successful? And how did we get to our present (lack of) understanding about the universe? As we will see in the following sections, the evolution of a scientific cosmology necessarily involves asking questions and admitting the limits of one's knowledge. And despite the unanswered questions, the scientific approach has given us a nearly comprehensive understanding of the formation and composition of the Earth, solar system, and galaxies around us. The process of this discovery is a very human process, which involves a "scientific community" of all of the nations of the Earth.

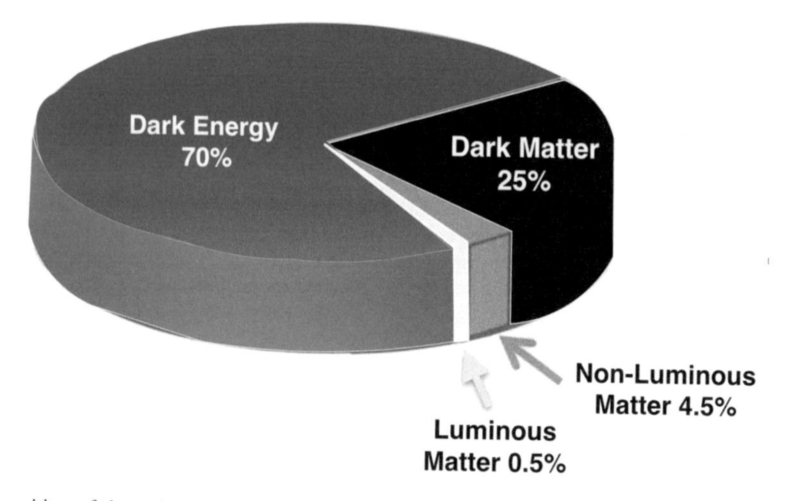


**Fig. 9.2** Composition of the universe, as determined by twenty-first century astronomy and cosmology: 96% of our universe is composed of unknown materials, such as "dark energy" and "dark matter" while only 5% is made of ordinary matter (protons, neutrons, and electrons) and only 0.5% coming from visible matter from the stars and galaxies that we see (*image by the author; based on* Wikimedia commons—https://commons.wikimedia.org/wiki/File:Matter_Distribution.JPG)

## Dark Matter Experiments

A prime example of the global community of scientists is the search for the nature of dark matter, which includes dozens of massive experiments. The presence of dark matter in outer space suggests that billions of the particles pass through every square meter of Earth each second, and these experiments hope to find these particles before they return to space. Scientists have been looking in laboratories on the surface of the earth, and kilometers beneath the surface for these particles. The LZ experiment, which uses 7 tons of liquefied Xenon, has been placed 4850 feet underground in a mine in South Dakota, USA, to detect interactions between dark matter and the Xenon nuclei. Xenon is chemically inert and has a large number of protons and neutrons in their nuclei, and so makes a good medium for detecting the elusive dark matter. This experiment involves 190 scientists in 32 institutions and links together the USA, UK, Portugal, Russia, and China in a search for dark matter (http://lz.lbl.gov/laboratory/). Dark matter could also be composed of a new particle known as the axion. To detect axions, scientists have built one of the most sensitive electromagnetic receivers in the world, known as the ADMX experiment, that creates a huge magnetic field which fills a chamber bathed in microwave radiation. Interactions between axions and the electromagnetic field can be detected by a tiny deposition of energy into the field—at a level of 1 *yoctowatt*, or $10^{-24}$ watts. This project involves 30 scientists from eight institutions, and has already placed limits on the mass of the axion (http://depts.washington.edu/admx/). A third underground experiment just starting is the Cryogenic Dark Matter Search (CDMS), which includes super-pure Geranium disks cooled to 0.040 K, placed 2 km below the earth. The CDMS can detect any particle that crosses from the "phonon" or sound wave produced by the dark matter bouncing off the atomic nuclei (Fig. 9.3).

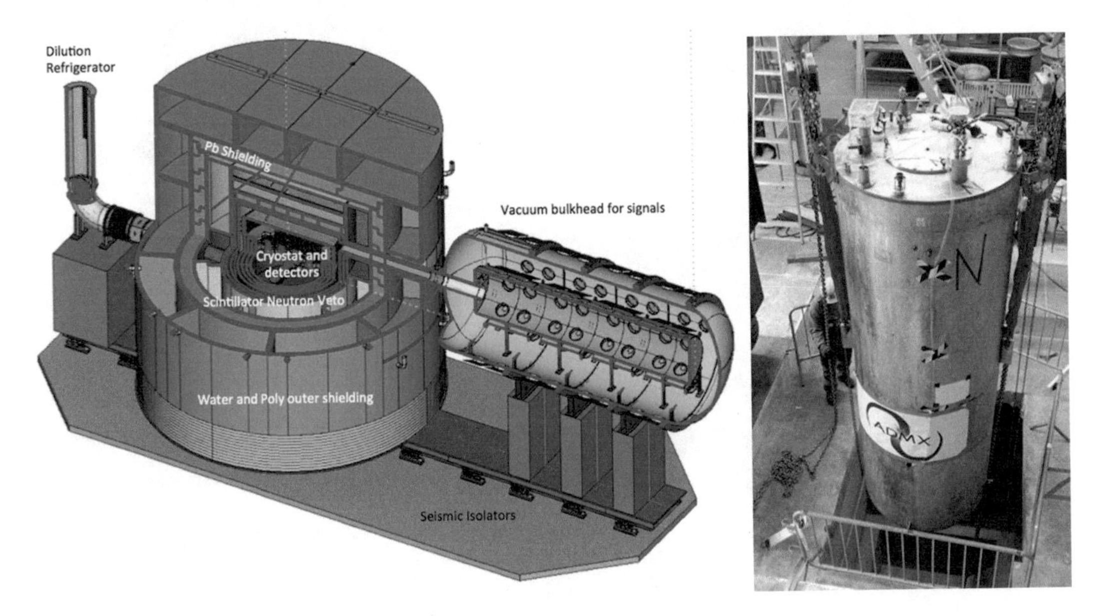


**Fig. 9.3** Two new dark matter telescopes—(*left*) the CDMS experiment, which is hoping to detect cosmic dark matter from over a mile beneath the earth, and the ADMX experiment, which uses super-cooled Germanium disks to detect interactions with dark matter (figure credits: http://cdms.berkeley.edu/Pictures/Source/supercdms_snolab_layout.png)

Dark matter can also be detected from observations of galaxies and galaxy clusters in space. The earliest detections of dark matter were based on motions of stars and galaxies that suggested additional mass beyond what could be accounted for by stars. The latest generation of dark matter observations can not only detect the amount of dark matter but also map the distribution of this invisible material. The dark matter can be detected by its effect on the light of galaxies, which is bent by the distortions of space-time. These distortions cause galaxies to appear slightly elongated, and a technique known as Gravitational Microlensing can detect these asymmetries as a function of position on the sky and distance—creating a three-dimensional map of cosmic dark matter. These enormous blobs of dark matter extend of 80 million light years across space, and enclose over ten trillion solar masses spread out over ten times the extent of a typical galaxy cluster (Massey, R., Rhodes, J. et al. 2007). Follow-up studies over even larger volumes of space studying tiny distortions in over two million galaxies in a 139 square degree field have enabled astronomers to measure the clustering of dark matter. The assembly of the large blobs of dark matter is being compared to superclusters of galaxies, which gives clues about the mass and temperature of the dark matter particles. These studies will explore even larger volumes to map the invisible dark matter across billions of light years (Vikram et al. 2015, http://arxiv.org/pdf/1504.03002v3.pdf) (Fig. 9.4).

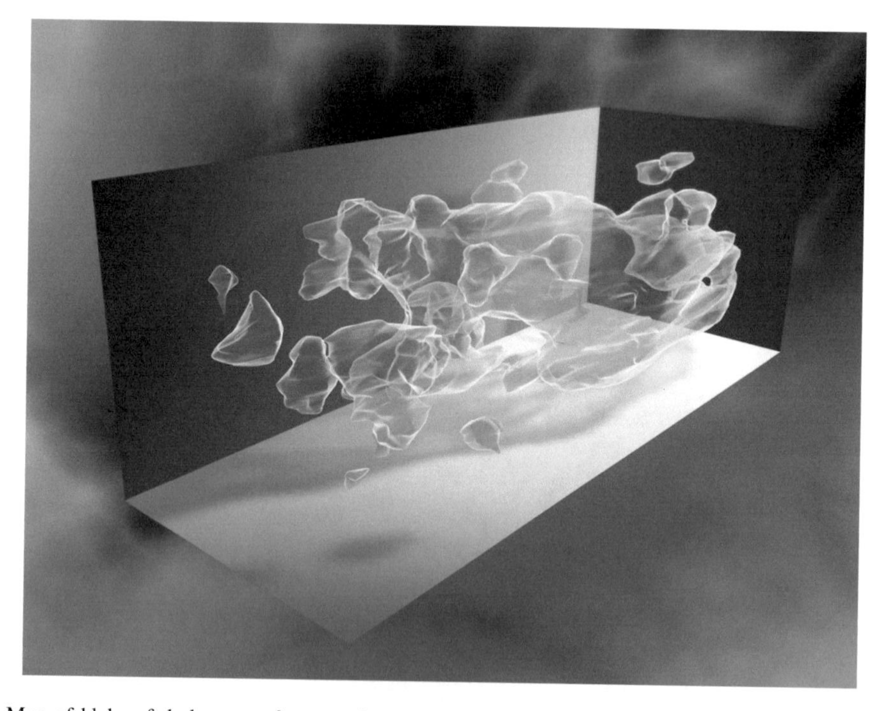

**Fig. 9.4** Map of blobs of dark matter from studies of tiny distortions of galaxy clusters in a technique known as Gravitational Microlensing. These blobs extend over 1 billion light years and each of the blobs contains over ten trillion solar masses of dark matter (from Massey, Rhodes et al. 2007)

## *The History of Scientific Cosmology*

The process of developing our modern scientific cosmology took over 2500 years and involved fascinating individuals who asked the right questions, admitted when their theories or observations fell short, and then applied the latest tools to fix gaps in their knowledge. These individuals are products of the scientific community of their time and depended on communication, competition, collaboration, and discussion to achieve their great works. Profiles of these leading "universe makers" can help us to understand how connections and collaborations can spark sudden discoveries that give our entire species new understanding into the workings of the universe. The works of Copernicus, Brahe, Kepler, Galileo, and Newton changed our view of the universe forever, as each of them created key elements of our first modern scientific or physical cosmology, which culminated in 1687 with Newton's *Principia*. We describe the important role of each in the following sections.

## *Nicholas Copernicus*

Nicholas Copernicus (1473–1543) (Fig. 9.5) is most famously known for his work *De Revolutionibus*, published in the year of his death, which presented his heliocentric theory and eventually convinced the world to abandon the earlier theories of Ptolemy, which had held sway for 1500 years. Copernicus was not the first to imagine a sun-centered or heliocentric universe. Scientists as far back as Aristarchus argued for a heliocentric theory, and authors such as Oresme in the fourteenth century suggested that the Earth might well be rotating about the Sun, hinting at the relativity of the motion between the Sun, stars, and the Earth (North, p. 304). But only Copernicus was able to offer a powerful case for heliocentric motion based on arguments that were easily tested. In his work he explains a heliocentric

**Fig. 9.5** Nicholas Copernicus

description of the apparent motions of the planets that included explanations for the appearance and disappearance of Mercury and Venus as a result of their movements between the Earth and the Sun. Retrograde loops of outer planets were described not as the result of epicycles, but instead as the natural consequence of the Earth passing the outer planets as they all orbited the Sun.

The acceptance of the Copernican theory is perhaps one of the best examples of Ockham's razor in action—the simplicity of having planets move around the Sun in simpler circular orbits was so appealing that it caused the earlier Ptolemaic system with its many epicycles, deferents, and eccentrics to quickly fall into disuse. Ironically, Copernicus tried to retain as much of the Ptolemaic theory as possible, which included uniform circular motion and the limited use of epicycles to provide the non-uniform component of heliocentric orbits.

Copernicus drew upon much improved data compared to what was available to Ptolemy, in the form of the Alfonsine tables developed between 1263 and 1272 under the support of king Alfonsine X in Toledo, Spain. The Alfonsine tables were based on earlier Arabic observations that were translated into Latin and then Spanish.

In order to match the data from the Alfonsine tables, Copernicus adopted simple Ptolemaic constructions for the planets, but with a center near the Sun for all of them, and in most cases a slight epicycle to account for variations in the observed speeds of the planets about the Sun. Circular orbits were proposed for the planets Mercury and Venus, but with a small offset of an epicycle near the Sun. The outer planets also retained some of Ptolemy's complicated machinery, in the form of an equant and epicycle. The overall effect of Copernicus' model was to simplify the motions of the planets and to shift our planet Earth into the role of one of many planets within the solar system.

Copernicus began the process of rearranging our cosmos to a simpler system with better agreement to the observations. However, he did so reluctantly and included a long disclaimer in his work to help reduce the risk of offense to the church. He also waited to publish his work until shortly before his death to reduce his own personal risk. Despite his caution, Copernicus led the way for later astronomers and set the stage for the oft-celebrated "Copernican Revolution" in cosmology.

In Copernicus' own words, from his work *De Revolutionibus*:

> And behold, in the midst of all resides the sun. For who, in this most beautiful temple, would set his lamp in another or a better place, when to illuminate all things at once? … Thus we discover in this orderly arrangement the marvelous symmetry of the universe.. this model permits anyone who is diligent to comprehend why the progressions and regressions of Jupiter appear greater than those of Saturn and smaller than those of Mars, and again greater for Venus than for Mercury… All these phenomena appear for the same reason: that the earth moves.
>
> (Danielson 2000, p. 117)

## *Tycho Brahe*

Tycho Brahe (1546–1601) (Fig. 9.6) became something of a celebrity after his discovery on 11, November 1572 of a new star, now known as the Tycho supernova. His fame was reinforced after his careful observations of a comet in 1577 showed that the comet moved in a non-circular orbit that reached much farther away than the Moon, based on its lack of observable parallax. Both discoveries were shocking at the time, since the prevailing model of cosmology held that the realm beyond the Moon was unchanging, made of ether, and in purely circular motion. The observations of a non-circular orbit for the comet and the appearance of a new star placed Aristotelian cosmology on its head, and also won Tycho much attention, eventually resulting in his commissioning a new royal observatory, which was named Uraniborg.

**Fig. 9.6** Tycho Brahe

Tycho's Uraniborg observatory was dedicated to Urania, one of the nine muses of astronomy, and outfitted by Frederick II of Denmark with the latest technology of the time. The observatory featured celestial globes, armillary spheres, sextants, octants, and azimuthal quadrants for sighting stars, including one quadrant that was 1.8 m in radius, by far the best in the world. Tycho's large ego assured the observatory was also equipped with abundant oil paintings of great astronomers from history, including several portraits of himself, and an imagined unborn descendant named Tychonides (North 2008, p. 325). Tycho showed more humility (and taste) in his observations, however, and carefully observed the stars with multiple instruments to assure accuracy. Tycho recorded (with flourish and with many assistants) subtle phenomena including atmospheric refraction, precession, and parallax of the Sun. In some cases his accuracy of angular measurements was less than 1 arc-min, better than at least a factor of 5 than earlier astronomers.

Tycho had to abandon several of the prevailing ideas about the heavens based on his new observations. These early ideas included the idea of an unchanging realm beyond the Moon, and of course the Ptolemaic universe with all the planets orbiting the Earth at the center of the universe. Tycho put some of his thoughts into the 1588 work "*De mundi aetherei recentioribus phaenomeni*," which gave details of intricate models of the lunar motion, and provided the best available data for future scientists in the motions of the planets. Tycho also developed a modified geocentric cosmology, shown in Fig. 9.7, in which the planets Mercury and Venus orbited the Sun, not the Earth.

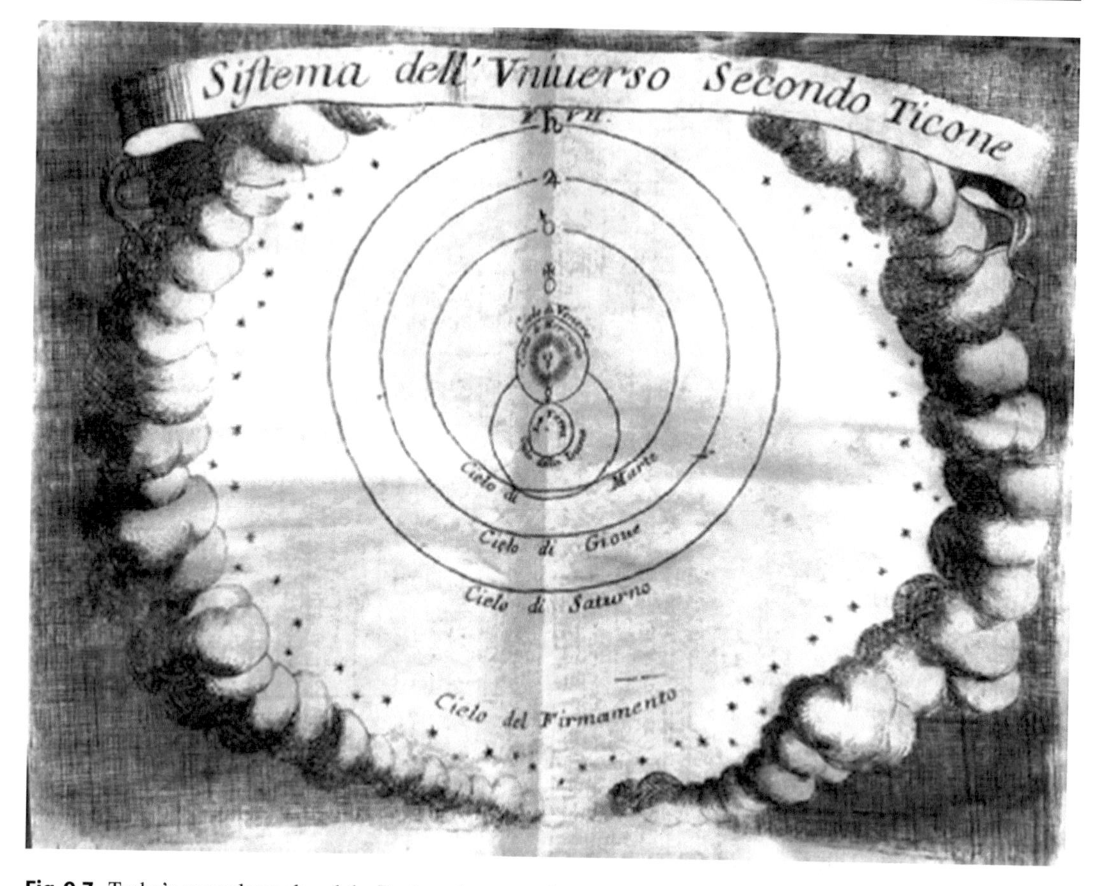


**Fig. 9.7** Tycho's cosmology placed the Earth at the center, but allowed for the planets Venus and Mercury to orbit the sun, which itself circled the Earth (image from Epitome Cosmographica, by P. Coronelli (1713); reprinted with permission from Wagner Collection of Claremont Colleges special collections)

Tycho's observations were superb; for example, his lunar observations recorded variations in the Moon's speed, the inclination of the Moon's orbit by 5.25° from the ecliptic, the precession of the Moon's orbit, and even detected subtle annual slowing and speeding up of the Moon's orbit. Unfortunately, Tycho did not live to see the results of his life's work, as he died suddenly in 1601. Tycho's life work was in the hands of the brilliant, mystic, mathematician, and astrologer Johannes Kepler, who would put Tycho's observations to good use.

Tycho's breathless wonder in witnessing the new star (now known as the "Tycho supernova") is described in his own words from his 1573 work, *De Nova Stella*:

> Last year [1572], in the month of November, on the eleventh day of that month, in the evening, after sunset, when according to my habit I was contemplating the stars in a clear sky, I noticed that a new and unusual star, surpassing the other stars in brilliancy, was shining almost directly above my head. And since I had almost from boyhood known all the stars of the heavens perfectly.. it was quite evident to me that there had never before been any star in that place in the sky… I was so astonished at this sight that I was not ashamed to doubt the trustworthiness of my own eyes… the observations of all the founders of science, made some thousands of years ago, testify that all the stars have always retained the same number, position, order, motion, and size as they are found, by careful observation on the part of those who take delight in heavenly phenomena, to preserve even in our own day.
>
> (Danielson 2000, p. 129)

## Johannes Kepler

The life of Johannes Kepler (1571–1630) (Fig. 9.8) was difficult from the day he was born in a small town near Stuttgart in 1571 to a "criminally inclined" mercenary soldier and a "bad-tempered" woman later accused of witchcraft (the descriptions come from Kepler himself!) (North 2008, p. 340). Kepler struggled to make a living between 1596 and 1624, which was much of his adult life. During this period he was expelled from Graz in 1598 as part of a sectarian purge, endured a stressful witchcraft trial for his mother in 1617–1620, and had only irregular employment as an astrologer and school teacher.

Kepler described himself as a "Lutheran astrologer" and was enthralled by investigations of mysteries such as the "Music of the Spheres," an imagined harmony of sound produced by planetary orbits, astrology, and concordances between planetary orbits and the Platonic solids (cube, tetrahedron, octahedron, dodecahedron, and icosahedron). His first work was the appropriately named *Mysterium Cosmographicum* (1596) and presented a solid framing of the Copernican system that included a basis in these mathematical shapes. An epiphany by Kepler gave him the central idea of the work—the orbits of the planet fit a progression described by a set of nested Platonic solids. A second book *Harmonice Mundi* (1619) (*Harmonies of the World*) explored the musical aspects of planetary orbits and including many of Kepler's mystical ideas about the universe.

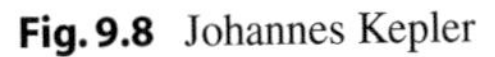

**Fig. 9.8** Johannes Kepler

Kepler began working with the astronomer Tycho Brahe in 1598 and was assigned the task of developing a theory to describe the irregularities in the orbit of Mars. After Tycho's death, Kepler was able to access the entire set of records, which included the next best set of data on planetary motion known as the Rudolphine tables. In these tables were the seeds of destruction of some of Kepler's cherished ideas, in the form of tiny discrepancies ranging from 2 to 8 arc-min between Kepler's models and Tycho's carefully observed positions.

Kepler worked on these matters persistently until he was able to construct a theory involving elliptical orbits that dispensed completely with the Ptolemaic equants and epicycles and provided a complete and more accurate description of planetary motions using just a simple equation of elliptical motion. These observations were published in his book *Astronomia Nova* in 1615 and provide the basis of the three well-known Kepler's laws:

1. Planets move on elliptical orbits
2. Planets sweep equal areas in equal times
3. The distance of the planet from the Sun ($A$) and its orbital period ($T$) is related by the equation $T^2 = A^3$.

For Kepler, achieving this scientific breakthrough required him to abandon his cherished ideas of nested planetary solids, and all of the epicyclical Copernican theories he had painstakingly constructed in his early years (Fig. 9.9). But the requirement of science to discard theories which did not agree with data, and the appeal of the simpler elliptical orbits made this sacrifice necessary, and set the stage for a physical theory of planetary motion.

In his final days, Kepler continued to investigate the "Music of the Spheres," and a quote from his book *Harmonies of the World* gives a sense of his inspired approach to finding the mysteries of the universe:

> The movements of the heavens are nothing except a certain ever-lasting polyphony (intelligible, not audible).. Hence it is no longer a surprise that man, the ape of his Creator, should finally have discovered the art of singing polyphonically, which was unknown to the ancients, namely in order that he might play the everlastingness of all created time in some short part of an hour by means of an artistic concord of many voices.
>
> (Ferris 1988, p. 77)

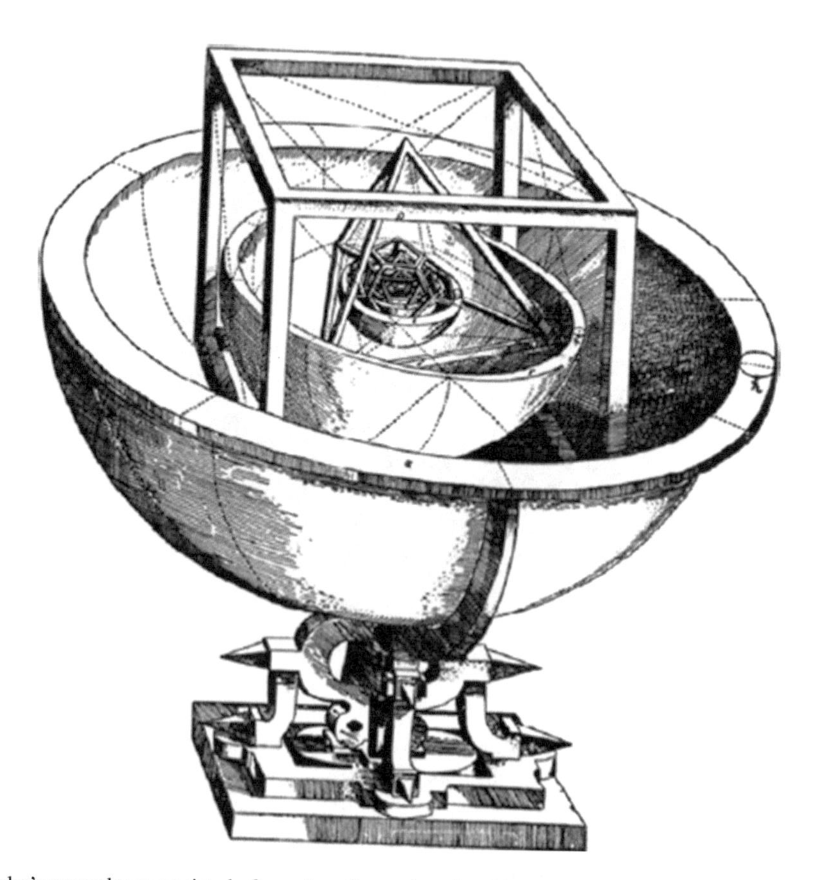

**Fig. 9.9** Kepler's cosmology consisted of a series of nested perfect Platonic solids, which appeared to match in their separations in the same ratio as the separation of the planets. However, more exact observations from Tycho Brahe forced Kepler to abandon this conception for the more modern idea of planets following elliptical orbits (*image source*: http://commons.wikimedia.org/wiki/File:Kepler-solar-system-1.png)

## *Galileo Galilei*

The ebullient, charismatic, and energetic nature of Galileo (1564–1642) (Fig. 9.10) jumps from the pages of his notebooks and published works and reflects the energies of the dynamic and prosperous Italian principalities of Venice, Padua, and Pisa where Galileo did most of his work. Galileo's work was distinctive in its ability to convey the excitement of entirely new telescopic views of the Moon, Venus, Jupiter, and star clusters and gave to the entire world its first close look at the world beyond the Earth.

**Fig. 9.10** Galileo Galilei

Galileo is famous for developing the telescope and applying it to astronomical observations, but actually was not the first to do either of these things. Kepler himself had written about refraction in one of his many works and described how to construct a telescope using lenses. The first telescopes had been made by 1608 by Hans Lippershey, an obscure Dutch lens maker, and copies of the device were made soon afterward and were on sale in Spain and France a year later. Thomas Herriot in England reported drawings of the Moon based on telescopic observations with a 6× magnification telescope in 1609.

Galileo's unique contributions to astronomy came from refining and applying the technology of the telescope and communicating his observations with clarity and excitement. In Galileo's hands, the power of the telescope increased tenfold, as he lengthened the focus of the telescope and increased its power from a paltry 3–6× magnification to 20× and then 30× in successively better instruments. Galileo's new instrument was then capable of discovering new things with each pointing, and he quickly published his exciting views of the planets, stars, and the Moon in his work *Sidereus Nuncius* (*The Sidereal Messenger*) in 1610.

Previous theories of the stars, planets, and the Moon came crashing down with the irresistible power of Galileo's telescope and writing. The Moon had previously been thought to be solid and smooth, perhaps made from water, but was shown to be rocky, arid, and mountainous. Venus, still believed by some to orbit the Earth, was clearly shown to have crescent and full phases only possible in a Copernican cosmology where it rotates about the Sun. Jupiter, perhaps most miraculously, was found to have its own set of stars, named by Galileo the "Medician Stars," that danced about it in what appeared to be orbits.

The discoveries continued at every turn; Orion's nebulosity and star clusters, the Milky Way was not smooth but composed of innumerable stars, Saturn's mysterious lobes (known now to be "rings" of icy particles), and the actual sizes of Jupiter and Mars could be measured, with features visible on their surfaces. Galileo had mastered observational astronomy and shared his exciting results with the world, making him a worldwide celebrity. The excitement also had a cost, which was his eventual house arrest in 1633, with a pardon (and apology) only coming in the year 2000, over 360 years later!

However, Galileo's ideas and writings were irrepressible, and speak to us with vigor, even centuries later. Galileo was something of a populist and wrote many of his most important works (such as his 1632 "Dialogue Concerning the Two Chief World Systems") in Italian (Fig. 9.11).

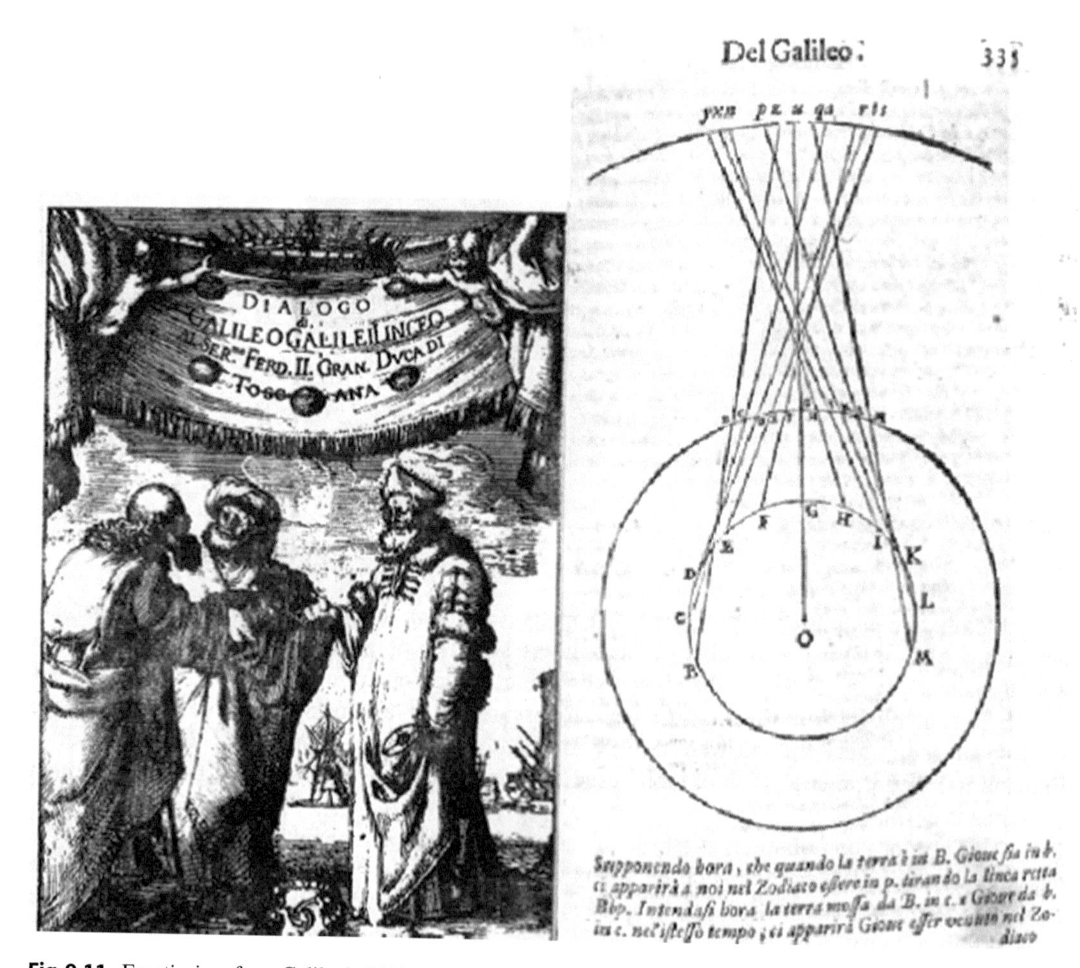


**Fig. 9.11** Frontispiece from Galileo's 1632 work "Dialogue Concerning the Two Chief World Systems," (*left*) in which Galileo puts forth arguments for the heliocentric system, in common Italian, as well as a complete description of the retrograde loops observed for planets (*right*). Galileo's attempts to communicate the findings of science and its direct appeal to the common sense of the reader are in the truest spirit of the scientific method and did much to help establish the heliocentric scientific cosmology in Europe (image courtesy of Claremont Colleges Special Collections)

In Galileo's *Sidereus Nuncius*, we can relive his excitement in discovery:

> To this day the fixed stars which observers have been able to view without artificial powers of sight can be counted. Therefore it is certainly a great thing to add to their number and to expose to our view myriads of other stars never seen before and outnumbering the old, previously known stars more than ten to one.
>
> Again, it is a most beautiful and delightful thing to behold the body of the moon.. so that the diameter of the moon appears about thirty times larger, its surface about nine hundred times, and its solid mass nearly 27,000 times larger than when it is viewed only with the naked eye…
> In addition to this, to trace out with one's finger the nature of those stars which all astronomers until now have called nebulous, and to demonstrate that it is very different from what has hitherto been believed – this also will be thrilling and beautiful. But what is most exciting and astonishing by far, and what particularly moved me to address myself to all astronomers and philosophers, is this: I have discovered four planets, neither known nor observed by anyone before my time.
>
> (Danielson 2000, p. 146)

We can also sense Galileo's amazement at discovering hundreds of new stars (Fig. 9.12)—the same excitement now felt by astronomers as they discover new galaxies everywhere they look using the world's largest telescopes:

> In the first I had intended to depict the entire constellation of Orion, but I was overwhelmed by the vast quantity of stars and by limitations of time, so I have deferred this to another occasion. There are more than five hundred new stars distributed among the old ones within limits of one or two degrees of arc. Hence to the three stars in the Belt of Orion and the six in the Sword which were previously known, I have added eighty adjacent stars discovered recently, preserving the intervals between them as exactly as I could. To distinguish the known or ancient stars, I have depicted them larger and have outlined them doubly; the other (invisible) stars I have drawn smaller and without the extra line. I have also preserved differences of magnitude as well as possible…. I have observed the nature and the material of the Milky Way.

**Fig. 9.12** Reproduction of Image of Orion region from Galileo's notebook. The sketch records dozens of new stars, and Galileo (like modern scientists today) was amazed to discover hundreds of new objects everywhere he pointed his new telescope—at the time the world's largest (from Galileo, 1610, *Sidereus Nuncius*, accessed at http://people.reed.edu/~wieting/mathematics537/sideriusnuncius.pdf)

> With the aid of the telescope this has been scrutinized so directly and with such ocular certainty that all the disputes which have vexed philosophers through so many ages have been resolved, and we are at last freed from wordy debates about it. The galaxy is, in fact, nothing but a congeries of innumerable stars grouped together in clusters. Upon whatever part of it the telescope is directed, a vast crowd of stars is immediately presented to view. Many of them are rather large and quite bright, while the number of smaller ones is quite beyond calculation.
>
> (Galileo, 1610, *Sidereus Nuncius*, p. 32)

## *Isaac Newton*

Galileo and Kepler set the stage for observational astronomy, and refinements in telescopes enabled Huygens to publish amazing observations of Saturn in his *Systema Saturnium* of 1659. Soon afterwards, Hevelius published detailed atlases of lunar features, measured the librations of the Moon, and even attempted to construct a 150-foot long telescope in 1679. However, the linchpin of cosmological progress during the period came from the work of Isaac Newton (1642–1727) (Fig. 9.13), working quietly in faraway England.

**Fig. 9.13** Isaac Newton

Newton is famous for inventing the reflecting telescope, based on a "diversion" in his spare time, which he presented to the Royal Society in 1672. Newton was most interested, however, in the mathematical machinery that determines the motions of planets, and therefore focused on refining his new technology of calculus to begin to tackle the mathematics of gravity. Like Galileo's newly developed telescope, Newton's calculus enabled new discoveries at every turn.

Unlike Kepler, Newton was not prone to far flung fancies of the imagination, but instead was motivated by an austere and minimalist philosophy. Unlike Galileo, Newton communicated his discoveries only reluctantly. Newton's theories of calculus, gravitation, optics, and motion were only published after repeated urgings from his friends, such as the Royal Society president Edmund Halley. Newton's unique self-reported ability to "think without ceasing" upon the best available data on the planets, and other astronomical phenomena. Newton's own summary of his life work is appropriately succinct and modest:

> If I have made any valuable discoveries, it has been owing more to patient attention, than to any other talent.

Newton began by realizing that the elliptical motion of the planets was caused by a force that varied by the inverse square of the distance between the Sun and the planet, his famous Universal Law of Gravitation. With this law, Newton could also begin to explain unexplained astronomical mysteries measured by Tycho, and later by Flamsteed, such as the nutation (or top-like wobbling) of the Earth's spin and the irregularities in the Moon's motion. Both motions could be explained by Newton as natural products of the tugging force from the Sun as the Moon orbited the Earth.

Newton also put his mind to explaining cometary orbits, both in theory and observation, to coincide with the appearance of the "Great Comet of 1680." In collaboration with the first Astronomer Royal, John Flamsteed, and with encouragement from Edmund Halley, Newton solved an equation of motion that completely determined the comet's orbit. Newton showed that the comet traveled along a parabolic path (Hoskin 1997; North 2008). In his studies of comets, Newton even managed to take new observations of the comet of 1680–1681 with his own 7-foot telescope. Perhaps most importantly, Newton (with collaboration from Halley) was able to show that the comet was not an aberration of the sky, but instead an object in a physical orbit, that could reappear on a predicted schedule based on the laws of physics.

Newton's work on gravity and planetary motions was incorporated into his more general work, *the Principia*, in 1687, along with his famous laws of motion. Edmund Halley brought the *Principia* to print in 1687 using some of his own funds after the Royal Society exhausted its limited funds on an unsuccessful book on the *Biology of Fishes*. Halley also spread copies of the work to the wider learned community of Europe. From this work, perhaps the most highly regarded work of science in the past 1000 years, the theoretical framework for a physical universe governed by forces at a distance was set. The motions of planets, comets, and the Earth were all completely determined by a single force law, which acted the same across all of space. Newton was able to integrate breakthroughs in physics and astronomy and create the first purely physical cosmology.

One famous quote from Newton describes his approach to discovery:

> I do not know what I may appear to the world; but to myself I seem to have been only like a boy playing on the seashore, and diverting myself in now and then finding a smoother pebble or a prettier shell than ordinary, whilst the great ocean of truth lay all undiscovered before me.
>
> (Brewster 1860)

## The Dawn of Astrophysics

After Newton, the next qualitative breakthroughs in cosmology depended on extensive observations of the stars, nebulae, and the emergence of astrophysics, in which the larger universe can be used as an extensive physical laboratory by taking careful observations of the positions, shapes, and spectra of the objects.

One key element enabling more precise measurements to be possible was the construction of increasingly more powerful telescopes (Fig. 9.14). During the eighteenth and nineteenth centuries, telescopes grew in size from Galileo's 4′ refractor or Newton's 7″ reflector to the enormous telescopes of Lick and Mt. Wilson observatory, which ranged in size from 36″ to 100″, greatly extending the distance at which astronomical objects could be detected and studied. Perhaps equally as significant as the development of large telescopes was the application of these telescopes with systematic observations, to compile catalogs of objects, and to begin to amass data on the composition of the telescopic universe.

**Fig. 9.14** The development of modern astrophysics went hand in hand with the development of more powerful telescopes and scientific instruments (image from Epitome Cosmographica, by P. Coronelli (1713); reprinted with permission from Wagner Collection of Claremont Colleges special collections)

Three key figures in creating the modern science of astrophysics are William Herschel, George Ellery Hale, and Edwin Hubble, who played perhaps the most important roles in applying and developing large telescopes to enable the development of our modern astrophysical cosmology. We briefly describe each of their contributions below.

## *William Herschel*

Astronomy was a family business for William Herschel (1732–1822) (Fig. 9.15), his son John (1792–1871), and his sister Caroline (1750–1848). Together they pursued astronomical observations as a life's work with single-minded determination and systematic effort that enabled many discoveries. Born in Hanover, Germany, William Herschel's father was in a military band, which William joined at age 15 as an oboist. Herschel left the military band and the carnage of the Seven Years War to settle in England 4 years later, and made a living as an organist in Bath teaching and performing music. His astronomical career began soon afterward, as he found that his considerable talent with his hands enabled him to make the highest quality telescopes in Europe (Fig. 9.16). What began as a self-taught hobby in the 1760s became a profession, as William built a 12-in diameter, 20-foot long reflecting telescope in 1776, followed by an 18.7-in. diameter telescope with a 20-foot long focal length in 1782, and an enormous 49.5-in. aperture telescope with a 40-foot focal length in 1789.

**Fig. 9.15** William Herschel

**Fig. 9.16** William Herschel's 10-foot telescope from 1790, which featured a 9-in. diameter mirror and controls to move the telescope vertically and laterally. Herschel made this telescope himself, and used it to discover hundreds of new objects, including galaxies and comets with his sister Caroline (photograph by the author, from Cambridge University Whipple Museum of Science)

At each stage, William's telescopes were shown to be finer in quality and more capable than any of the telescopes at the professional observatories. Herschel describes his process of building telescopes:

> I have tried to improve telescopes and practiced continually to see with them. These instruments have played me so many tricks that I have at last found them out in many of their humours and have made them confess to me what they would have concealed, if I had not with such perseverance and patience courted them.
>
> (Ferris 1988, p. 154)

Herschel's career-making discovery occurred in March of 1781, when Herschel viewed in his telescope an object which looked like a small comet only 5 arc-s in diameter. The discovery of the object could not be easily confirmed by the royal astronomer Nevil Maskeleyne, as the reference stars Herschel observed were too faint for the Greenwich telescope to view with its micrometer, and the "comet" itself was too faint to see at the Oxford Radcliffe Observatory (North 2008, p. 439). Later in 1781, the object's orbit was determined and found to be consistent with a new planet, which Herschel wanted to name the planet *Georgium Sidum* (George's Star) after the king of England. Herschel's newly discovered planet was eventually named Uranus, after the father of Saturn and grandfather of Jupiter (although some of the astronomical communities wanted to name the planet after Herschel himself).

Herschel's accomplishments were noticed and gave him professional status and a small royal pension. The money enabled him to move into comfortable quarters near Windsor, and he supplemented his income by making telescopes for sale. His largest construction, the 40-foot focal length telescope, was completed in 1789 with funding from King George III. The 40-foot telescope was so large it required enormous scaffolding, pulleys, ropes, and a team of men to turn the entire structure, which was as tall as a four-story building.

With the increased power of Herschel's telescopes, many new comets, nebulae, and stars were immediately visible. The 40-foot reflector quickly discovered the sixth and seventh moons around Saturn (Enceladus and Mimas) and observed many nebulae for the first time. Before Herschel, astronomers such as Cheseaux, le Gentil, Lacaille, and Messier had observed about 90 different nebulae. When Herschel was done cataloging nebulae, he had observed over 2500! Similar increases in the number of known binary stars, faint star clusters, and cataloged stars were obtained by the tireless efforts of William Herschel, and his sister Caroline, who conducted many of the observations and kept immaculate records of their discoveries. Caroline herself discovered at least eight new comets between 1786 and 1797 (North 2008, p. 438), and together William and Caroline published an updated version of Flamsteed's star catalog and "swept" the sky for new nebulae and star clusters.

One unique aspect of Herschel's research was the way he applied scientific techniques of surveying to large fields of stars, instead of being content to view select small fields through his telescopes. Herschel would use star counts in "sweeps" of different directions to estimate the number of stars in the galaxy. Herschel (both William and Caroline) would observe on virtually every clear night for decades, starting when he was still a musician at Bath (where he said to have observed even during his intermissions of concerts). Their practice of "sweeping" the sky involved viewing with a black hood to preserve their night vision, and sweeping 10–30 times across a segment of sky, recording the results in a "*Book of Sweeps*" (Burnham and Rexroth 1978; Ferris 1988; North 2008).

Herschel also had many impressive insights into the nature of the objects he was observing. The Orion nebula was described by Herschel to be "the chaotic material of future suns" and the Andromeda galaxy was recognized by Herschel to be composed of "the united lustre of millions of stars" (Ferris 1988, p. 157). Both insights were astrophysically correct, but would not be proven for centuries. Herschel described his approach to research in a paper delivered to the Royal Society in 1811:

> A knowledge of the construction of the heavens has always been the ultimate object of my observations, and having been many years engaged in applying my forty, twenty, and large ten feet telescopes, on account of their great space penetrating power, to review the most interesting objects discovered in my sweeps.
>
> (Danielson 2000, p. 282)

William Herschel was married in 1788 to Mary Pitt, a recently widowed friend, and their one son, John Hershel, followed in the family business, eventually becoming the Astronomer Royal of England and a founding member of the Royal Astronomical Society (Fig. 9.17). John Herschel continued his family's work by conducting original observations from Cape Town with his aunt Caroline's old 20-foot reflecting telescope, which discovered countless binary stars and helped locate the Sun's position within the Milky Way galaxy for the first time. John Herschel's impressive list of discoveries include an 1833 catalog of 1307 nebulae and star clusters, an enormous catalog of 10,300 binary stars (published posthumously in 1874), and a catalog of 1847 which listed new discoveries of 1269 Southern nebulae and star clusters observed from South Africa between 1834 and 1838.

**Fig. 9.17** Caroline Herschel, later in her life in 1829 (*left*) and John Herschel, as a young man (*right*) as portrayed in 1829 (*image sources*: http://commons.wikimedia.org/wiki/File:John_Herschel00.jpg, http://commons.wikimedia.org/wiki/File:Herschel_Caroline_1829.jpg)

For the first time, the Herschels were able to show the Sun to be located within the Milky Way, itself a large bunching of stars. In William Hershel's own words, in his 1785 work *On the Construction of the Heavens*, "our Milky Way is a most extensive stratum of stars.. a very extensive, branching, compound congeries of many millions of stars" (Hetherington 1993, p. 274). Previous authors had speculated on the Milky Way as being an "island universe" (Kant), or part of a set of shells of stars (Thomas Wright), or just one of "innumerable Milky Ways" (Lambert), but the Herschels were able to provide actual data on the numbers, positions, and brightnesses of stars that enabled the first scientific cartography of the Milky Way.

The Herschels showed that the asymmetry in nebulae between northern and southern sky is caused by the position of the Sun slightly north of the mid-plane of the Milky Way. From the work of the Herschels comes what became a second major shift in the perspective of human consciousness—the awareness that Earth is not only just one planet orbiting the sun, but also the Sun itself is just one of countless stars within the enormous "congerie" of stars that is the Milky Way!

Herschel summarizes his conception of the Sun's place in the Milky Way in his statement to the Royal Society in 1785:

> That the Milky Way is a most extensive stratum of stars of various sizes admits no longer of the least doubt; and that our sun is actually one of the heavenly bodies belonging to it is evident …
>
> (Danielson 2000, p. 277)

## Astronomical Spectroscopy

The emergence of the science of astrophysics was made possible with the combination of astronomy and spectroscopy, enabling detailed observations of the physical properties of nebulae and stars. The first astronomical spectroscopy was performed by Newton himself, in his experiments on color with sunlight. Newton famously demonstrated the "refrangibility" of light by using a prism to break sunlight into its constituent colors, and then reassemble them, showing that the colors were an intrinsic property of "white" light and not something imparted to the light by the prism. In the words of Newton (Newton 1671), white light is "ever compounded, and to its composition are requisite all the aforesaid primary colors, mixed in due proportion."

Further experiments with the solar spectrum in 1802 by Wollaston and in 1814 by Joseph von Fraunhofer showed that the light of the Sun was not just of "primary colors" but appeared to be broken into hundreds of bands of colors and separated by vertical lines (Fig. 9.18). In Fraunhofer's words:

> I found with the telescope almost countless strong and weak vertical lines, which however are darker than the remaining part of the colour-image; some seem to be nearly black.
>
> (Hearnshaw 1986, p. 27)

**Fig. 9.18** Joseph Fraunhofer's original solar spectrum, the first astrophysical spectrograph, which was commemorated in a postage stamp from the German post office (*image source*: http://commons.wikimedia.org/wiki/File:Joseph_Fraunhofer_(timbre_RFA).jpg)

Fraunhofer's spectrograph was an improvised construction using a slit in the wall of his glassworks warehouse window, a small prism, and a telescope (fashioned from a surveyor's theodolite), and with it he took the first detailed astronomical spectrum. Fraunhofer described the locations of a set of strong lines in the Sun (which he labeled A–G) and 574 fainter lines. Fraunhofer also observed lines in the spectrum of Venus, Sirius, and several bright stars as well and was able to show variations in the strengths of the largest lines (A–G) in these sources.

John Herschel in 1823 suggested the possibility of connecting the mysterious dark bands in the Sun with similar bands seen in the spectra of flames in which various elements or "bases" were added:

> The colours thus communicated by the different bases to flame afford, in many cases, a ready and neat way of detecting extremely minute quantities of them.
>
> (Hearnshaw 1986, p. 30)

By 1840, careful physical experiments by Bunsen and Kirchhoff in Heidelberg were able to realize Herschel's suggestion, and they showed that the dark bands first seen in 1802 were in fact the result of "reversed" lines of elements in the surface of the Sun. Kirchhoff and Bunsen's laboratory work could produce bright emission lines from heated salts and also dark "reversed" lines by placing clouds of atoms in front of a brighter source.

The conclusion Kirchhoff reached was that the bands of the Sun could be formed from the specific lines from the elements of lithium, sodium, potassium, strontium, and barium seen in the laboratory. The larger implication was that the Sun, long thought by ancient people to be made of divine substance (ether or other materials not seen on the Earth), could contain ordinary elements found on the Earth. With these advances it now became possible to perform a chemical analysis of the Sun or stars. In the words of Kirchoff and Bunsen, from their 1860 paper:

> "we may conclude that the solar spectrum, with its dark lines, is nothing else than the reverse of the spectrum which the sun's atmosphere alone would produce. Hence, in order to effect the chemical analysis of the solar atmosphere, all that we require is to discover those substances which, when brought into the flame, produce lines coinciding with the dark ones in the solar spectrum."
>
> (Hearnshaw 1986, p. 44)

The process of identifying the source of the absorption bands in the Sun and stars took most of the nineteenth century and from this hard work of astronomers across the globe, the science of astrophysics was born. Early drawings of stellar spectra were provided by G.B. Donati in Florence, and supplemented by spectacular spectra taken by A. Sechhi in Rome, using a 24-cm refractor. Spectral atlases by Secchi published in 1867 began a classification of stars into classes based on their visible bands and provided excellent drawings of the spectra of bright stars. One key part of the puzzle was provided by William Huggins, a British amateur astronomer, who built a telescope and astrophysical laboratory after attending a meeting of the Pharmaceutical Society in 1862 where he learned about Kirchhoff's results and was spurred to look for elements in the stars. In Huggin's words:

> The news reached me of Kirchhoff's discovery of the true nature and chemical composition of the sun from his interpretation of Fraunhofer lines. This news was to me like the coming upon a spring of water in a dry and thirsty land.
>
> (Hearnshaw 1986, p. 67)

Huggins set to the task and soon built a pair of telescopes, including a 5″ and 8″ telescope, in his garden in the outskirts of London. The observatory was equipped with a spectrograph that included glass prisms containing the poisonous gas hydrogen sulfide (which dispersed light better than solid glass) and a device for creating sparks and flames of various elements to compare with the stellar spectra. With his device, Huggins published atlas of spectra in 1864 that included spark spectra of nitrogen, oxygen, and 25 other elements, with his custom-built six-prism spectroscope (Fig. 9.19).

**Fig. 9.19** Heroes of early astronomical spectroscopy. Joseph Fraunhofer (*left*) demonstrating an early spectroscope, in a 1900 painting by Richard Wimmer, and William Huggins later in life, taken in 1910 (*image source*: wikimedia.org)

After 1875, Huggins' wife, the former Margaret Lindsay Murray, joined him in what became a lifelong collaboration in astronomy. Huggins continued in his work until he was in his eighties and together they located hundreds of lines in stellar spectra that corresponded to the ordinary elements found on the Earth, such as calcium (the source of the Fraunhofer H and K lines) and sodium (the source of the D lines). Additional lines were found in nebulae, including the lines of hydrogen, which Fraunhofer had labeled as the C and F lines (now known to astronomers as Hα and Hβ).

From the work of Huggins, most of the lines within the stars were identified to correspond to known elements. However in some cases, no known element could explain lines in the Sun and stars. Huggins himself discovered two greenish lines in the planetary nebula NGC 6543, which he called "Nebulium" (this was later shown to be from ionized oxygen). Another example was a new line discovered in a solar spectrum by Lockyear in 1868, which did not have any corresponding elemental lines in spark or flame spectra. Lockyear named this line D3, as it was close to the D1 and D2 lines, and Lockyear named the element "helium" as he suspected it was only found in the sun. However, experiments by Ramsay in 1895 duplicated the mysterious D3 line, after a mineral known as cleveite (containing small amounts of helium) was evaporated and measured with a spectrograph. Clearly the Sun, and by implication the stars and nebulae, was composed of ordinary matter, and chemical and physical analysis of the distant universe was now possible using spectroscopic techniques (Hearnshaw 1986, p. 84).

The next phase of astronomical history awaited the Modern Observatory, most superbly exemplified by the telescopes of Mount Wilson and Palomar Observatories, made possible by the tireless efforts of George Ellery Hale and the observations of his pupil, Edwin Hubble.

## George Ellery Hale and the Modern Observatory

George Ellery Hale (1868–1938) (Fig. 9.20) is appropriately known as the father of the modern observatory, and was responsible for raising funds, and administering the creation of the Yerkes, Mount Wilson, and Palomar observatories. Each of these observatories between 1896 and 1960 had the distinction of owning the largest and finest telescopes in the world, and from these telescopes the best scientists in the world were able to construct our modern cosmology from the new images and spectra of galaxies and distant stars. Many other distinctions belong to George Ellery Hale, as he was also

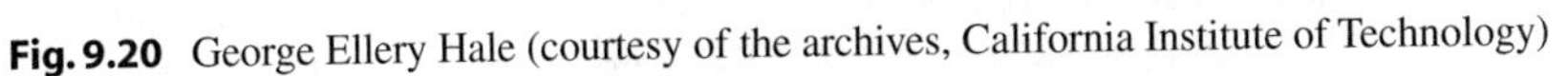

**Fig. 9.20** George Ellery Hale (courtesy of the archives, California Institute of Technology)

instrumental in developing partnerships between modern industrial leaders and scientific research, and in developing professional research institutes that offered groups of brilliant individuals the opportunity to work together and share their ideas in serene and well-appointed academic environments. He was also a key figure in founding the American Astronomical Society, the founder of the *Astrophysical Journal*, and played a key role in the early days of the California Institute of Technology and the Huntington Library.

This one remarkable man ushered in a new era of science, which brought together complex equipment, technicians, teams of scientists, and funding partners in new institutions and partnerships. As such, the twentieth century became the era of "Big Science" where the intellectual and financial resources needed to conduct research necessarily were beyond the means of any individual and required coordinating large groups of people toward the same goal. Bringing all of these resources together took a toll on Hale himself, who suffered throughout his life from mental illness. However, the success of his work and his vision is responsible for the sublime view offered by modern astronomy of the origin and evolution of the universe.

From his childhood, George Ellery Hale developed his talent for fundraising and building complex instruments for scientific research. His first exercise in this activity occurred in December 1882, when Hale was just 14-year old, Hale acquired funding for a fine 4″ refracting telescope from Alvan Clark and Sons to observe a transit of Venus on December 6. The funding source for this (and many of his early researches) was his father, who became wealthy providing elevators for the many new skyscrapers of nearby downtown Chicago. By the time Hale was 22, he had completely outfitted a research laboratory for solar research at the family mansion which included a 12″ telescope, a research grade spectrograph and a newly developed instrument known as a spectroheliograph invented by Hale during his college years at MIT for studying the surface of the Sun.

After graduation from MIT, Hale moved back to Chicago with his new bride and supervised the construction of a new building for the Kenwood Physical Observatory, Hale's personal observatory in his parent's backyard. The research of this observatory was of such high quality that it attracted the attention of the first president of the newly founded University of Chicago, William Rainy Harper. Hale was given an offer to join the faculty of Chicago, both based on his great skill as an instrument builder and astronomer (Osterbrock 1997, p. 8). Hale accepted Harper's offer and used his new position at Chicago to begin his lifelong quest to build the world's largest telescopes.

The first of Hale's giant telescopes was the Yerkes refractor, to be based in Williams Bay, Wisconsin. Hale was able to build this telescope by buying a pair of giant 40″ diameter lenses from a struggling University of Southern California in 1892, with the help of Chicago industrialist Charles T. Yerkes. Yerkes gave Hale the charge to build "the largest and best.. telescope in the world.. and send the bill to me." The project took 5 years and a cost of over $500,000—over 10 million of today's dollars (Osterbrock 1997). Hale's unique first-hand experience in spectroscopy convinced him to dedicate the new observatory for *astrophysical* research, which required not just the telescope, but spectrographs, photographic equipment and darkrooms, and well-equipped optical and mechanical shops. In Hale's own words:

> One of the principle aims of the Observatory is to bring together the physical and astronomical sides of the work.. The Observatory will be in reality a large physical laboratory as well as an astronomical establishment. All kinds of spectroscopic, bolometric, photographic and other optical work will be done in these laboratories under much better conditions than those that prevail in cities.
>
> (Wright 1966, p. 108)

Hale also was able to equip the observatory with some of the best astronomers in the country, including many that had experience with the previously largest telescope, the Lick 36″ telescope in California. The Yerkes telescope was completed in 1897 and began its work by immediately by making startling new discoveries. While testing the new optics, Barnard discovered a faint star near Vega, and soon afterward the Yerkes refractor discovered absorption lines of carbon in the Sun and stars, the first detection of the "stationary" lines of calcium atoms. The discovery began the first spectroscopic studies of the interstellar medium, and soon the Yerkes telescope also pioneered in the precise measurement of star positions, parallax, and proper motion, which was possible with the long focal length refractor. These observations were crucial for measuring distances and masses of stars.

Despite the success of the Yerkes refractor, Hale had his mind on even bigger "schemes" (a term he used for long-term strategies). A reflecting telescope, on a mountain site in the Western United States, would gather clearer images on many more nights and would be free from "chromatic aberration"—the color given to light from a lens-based telescope (Fig. 9.21).

Hale had already obtained some key components of the future Mt. Wilson observatory as early as 1900, while he was at Yerkes. These parts included a horizontal solar telescope, which used two mirrors to gather the Sun's light and an enormous 60″ mirror, which would become the core of the Mt. Wilson telescope. Hale was careful in his work to retain ownership of the 60″ mirror by purchasing the glass for the mirror with his father's money, and by paying part of his technician's salary with his father's and his own fortunes. Hale's long-awaited opportunity to build the Mt. Wilson 60″ telescope came in the form of the Carnegie Institute, which was newly constituted and empowered to fund research in astronomy. Hale directed the institute toward his thinking about a Mt. Wilson observatory, and managed to cleverly implement nearly every aspect of his "scheme" to build a California solar observatory, which included a 60″ telescope with Carnegie money.

Hale left for California in 1903 and by 1904 had obtained funding for the observatory that included $65,000 a year for 5 years (Osterbrock 1997, p. 45). By 1905, Hale also had convinced many of his former associates from Yerkes to move west to help found the observatory. His associates from Yerkes came to California loaded with equipment: Adams, Ritchey, and Barnard brought to Mt. Wilson a wide-field camera (the Bruce photographic telescope), parts for the horizontal solar telescope (the Snow telescope), and of course the 60-in. mirror. Together Hale and his newly arrived colleagues helped build the future center of astronomy—the Mt. Wilson observatory.

When the Mt. Wilson 60″ and 100″ telescopes were completed, they completely outclassed all other telescopes in the world both in the sheer size of the mirror, which enabled fainter objects to be seen and in the pristine quality of the site, which allowed clear and sharp images to be taken on most nights of the year (Fig. 9.22).

**Fig. 9.21** The Yerkes 40″ refractor, Hale's first largest telescope in the world, completed in 1897, and still the world's largest refracting (lens-based) telescope. The telescope was famous enough to attract a very well-known visitor, Albert Einstein, shown in the center of this 1921 photo (*image source*: http://commons.wikimedia.org/wiki/File:Yerkes_Observatory_Astro4p7.jpg)

**Fig. 9.22** The Mt. Wilson 100″ Hooker telescope (*left*) and the 60″ telescope (*right*). These two telescopes were the world's largest for the first half of the twentieth century, and were used by Edwin Hubble to discover the cosmological expansion of the universe (*image sources*: http://commons.wikimedia.org/wiki/File:100inchHooker.jpg; *Photo of 60″ by the author*)

Perhaps more importantly, the Mt. Wilson telescopes were outfitted as complete astrophysical laboratories. The reflecting optics were far more efficient than the lens-based systems at Yerkes and Lick Observatories, and the Mt. Wilson telescopes were also arranged to enable the light to be directed into a focal point along the polar axis of the telescope, in what is known as the "coude" focus. The "coude" focus enabled the light to be analyzed in a temperature-controlled laboratory, where a huge spectrograph was constructed to split the light into a very highly dispersed spectrum.

Hale had a particular interest in spectroscopy, having founded the American "Astrophysical" Society (the "Astrophysical" emphasized spectroscopy) and having spent his youth and early career acquiring and building spectrographs. In Hale's own words:

> The star light passes through no glass, but after entering the open tube falls directly upon the silvered upper surface of a mirror which lies at the bottom of the tube. The image formed by this mirror after reflection to one side at the upper end of the tube, falls directly upon a photographic plate. There through the use of suitable mechanisms, it is maintained in a fixed position for as many hours as is needed.
> George E. Hale, from a statement written for donors to the Mt. Wilson Observatory, c. 1902 (Wright 1966, p. 161)

> The spectra of stars might be photographed on so large a scale as to permit the study of their chemical composition, the temperature and pressure in their atmospheres, and their motions with that high degree of precision which can now be reached only the case of the sun.
> George E. Hale, in a letter to W.R. Harper, president of the University of Chicago, c. 1898 (Wright 1966, p. 155)

Hale's success in building the giant observatories had perhaps as much to do with his tireless efforts in building networks of scientists and philanthropists, as it did with bringing together the huge components of steel and glass. Hale's network included local magnates such as Henry E. Huntington and John D. Hooker, and the board of the newly developed Throop Institute of Technology. Hale also was instrumental in building societies of scientists and had a founding role in the most important institutions of modern astronomy and science research, including the National Research Council, the International Astronomical Union, the American Astronomical Society, and the International Astronomical Union. These diverse contacts gave Hale unique access to both practicing scientists and directors of government and private agencies that helped him bring together expertise and funding. Hale's efforts transformed both astronomy and southern California, as he played a key role in founding the Huntington Library, the California Institute of Technology, the Pasadena Civic Center, in addition to his better-known role in founding Mount Wilson and Palomar Observatories.

With funds in hand from the Carnegie Foundation, Hale began Mt. Wilson as the Mt. Wilson solar observatory, which grew from a re-constructed Snow horizontal solar telescope to include a 60-foot solar tower, and later a 100-foot solar tower, equipped with the latest, highest power spectrographs available. Hale's 60″ telescope and 100″ telescope came from funds acquired by the Carnegie Institute, although both telescopes made use of glass blanks paid for by Hale's father and by John D. Hooker, respectively. In a short period of only 8 years, Hale built Pasadena into the world center of astronomy.

Hale's fame and results with the solar telescopes quickly extended around the world, as he was acquiring the world's sharpest images of sunspots and the first measurements of their magnetic fields. From the 60″ and 100″ telescopes came an even more exciting legacy, which included high-quality spectra of stars and galaxies that revealed the temperatures and densities of stellar atmospheres for the first time and enabled the discovery of the expanding universe by Hale's young associate Edwin Hubble and his assistant Milton Humason. The Palomar Observatory, Hale's final achievement, was dedicated in June of 1948, 10 years after Hale's death, and included a 200″ Pyrex disk and even more powerful spectrographs capable of measuring individual stars in distant galaxies (Fig. 9.23). The 200″ Hale telescope at Palomar was able to fine-tune the exciting discoveries begun by George Ellery Hale's telescopes at Yerkes and Mt. Wilson. The suite of telescopes built by George Ellery Hale gave humanity a window to the universe that changed our perspective forever.

**Fig. 9.23** The final legacy of George Ellery Hale, the Palomar 200″ telescope, known as the "Perfect Machine" and perhaps the most impressive embodiment in the world of the modern astrophysical observatory (image courtesy of Dr. Thomas Jarrett, Caltech from http://web.ipac.caltech.edu/staff/jarrett/index-2.html)

As Hale himself said in one of his many popular articles on astronomy (this one to Harper's Magazine in 1928):

> Like buried treasures, the outposts of the universe have beckoned to the adventurous from immemorial times. Princes and potentates, political or industrial, equally with men of science, have felt the lure of uncharted seas of space, and through their provision of instrumental means the sphere of exploration has rapidly widened…
>
> Each expedition into remoter space has made new discoveries and brought back permanent additions to our knowledge of the heavens. The latest explorers have worked beyond the boundaries of the Milky Way into the realm of spiral "island universes".. While much progress has been made the greatest possibilities lie in the future. (Hale 1928)

## Edwin Hubble and the Expanding Universe

Edwin Hubble (Fig. 9.24), namesake of the Hubble Space telescope, ushered in a new era of cosmology based on his discovery of an expanding universe. Hubble was one of many of Hale's "future" astronomers who were able to use Hale's new telescopes to their limits with imaginative and patient observations of the most distant galaxies. Hubble himself was educated in mathematics and astronomy at the University of Chicago and spent time as a young student at Yerkes Observatory with Hale. His career in astronomy did not begin without many interesting detours which included a year studying Law at Oxford as a Rhodes scholar in 1912, brief stints teaching high school in Indiana, practicing law in Kentucky, a short stint as a prizewinning boxer in Europe, and a 2 year period serving in World War I as an infantryman (North 2008, p. 587). However, Hubble did manage obtain his Ph.D. in astronomy at Yerkes observatory in 1917 and was invited by Hale to join the staff at Mt. Wilson, which Hubble did after his tour in the US Army.

Soon after Hubble arrived at Mt. Wilson in 1920, the perspective of our place in the universe underwent two profound and permanent shifts. First, Harlow Shapley, by observing the positions and distances of the many globular clusters orbiting around the Milky Way, was able to show that our Sun sat at the edge and not the middle of this swarm. This presumably meant that the Earth and the Sun were at the edge of the galaxy, in a not particularly interesting "suburb" of the galactic disk. Soon afterward Hubble was able to show an even more dramatic shift in perspective.

**Fig. 9.24** Edwin Hubble (image from the Huntington Library; reprinted with permission)

With his assistant, Milton Humason, Hubble measured the Doppler shifts from dozens of galaxies and found that these shifts seemed to increase in proportion to the distance to the galaxy. The rate of increase, expressed in units of velocity per unit distance, is now known as the Hubble Constant in his honor and is one of the primary pieces of evidence for the expanding universe. Each galaxy cluster was observed using the "Cepheid variable" stars within them as a cosmic distance indicator. The variable stars had periods proportional to their luminosity and so were excellent for measuring distances to galaxies.

Hubble extended the Cepheid distance measurements for use in distant external galaxies by using the calibration of the Cepheid Period Luminosity relation developed by Henrietta Leavitt from observations of the nearest galaxy to ours, the Large Magellanic Cloud. Harlow Shapley had already used these stars to measure the distances of globular clusters, which were over 30,000-light years away. Hubble took the technique even further, by showing that the Cepheids in the Andromeda galaxy implied an enormous distance of over 1 million light years—a huge factor greater than the presumed size of the universe. Hubble in 1924 wrote excitedly to Shapley:

> You will be interested to hear that I have found a Cepheid variable in the Andromeda Nebula (M31). I have followed the nebula this season as closely as the weather permitted and in the last five months have netted nine novae and two variables.. The distance comes out something over 300,000 parsecs. [300,000 parsecs = 1,000,000 light years].
> (Hetherington 1993, p. 287)

Hubble continued to expand the known universe by systematically applying the Cepheid technique to more distant galaxies. His work showed that Andromeda was perhaps the closest of the major galaxies, and soon he was measuring Cepheids in galaxies within the Virgo and Perseus clusters. Each more distant galaxy cluster showed smaller diameter galaxies and fainter Cepheid stars. To his amazement, Hubble discovered from his spectroscopy with the Mt. Wilson telescopes that more distant galaxies receded from our view at ever increasing rates of speed.

At first glance, the Milky Way seemed to be in a privileged central position of this expansion. Further consideration of the evidence in the context of the newly developed cosmology from Einstein and Dutch theorist De Sitter showed that the expansion viewed by Hubble was completely consistent with an overall expansion of space. Einstein's Relativity Theory predicted that each part of space (and its billions of galaxies) would observe all other parts of space expanding. The Milky Way was part of a much greater cosmic expansion, which also implied that the universe had a beginning at a much higher density and temperature. This picture, first developed by Hubble and Humason, and later refined by Hubble's student Alan Sandage, is now known as the Big Bang. The Big Bang model predicts that the universe begins in an unimaginably hot singularity, and expands forever, slowly decelerated by the gravity of its galaxies, of which the Milky Way is just one of billions.

Hubble's work brought about a complete revolution in astronomy. No longer were galaxies "nebulae" or "island universes" but instead were large swarms of stars just like the Milky Way. Hubble organized sequences of galaxies into classes of spirals, barred spirals, and ellipticals and argued that their enormous distances were supported both by angular sizes and by Cepheid variable stars. Some resistance to this picture included evidence of very rapid internal motions within galaxies from plates taken by the Mt. Wilson astronomer Van Maanen, which was used by Harlow Shapley to argue that the galaxies had to be closer to prevent motions at their edges faster than the speed of light.

In the end, the excellent data from Hubble was irrefutable. Shapley had to concede that Hubble's observations confirmed that the nebulae were indeed at great distances, making them full-sized external galaxies like the Milky Way. Shapley received a letter from Hubble describing his discoveries and commented (Hetherington 1993, p. 295), "Here is the letter that has destroyed my universe." Later re-analysis of Van Maanen's data by the astronomer Knut Lundmark in 1927 showed that Van Maanen had erred by a factor of 10 in the velocities, which further strengthened Hubble's conclusions (North 2008, p. 577).

The first diagram Hubble constructed of velocity and distance was soon refined, using newer and better calibrations of the Cepheid distance relationship. The Hubble Constant, the slope of the linear relation between velocity and distance, has recently been accurately measured using the Hubble Space telescope, in what is called the "Hubble Key Project," a multinational effort of nearly 150 astronomers, led by Wendy Freedman, director of the Carnegie Observatories in Pasadena (Freedman 2001). The realization of the modern astronomical establishment is perhaps embodied in this collaboration, which makes use of the ground-based telescopes (such as those built by George Ellery Hale) and applies the techniques of Edwin Hubble to ever more distant galaxies to measure the speed of recession to only a few percent. Hale would have appreciated the international cooperation involved in the effort—a truly worldwide campaign of using telescopes in Chile, Hawaii, California by astronomers based in 12 countries, and making use of several of the "Great Observatories" in space, most especially the crown jewel of the Hubble Space telescope (Figs. 9.25 and 9.26).

Hubble's summary of the results of science in building our knowledge of the universe can be found in his work "*The Realm of the Nebulae*," from which an excerpt is below:

> The exploration of space has penetrated only recently into the realm of the nebulae. The advance into regions hitherto unknown has been made during the last dozen years with the aid of great telescopes. The observable region of the universe is now defined and a preliminary reconnaissance has been completed… Mathematics, it has been said, deals with possible worlds – logically consistent systems. Science attempts to discover the actual world we inhabit. So in cosmology, theory presents an infinite array of possible universes, and observation is eliminating them, class by class, until now the different types among which our particular universe must be included have become increasingly comprehensible.
>
> (Hubble 1936)

**Fig. 9.25** The Hubble Deep Field and Hubble Deep in the Field. (*Top*) Edwin Hubble "Deep in the field" in 1914 (*center, top*), with friends Roe and Lydia Roberts (*front left*) and Davis and Earl Hale (*upper right*) while visiting Indiana; and the Hubble Deep Field (*bottom*), which is one of the "deepest" images ever taken by modern astronomers, which made use of the Hubble Space telescope and discovered hundreds of previously undiscovered galaxies in a patch of sky the size of a grain of sand held at arm's length (*image source*: http://www.astro.louisville.edu/education/hubble_in_louisville/hale/index.html, R. Williams (STScI), the Hubble Deep Field Team and NASA)

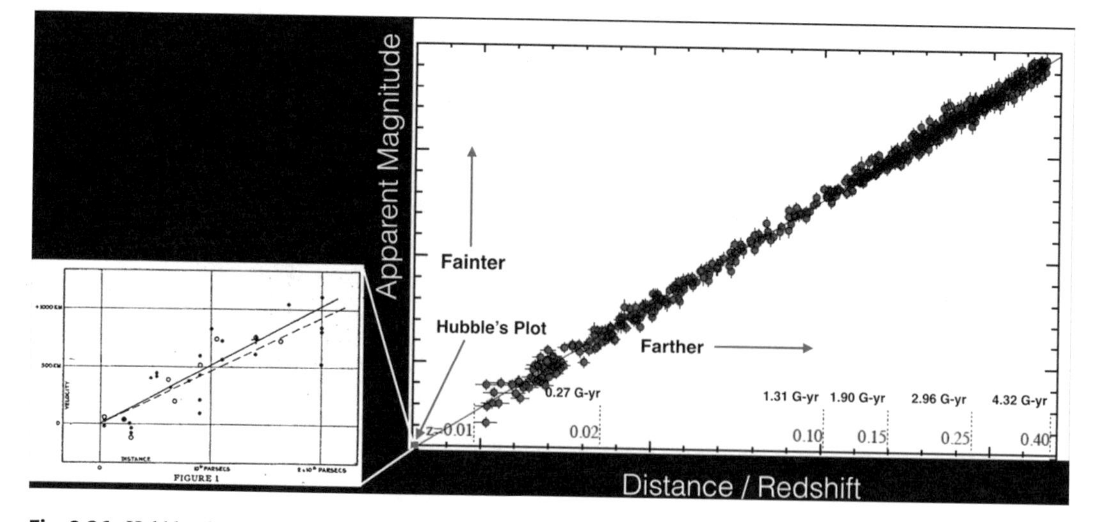


**Fig. 9.26** Hubble plots, old and new. Hubble's original 1929 plot of recession velocity of galaxies against distance (*left*). Also shown (*right*) is a modern plot of galaxy brightness against distance in billions of light years (G-year) from a recent paper using distant supernovae to measure distances (Riess et al. 2016). The most distant data points, which look back over 4.3 billion light years, show the evidence for "Dark Energy" which results in a slightly greater rate of expansion than expected at large distances (*figure adapted from Riess et al. 2016)*

## A Census of the Modern Universe

Since Hubble's time, modern astrophysical discoveries are being made by an increasingly international astronomical community. Hundreds of scholars working at observatories and universities around the world contributed to an accelerating pace of discovery during the middle and end of the twentieth century. The amazing convergence of computer technologies, sensor development, and large telescopes enabled the decades of the 1990s and 2000s to be known as "the Golden Age of Astronomy." From these discoveries we have been able to put together a coherent picture of the local universe, with a nearly complete census of the stars in our part of the Milky Way, and huge catalogs of galaxies that extend to nearly the edge of the observable universe.

Large telescopes on the Earth are joined by a suite of orbiting Great Observatories to peer into the darkness of space. Together this fleet of telescopes extends our view of space beyond the optical wavelengths of our eyes into radio, infrared, ultraviolet, X-ray and gamma-ray wavelengths. From all of these tools, which have been assembled from the breakthroughs of twentieth century science and technology, astronomers are able to assemble a complete picture of galaxies. Astronomers can now study both the visible and invisible parts of galaxies, to understand the life cycles of stars within galaxies, from their first formation to their violent deaths in novae, supernovae, and gamma-ray bursts. A brief summary of the "ingredients" of the universe follows, along with a timeline that describes the formation and evolution of a star and a galaxy. Each small part of the story may represent the life's work of one of the hundreds of scientists, and we are indebted to their hard work for the vivid and detailed picture we enjoy today. The summary follows in order of size from small to large.

### *Stars*

Stars are the "atoms" of the large-scale universe. They make up the majority of the visible sources of matter in our universe and produce light from nuclear fusion in their cores (Fig. 9.27). A star can be described as a thermonuclear explosion held together by its own gravity. Beginning from large clouds

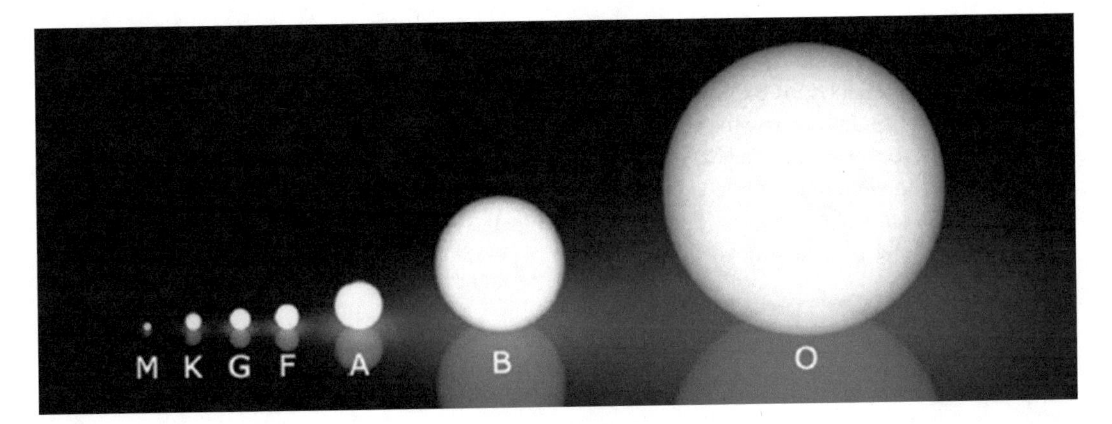


**Fig. 9.27** An image showing a schematic sample of stars, showing the early classification system known as the Morgan-Keenan system, which assigns letters to stars based on their temperature, as determined from stellar absorption lines. The M stars (*left*) are the smallest and coolest of the stars, while G stars (*middle*) are like our Sun. The O stars (*right*) are the hottest, brightest, and shortest lived of the stars and are found near star-forming regions such as the Orion nebula (*image source*: http://commons.wikimedia.org/wiki/File:Morgan-Keenan_spectral_classification.png)

of interstellar dust and gas, a proto-star can collapse according to gravity to form the higher density and temperature needed to initiate nuclear reactions. The process of forming stars has been witnessed using infrared and space telescopes, which are needed to peer through the dense clouds that surround the infant star. With the onset of nuclear reactions a star provides powerful winds that clear out light atoms and ices close to the star, leaving behind rocky iron-rich elements that can form terrestrial planets. Once the star is formed, it settles into a stable gravitational arrangement that can last for billions of years, as the core of the star is able to fuse Hydrogen into Helium releasing the light that we see as daylight from the sun. Eddington describes the interior of a star and the journey of light (or "aether waves" in his 1925 terminology) very eloquently:

> The inside of a star is a hurly-burly of atoms, electrons and aether waves. We have to call to aid the most recent discoveries of atomic physics to follow the intricacies of the dance. We started to explore the inside of a star; we soon find ourselves exploring the inside of an atom. Try to picture the tumult! Disheveled atoms tear along at 50 miles a second with only a few tatters left of their elaborate cloaks of electron s torn from them in the scrimmage. The lost electrons are speeding a hundred times faster to find new resting-places. Look out ! there is nearly a collision as an electron approaches an atomic nucleus; but putting on speed it sweeps round it in a sharp curve. A thousand narrow shaves happen to the electron in 10~10 of a second; sometimes there is a side-slip at the curve, but the electron still goes on with increased or decreased energy. Then comes a worse slip than usual; the electron is fairly caught and attached to the atom, and its career of freedom is at an end. But only for an instant. Barely has the atom arranged the new scalp on its girdle when a quantum of aether waves runs into it. With a great explosion the electron is off again for further adventures. Elsewhere two of the atoms are meeting full tilt and rebounding, with further disaster to their scanty remains of vesture. As we watch the scene we ask ourselves, Can this be the stately drama of stellar evolution? It is more like the jolly crockery-smashing turn of a music-hall.
>
> A.S. Eddington, *The Internal Constitution of the Stars,* p. 19-20 Cambridge University Press 1926

The further evolution of a solar system involves accretion from these rocky "planitesimals" which can consolidate millions of the fragments into single planets that share the same plane and direction of orbital motion. After formation, the stars burn from nuclear reactions in their cores for a duration that depends on the available fuel supply (proportional to the mass of the star) and inversely on the rate of fuel consumption or luminosity (proportional to the mass of the star, the third or fourth power). The result is that massive stars burn out quickly and smaller stars burn slowly and faintly.

When fuel is exhausted in a star, the star collapses under its own weight as the radiation from the core is no longer pushing out the atmosphere. The collapse can be violent for a massive star, and the explosive pressure after the collapse produces a powerful supernova. A star like the Sun ends in a gentler explosion and will produce a nova and a planetary nebula in about 5 billion years.

After the explosion, a fraction of the mass is left behind in a compressed state, which can be a black hole, neutron star, or white dwarf, depending on the initial mass of the "progenitor" star. The explosions of stars release most of their mass to the surrounding galaxy, where they can trigger molecular clouds into collapse. In this way the supernova completes a "life cycle" of stars where the elements of the universe are recycled into the next generation of stars and planets. All of the elements in our bodies besides hydrogen were once in the core of a long-lost star, giving us a direct connection to the stars for the very elements of our bodies. The oxygen we breathe and all of the heavy elements in our bones were forged in long-past supernovae.

## *Star Clusters*

Stars usually form in batches, ranging in size from a pair of binary stars, to mid-size groupings of dozens of stars known as "open clusters," to enormous "globular clusters" of over 100,000 stars (Fig. 9.28). In the early part of the universe even larger groupings of millions of stars known as "dwarf galaxies" can be formed. The size of a batch is determined by the temperature, density, and size of the initial cloud of interstellar medium that collapsed to form the star. Since each group of stars is formed at the same time, it locks in the mixture of chemical abundances of elements appropriate for the time and place of their formation. Very old star clusters formed in a time when the universe was "pristine" and contained very little heavier elements such as oxygen, silicon, and iron. Since the Big Bang is expected to only produce hydrogen, helium, and traces of very light elements, the earliest stars would be free of heavy elements. Yet even the most ancient stars known all contain traces of heavy elements. The pattern of elements in these first batches of stars, and also in the very distant interstellar medium, contains the mix of elements released by a short-lived and massive group of stars that are thought to have formed and died very soon after the Big Bang.

**Fig. 9.28** Star clusters can include "open clusters" such as the Pleiades (*left*), which is a young (~100-million-year old) star cluster just emerging from the wispy interstellar medium that formed it, to ancient clusters such as the "globular" cluster M80 (*right*), which contains over 100,000 stars, and was formed within the first billion years of the 13.7-billion-year history of the universe (*image source*: http://commons.wikimedia.org/wiki/File:M80.jpg, http://commons.wikimedia.org/wiki/File:Pleiades_large.jpg)

Later batches of stars are enriched in heavy elements, which accumulate gradually in the course of cosmic time as each generation of stars evolves and dies, and release additional heavy elements into the universe. Much of what we know about the history of star formation in the universe comes from the "cosmic archeology" of stars.

Much as archeology constructs the history of time on the earth from strata of accumulated layers of dirt and debris, "cosmic archaeology" pieces together the history of galaxies and the universe from studying the motions, positions, and numbers of different ages of stars. In this way, astronomers have shown that the Milky Way galaxy probably collapsed from a very large nearly spherical cloud 10–12 billion years ago, and then gradually condensed and flattened to form the disk shape over the subsequent 3–4 billion years.

Since many of the oldest stars are found in large globular clusters and dwarf galaxies (Fig. 9.29), the motions and compositions of these stars are some of best clues about the early universe. In some sense, these old star clusters are much like a "time capsule" carrying within their atmospheres a sample of the early universe, which astronomers can study in detail using large telescopes and spectrographs.

**Fig. 9.29** A pair of "dwarf galaxies"—the Sagittarius dwarf galaxy (*left*), taken with the Hubble Space telescope (*right*), an old galaxy seen near the galactic center, and the dwarf galaxy IC10, showing *red* glowing gas where stars are forming (*image source*: http://commons.wikimedia.org/wiki/File:Sagittarius_dwarf_galaxy_hst.jpg, http://commons.wikimedia.org/wiki/File:IC10_BVHa.jpg)

## *"Normal" Galaxies*

From the earliest telescopic astronomy of Herschel, Messier, and others, the "Nebulae" were a source of great fascination. We know now that many of these nebulae discovered by 17th- and 18th astronomers were comets, interstellar clouds, or glowing clouds of gas surrounding dying stars, either in the form of "reflection nebulae" or "planetary nebulae." However, many of the nebulae are themselves complete galaxies containing many hundreds of billions of stars. The first glimpses of these galaxies, made possible with large telescopes such as those at Mt. Wilson and Palomar, enabled mankind to look beyond our Milky Way across millions of light years of space into the cosmic deep. Recent observations of galaxies with the large Keck telescopes, or the Hubble space telescope, can peer across the universe and back in time to look 10–12 billion years into the past, at an era when the Milky Way itself was forming. The largest and most impressive galaxies were studied first, using the telescopes of Mt. Wilson. Soon afterwards a system of galaxy classification, known as the "Hubble sequence," was constructed to begin to understand the diverse forms seen within the "*Realm of the Nebulae*" (Hubble's title of an early work on the subject) (Fig. 9.30).

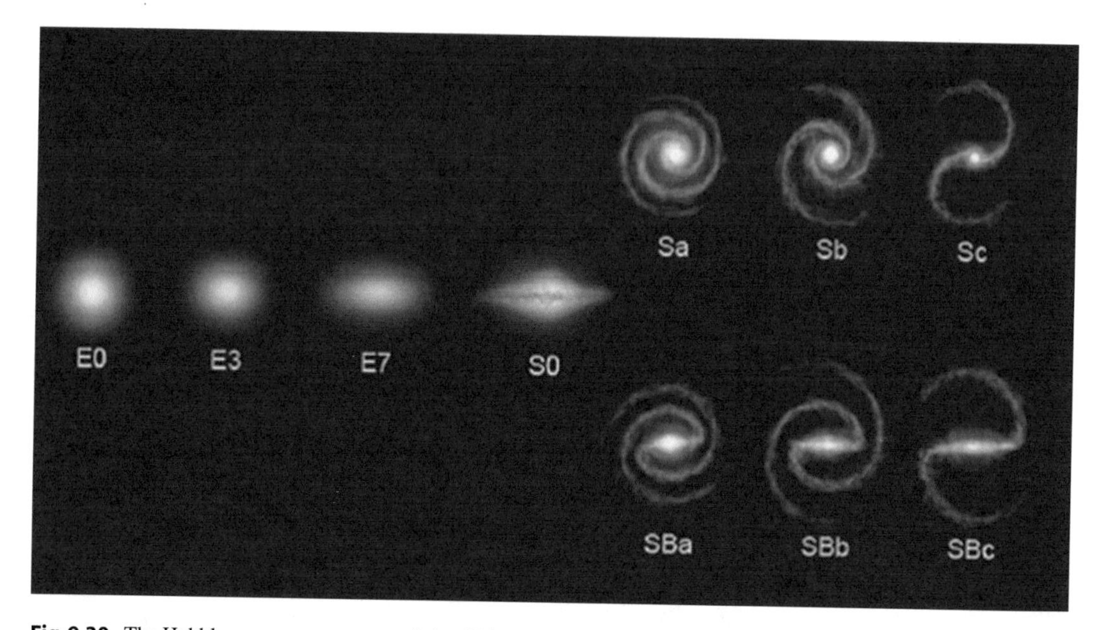


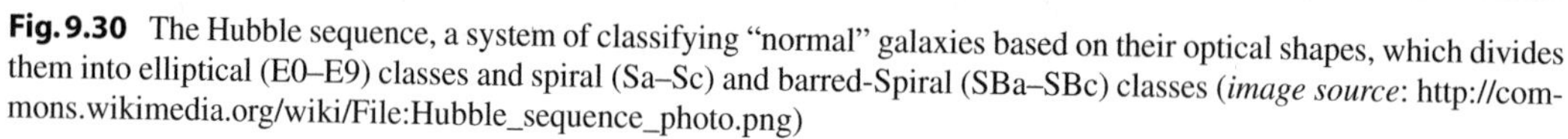

**Fig. 9.30** The Hubble sequence, a system of classifying "normal" galaxies based on their optical shapes, which divides them into elliptical (E0–E9) classes and spiral (Sa–Sc) and barred-Spiral (SBa–SBc) classes (*image source*: http://commons.wikimedia.org/wiki/File:Hubble_sequence_photo.png)

The Hubble sequence is now known to describe nearby and therefore relatively recent galaxies, in which there is sufficient time for the orbits of the billions of stars to form beautiful elliptical shapes, or the complete formation of disks and spiral arms. These "normal" galaxies are broken into groups based on their shapes, which include elliptical galaxies (usually the result of colliding galaxies which form a smooth swarm of stars in round or elliptical shapes), spiral galaxies such as the Milky Way (where spiral arms form from density waves which trigger star formation, giving the spiral arms their distinct, bright star populations), and barred-spiral galaxies (where a "bar" or horizontal group of stars appears due to the dynamic interaction between the orbiting stars and dark matter).

These "Hubble Sequence" galaxies are complete systems like the Milky Way with typically 100–200 billion stars, orbiting about a nucleus that can contain a supermassive black hole, ranging from 1–3 million solar masses (as in the Milky Way) to over one billion solar masses (as in M87). The entire assemblage of stars within a galaxy is one of the most beautiful spectacles of nature, and yet the bulk of the galaxy is largely invisible. Dark matter in galaxies is seen to comprise over ten times the mass of all the visible stars within galactic disks, making all of the structures of the Hubble Sequence that we would call galaxies just tracers of a much larger underlying dark matter mass. As such the stars we see in a galaxy are much like the tips of icebergs of dark matter or are like whitecaps on the ocean that sit on top of much larger "swells" (Fig. 9.31).

**Fig. 9.31** The galaxy M83 sometimes known as the "Southern Pinwheel" galaxy. This giant spiral galaxy shows beautiful structure in its knots of bright stars, and these luminous parts of the galaxy themselves are just the crests of much larger underlying distributions of dark matter (image by the author, based on images taken with the Carnegie Observatories Las Campanas 100″ DuPont telescope)

## *Irregular and Active Galaxies*

The exceptional nature of the galaxies Hubble used for his classification sequence defined what we call "normal" galaxies. These nearby galaxies, which were viewed in optical light of recent vintage (the past few million years), had a relatively small number of forms (spiral, barred spiral, and elliptical), which gives us the basis of declaring galaxies that do not share these shapes as "irregular." It is now known that in the early epochs of cosmic time, galaxies were colliding and merging frequently and slowly evolving into the more recent elliptical and spiral forms. Therefore most of the "irregular" galaxies are those that are caught in the midst of a violent collision or are still forming their disks, or otherwise are in a state of transition (Fig. 9.32). Irregular galaxies give us the basis of understanding the dynamics of galaxies, and from their arcs of stars, tidally stripped interstellar medium, and distorted shapes we can infer all manner of cosmic collisions which were much more common early in the universe's history.

Using supercomputer models, astronomers have recreated the sequence of events that form the wide variety of shapes seen in many of these irregular galaxies. Some of the types of irregular galaxies include tidal arcs, such as the "mice" or "antenna" galaxies, and amorphous blobs of stars and interstellar medium, such as the galaxy M87, which indicate a recent burst of star formation. Many of the "irregular" galaxies show a great sense of order, such as the "ring" galaxies, which are now known to be formed from a collision between a disk galaxy and a well-aligned dense galaxy that punches through the disk and sets up a shock wave that triggers a ring of star formation millions of years after the collisions. Sometimes the central black hole within a galaxy is absorbing millions of stars, giving rise to the "active" galaxy, which includes the class of galaxies known as quasars. These active galaxies can have cores that are in some cases thousands of times more luminous than the rest of the galaxy, and often appear as star-like points of light in the sky. Yet unlike

**Fig. 9.32** A sample of irregular galaxies, starting clockwise at *top left* with the "cartwheel" galaxy, the "mice" galaxies, the "tadpole" galaxy, and the "antenna" galaxies. The names are common names for the irregulars, which are in various stages of interacting with other galaxies, causing the distortion of the disk-like shapes seen in "normal" galaxies (image credit; NASA)

stars, these galaxies can be seen at distances of billions of light years, due to the enormous cataclysm of accreting matter within their nucleus. The quasars are very useful for studying the most distant realms of the universe, as their light can be absorbed by less luminous smaller galaxies and clouds of undifferentiated matter in the early universe. The process of studying the distant absorption from quasars helps astronomers understand the evolution and formation of galaxies that complement the study of ancient stars.

## *Clusters of Galaxies*

Like stars, galaxies are seldom formed in isolation. Groups of galaxies can be attracted to a central mass and describe large orbits about the center of a galaxy cluster. Each cluster of galaxies shares the same rate of cosmic expansion, and early work with the Virgo, Coma, and Perseus galaxy clusters was very helpful in measuring how the rate of cosmic expansion increases with distance. The measurement of the increase in apparent velocity with increasing galaxy distance is known as the "Hubble constant" and is one of the first indicators of the overall expansion of the universe. Some internal motions within galaxy clusters also suggest enormous clumps of dark matter within the center of the cluster, which exceeds the mass of stars by a factor of 10. Astronomers have extended our view of the sky out to billions of light years, and maps of the positions and distances of galaxies show that clusters of galaxies often clump into enormous "super clusters" separated by voids, giving us a glimpse of the "large-scale structure" of the universe. Explaining the clumpiness of the large-scale structure was a challenge, until the presence of dark matter provided an explanation. Supercomputer models of the universe have been constructed using literally trillions of particles and have supported a picture of "hierarchical galaxy formation" in which small clumps of galaxies under the influence of dark matter are guided along large filaments and sheets, providing the observed structure.

Even more distant galaxy clusters have been studied in recent years, by using the light from supernovae to trace the distance and motion of these more distant clusters. The very subtle variation in cosmic expansion with distance seen in these most distant clusters gives hints of a slight acceleration in the expansion of the universe, which has been attributed to the "dark energy." dark energy is the term used for an unknown repulsive force that pushes galaxies and clusters apart. If one converts the amount of energy needed into an equivalent mass (using Einstein's famous $E = mc^2$) we can determine that the dark energy exceeds in equivalent mass all the known sources of gravitation in the universe and comprises about 70% of the known mass in the universe, compared to about 27% for the dark matter, and about 3% for luminous matter such as stars and galaxies.

The discovery of dark matter, and now dark energy, is some of the most amazing achievements of modern astronomy and astrophysics, which also presents the most profound challenge to physics and astronomy in the coming decades (Fig. 9.33). We now know that our census of the universe, which up until very recently only included luminous stars and galaxies, was missing 97% of the mass of the universe! Completing this census will be the task of future telescopes, observatories, and physics experiments, which will most likely discover new types of particles and maybe even large clumps of dark matter and dark energy that could even form "dark galaxies" where billions of solar masses of dark matter gravitate and evolve separately from luminous matter.

**Fig. 9.33** *Left:* A pair of colliding galaxy clusters, together known as MACS J0025.4–1222, or the "bullet cluster," taken with the Hubble Space telescope and Chandra X-ray observatory. The huge *red* blob is X-ray emitting hot gas that traces the core of the ordinary "baryonic" matter, which includes electrons and protons. The *blue* blobs are computed densities of dark matter derived from a technique known as "weak gravitational lensing" (http://commons.wikimedia.org/wiki/File:Galaxy_Cluster_MACS_J0025.4-1222_Hubble.jpg). *Right:* Image of the galaxy cluster Abell 370, revealing the mix of hot x-ray emitting gas, made of protons and electrons (*blue*) and clouds of dark matter (*blue*). In the image are thousands of galaxies and arcs of light coming from gravitational lenses (http://chandra.harvard.edu/photo/2015/dark/)

## A Very Brief History of Time: Big Bang to the Present

If we put all of our observations of stars, galaxies, clusters of galaxies together with our knowledge of physics, it is possible to trace the physical state of the universe and its evolution from the very first instants of time to the present moment. The beginning of the universe, from our present-day knowledge of physics, starts at about $10^{-30}$ s after the Big Bang, when the first clump of matter began its explosive expansion that created the space and time that we now know as the universe.

It is important to understand that the universe is not expanding with a center and edge but that the space and time of the universe itself are created and grow with the Big Bang. Much as the surface of a sphere expands into three-dimensional space, without a center or edge within its two dimensions, the space and time of our universe expand within a four-dimensional space, without a center or edge. Because each part of space expands, any observer will see all other parts of space expanding away from them, giving the illusion of being in the center of a cosmic expansion.

The shift in perspective that came from the Big Bang was the most recent of the massive conceptual paradigm shifts provided by modern astronomy and astrophysics. Our Milky Way, itself an ordinary galaxy within an unremarkable cluster of galaxies known as the Local Group, is just one of millions of galaxies that mutually appear to be expanding from one another. This shift in perspective continues the earlier revelations from Copernicus, Herschel, and Hubble, which all further reduce the uniqueness of our Earth in the larger cosmic picture.

Turning the clock back on all of the expansion requires us to imagine the universe condensed into a single dense fireball, which in early times is expected to be increasingly small, dense, and hot. As we move time backward, the conditions of the universe in earlier times will resemble the conditions within a stellar atmosphere, the core of a star, and even hotter and denser conditions as we go back to the earliest times. Moving time backward draws on our knowledge of physics of stars, stellar interiors, nuclear physics, and particle physics, each one requiring a mastery of the physics of higher temperatures and densities. In this way, modern astrophysics and physics have reached in interesting meeting point. The study of the most distant and vast spaces of the universe, the earliest universe, necessarily involves a contemplation of the smallest and densest known physical phenomena seen within nuclear and particle physics. The history of the universe from the beginning of time is one of expansion and cooling, through different phases of matter.

In the first instants of time, the universe was hotter than the most powerful flashes of energy in particle accelerators, and it is thought that the laws of physics we now know were not yet in place. The universe in this time is sometimes known as the "Grand Unified Theory" epoch, since the four forces of our present universe (gravity, electromagnetism, strong and weak forces) have not yet had a chance to separate.

In the inferno of the early GUT universe, any of our known particles (protons, neutrons, electrons) would instantly be destroyed, and therefore the forces we use to describe the particles could not yet exist. However, within the first $10^{-24}$ s (or the first trillionth of the first trillionth of a second), space expanded to a point where temperatures had dropped to enable the first particles to "freeze out. By $10^{-12}$ s (or one trillionth of a second), exotic particles like the Higgs Boson "freeze out" as the temperatures drop below the energies found in the largest particle accelerators such as the CERN LHC (Fig. 9.34). By $10^{-6}$ s (one microsecond) temperatures drop below at a temperature of $10^{12}$ degrees, and the universe forms the first free protons and neutrons. At this point the universe has "cooled" to the temperatures found within the cores of the hottest known stars!.

**Fig. 9.34** The world's most powerful particle accelerators can re-create the conditions found in the early universe in the first one trillionth of a second. *Right:* Particles detected from the Large Hadron Collider (LHC) showing the fragments of exotic particles that emerge from colliding nuclei, which for a brief instant create a mini-Big Bang in which particles like the Higgs Boson can be detected. *Left:* The CERN Compact Muon Solenoid (CMS) detector, which like a telescope, allows us to study the extreme past of the universe, as it reveals exotic particles that were common during the early stages of the Big Bang (image credits: https://commons.wikimedia.org/wiki/Category:Higgs_boson#/media/File:CMS_Higgs-event.jpg; https://upload.wikimedia.org/wikipedia/commons/9/9c/CERN_CMS_endcap_2005_October.jpg)

Just as stars fuse nuclei together to form heavy elements, the early universe for a brief instant acted as an enormous stellar interior and fused the first nuclei of helium, beryllium, and lithium from about 1 s to 3 min, as temperatures cooled from $10^{10}$ degrees to a mere $10^6$ degrees. After 3 min, the cosmic element abundances of helium and other light elements are "locked in" since further nuclear reactions required higher temperatures found today only in the centers of stars (Fig. 9.35).

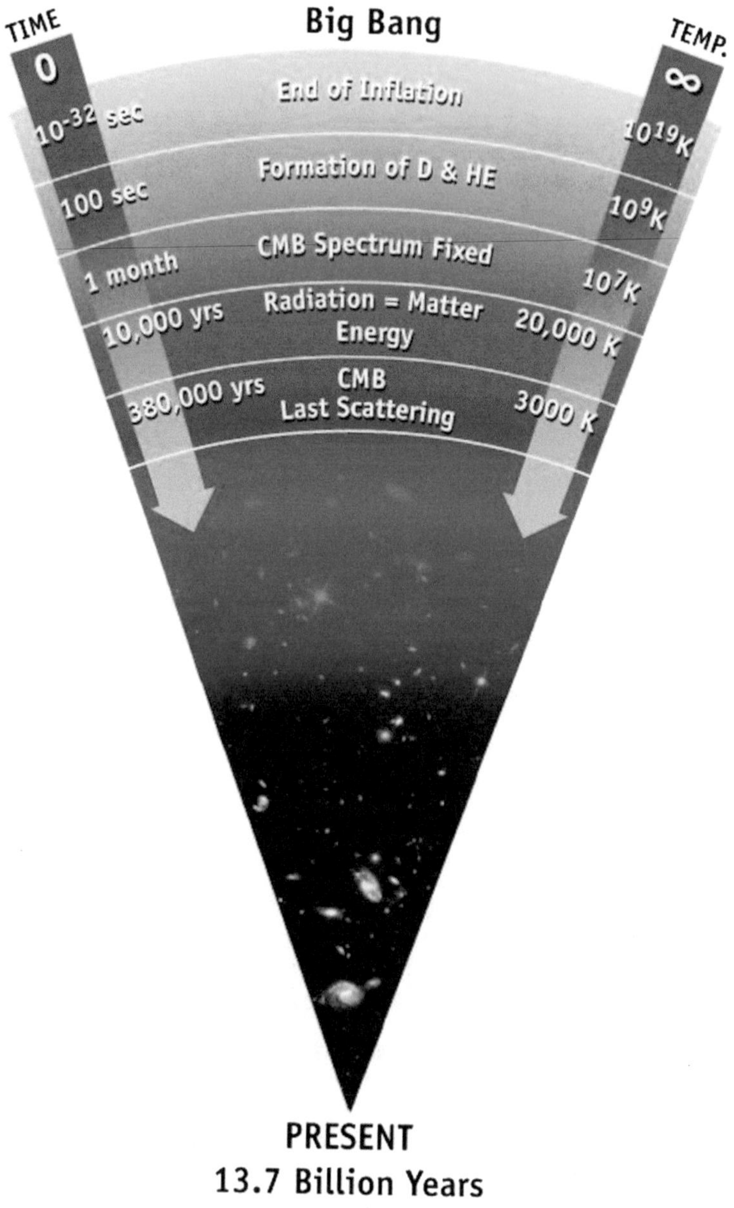

**Fig. 9.35** A timeline of the early universe, showing how space and time arose from the Big Bang, including "inflation" of space, the formation of the first nuclei of deuterium (D) and helium (He), within the first 100 s. Afterward the light and matter separate and are emitted as the Cosmic Background Radiation (CMB) (image courtesy of NASA/WMAP)

Further cooling over the next 400,000 years enabled the first atoms to form, as temperatures lowered below 10,000 K to allow electrons to attach to nuclei of hydrogen and helium. When the first atoms formed, the universe also became transparent for the first time, since the fog of electrons prior to this point was continuously absorbing and re-emitting light. In a very real sense, the formation of the first atoms was the separation of matter and light, which has been recalled poetically in the creation stories of many of the cultures mentioned earlier in the book. The light of the "recombination" forms the basis of the "cosmic background radiation," a sea of photons that pervades the entire universe. Since 400 of these photons occupy every cubic meter of space in the universe, we can get a direct sample of the early universe, even here on the Earth! When these photons emitted, they were much like the light leaving the surface of a hot star, in the blue and ultraviolet parts of the spectrum.

After billions of years of expansion (13.7 billion to be precise), photons acquire longer wavelengths and lower energies, shifting them to the wavelengths used for terrestrial radio and television broadcasting. It is still possible to detect these cosmic photons—an ordinary television set tuned to one of the UHF channels will show "snow" which has about 1% of the signal from cosmic background photons!

The light and matter which emerged from the recombination was very smooth, and the maps of the cosmic background radiation show that the density of the universe was constant over space to within one part in 100,000, as measured by the tiny bumps and wiggles in intensity of the CMBR across the sky. These small ripples in space from the early universe are enough to provide the initial input of energy to allow matter to collapse into the first generation of stars, some 400 million years after the Big Bang, which exploded and seeded the rest of the universe with trace amounts of heavy elements (Fig. 9.36).

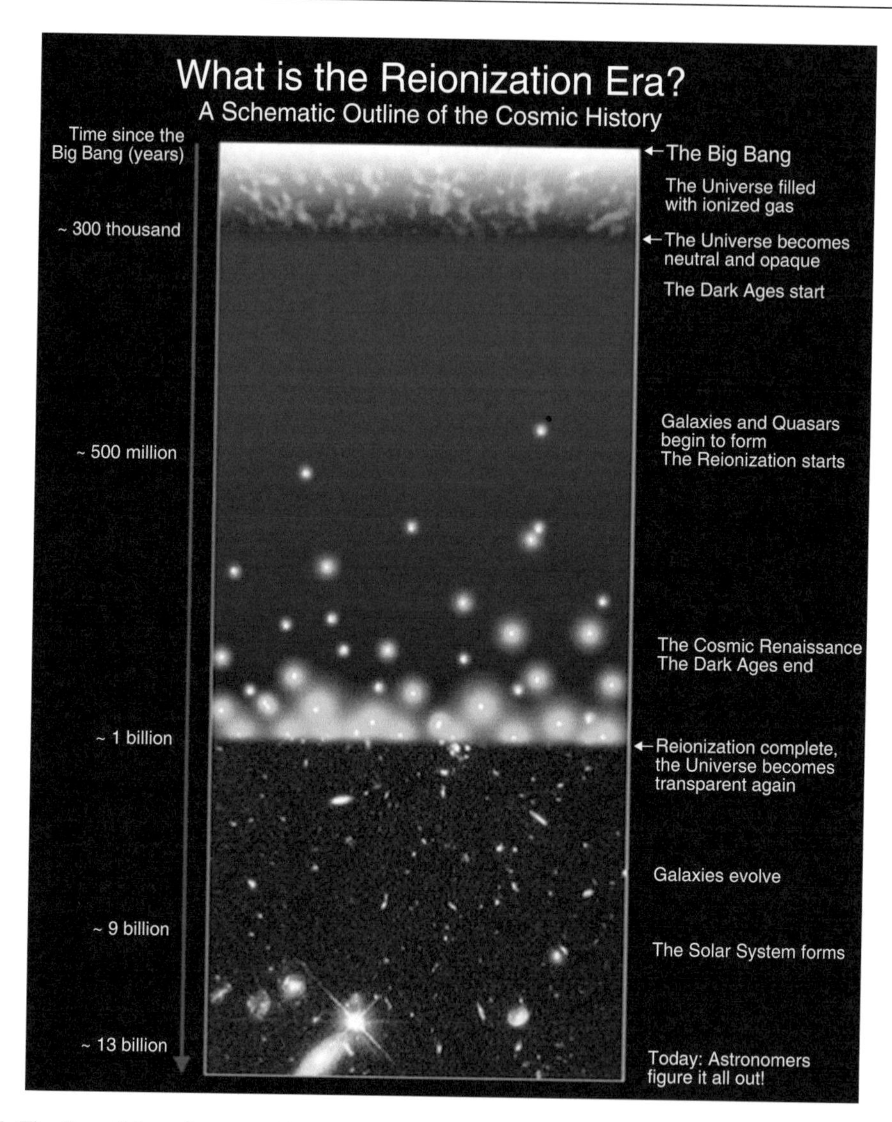


**Fig. 9.36** Timeline of the universe, starting with the "Big Bang," and then showing the epoch of recombination at 300,000 years and the formation of the first galaxies and quasars 500 million years after the big bang, after what cosmologists call the "Dark Ages." These first stars "reionize" the universe, but by 1 billion years the universe becomes transparent again and galaxies can form and evolve (image credit: S. G. Djorgovski et al., Caltech. Produced with the help of the Caltech Digital Media Center, and reprinted with permission)

These CMBR photons are the most distant form of light that we can see—as the universe farther and earlier than this point is filled with a fog of electrons that scatter light, much like the surface of a star is opaque. Seeing further and earlier than the CMBR requires future gravitational wave telescopes, which have the potential of peering deeper into the fog of the early universe, and revealing the distribution of mass before the first atoms were formed.

The next generations of stars formed in large clumps which today can be seen in the globular clusters and dwarf galaxies. Billions of years later further contraction gave rise to disk galaxies, which collided and merged and also began to form the large-scale super cluster, and clusters of galaxies observed today. The rest of cosmic history is possible to view directly with telescopes—different snapshots of galaxies at different distances give us a "time-lapse" picture of galaxy evolution and stellar evolution. As such the modern telescope is a very capable time machine able to look back in time to any era up to about 12 billion years in the past.

Our whirlwind tour of the universe has provided an overview of the fabric of the modern universe, with its dark matter, dark energy, stars, galaxies, and clusters of galaxies, as well as a view of the evolution of the universe from the earliest instants of time to the present day. The perspective we enjoy in the twenty-first century is one which earlier generations lacked. Ours is the first generation to ever peer directly across the dark expanse of space and the first ever to understand the various physical domains of the universe from the centers of stars and the beginning of the universe to the physical conditions of intergalactic space. Our telescopes, our astronomers, and our physicists have made this possible, by patiently applying the techniques of modern science and combining the life's work of some of the best minds over many generations (Table 9.1 and Fig. 9.37).

**Table 9.1** Notable telescopes of the world over the centuries

| Year | Telescope | Diameter (in.) | Diameter (m) |
|---|---|---|---|
| 1608 | Hans Lippershey | 3 | 0.08 |
| 1609 | Galileo (refractor) | 5 | 0.125 |
| 1670 | Newton (reflector) | 3 | 0.08 |
| 1789 | Herschel (reflector) | 49 | 1.24 |
| 1839 | Pulkovo Observatory | 15 | 0.38 |
| 1846 | Lassell (reflector) | 24 | 0.61 |
| 1850 | Lord Rosse Parsontown Observatory (reflector) | 88 | 1.8 |
| 1867 | Grubb Melbourne (reflector) | 48 | 1.2 |
| 1885 | Pulkovo (refractor) | 30 | 0.5 |
| 1887 | Lick Observatory (refractor) | 36 | 0.91 |
| 1897 | Yerkes Observatory (refractor) | 40 | 1.01 |
| 1906 | Mt. Wilson Observatory (reflector) | 60 | 1.5 |
| 1918 | Mt. Wilson Observatory (reflector) | 100 | 2.5 |
| 1950 | Palomar Observatory (reflector) | 200 | 5.2 |
| 1988 | Keck Observatory (reflector) | 393 | 10 |
| 2009 | Gran Telescopio Canarias | 409 | 1.04 |
| 2021 (est) | Giant Magellan Telescope | 1200 | 30 |
| 2024 (est) | Thirty-Meter Telescope (CIT) | 1200 | 30 |
| 2024 (est) | European Extremely Large Telescope | 1574 | 40 |

Telescope size appears to double approximately every 30 years. These powerful telescopes enable us to see farther into the universe and to reach back into the earliest times when galaxies and stars begin forming (data from North 2008, p. 491)

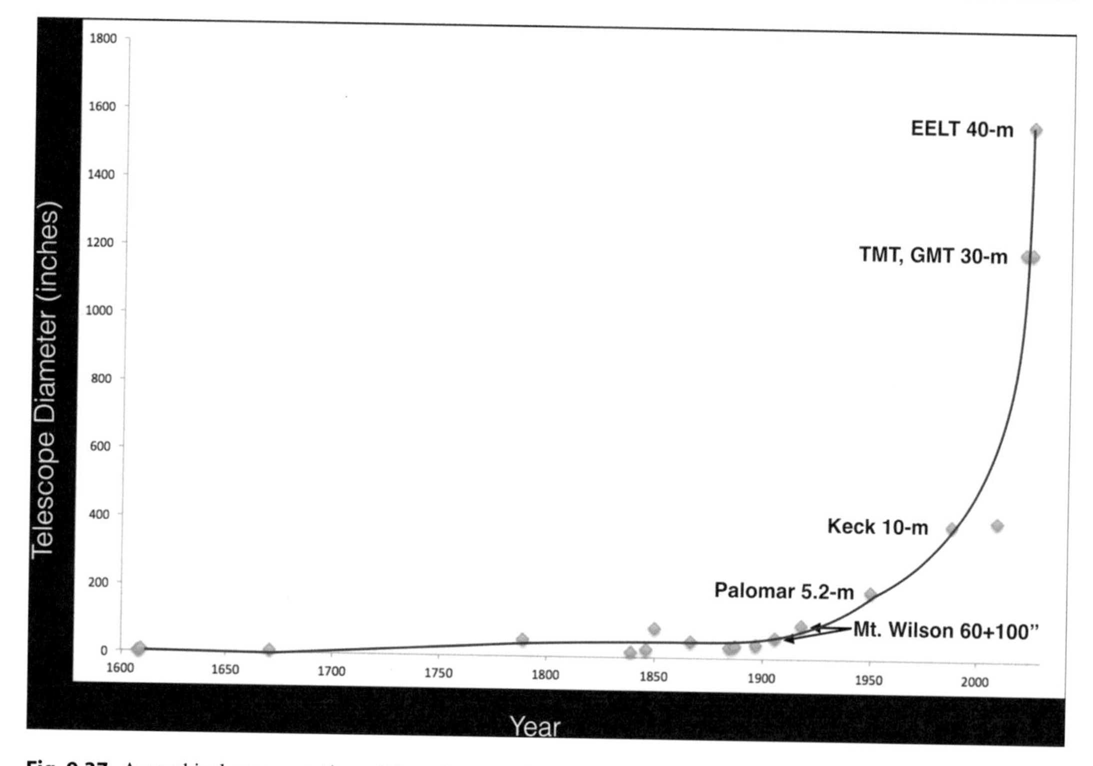


**Fig. 9.37** A graphical representation of the telescope sizes as a function of time, showing the dramatic geometric increase in size with time (figure by the author, with data from North 2008, p. 491)

We can also see how intimately we are connected with the larger universe—the starlight we see on any dark, still night can originate from before the time of our birth, and the atoms within our bodies themselves were born within the cores of long-exploded stars. Our Earth sits at the edge of a galaxy which itself expands within the larger expansion of space and is bathed with photons from nearby and distant galaxies, as well as light that came from a time before any galaxies (the CMBR). Humans have somehow learned all of this in the short expanse of time in the past few centuries, and we seem to be poised to answer many of the remaining questions, using the incredible resourcefulness of our species to construct new experiments, telescopes, and observations making use of all the evidence we have at our fingertips—both visible light and other invisible evidence of dark matter.

## References

Brewster, D. 1860. *Memoirs of the life, writings, and discoveries of Sir Isaac Newton*. Edmonston and Douglas: Edinburgh.

Burnham, R., and K. Rexroth. 1978. *Burnham's celestial handbook: an observer's guide to the Universe beyond the solar system*. New York: Dover.

Coronelli, P. 1713. *Epitome Cosmographica*. Venice: Poletti.

Danielson, D.R. 2000. *The book of the cosmos: imagining the universe from Heraclitus to Hawking*. Cambridge, MA: Perseus.

Einstein, A. 1933. Mein Weltbild. From http://en.wikiquote.org/wiki/Albert_Einstein.

Ferris, T. 1988. *Coming of age in the Milky Way*. New York: Morrow.

Feynman. 1970. Feynman lectures on physics. From http://en.wikiquote.org/wiki/Richard_Feynman.

Firestein, S. 2012. *Ignorance: how it drives science*. New York: Oxford University Press.

Freedman, W.L., et al. 2001. Final results from the Hubble space telescope key project to measure the Hubble constant. *Astrophysical Journal* 553(1): 47–72.

Hale, G.E. 1928. *The possibilities of large telescopes*. Harpers, p. 639.

Hearnshaw, J.B. 1986. *The analysis of starlight: one hundred and fifty years of astronomical spectroscopy*. New York: Cambridge University Press.

Hetherington, N.S. 1993. *Encyclopedia of cosmology: historical, philosophical, and scientific foundations of modern cosmology*. New York: Garland.

Hoskin, M.A. 1997. *The Cambridge illustrated history of astronomy*. New York: Cambridge University Press.

Hubble, E.P. 1936. *The realm of the nebulæ*. London: H. Milford; Oxford University Press.

Massey, R., J. Rhodes, et al. 2007. Dark matter maps reveal cosmic scaffolding. *Nature* 445: 286–290.

Newton, I. 1671. A letter of Mr. Isaac Newton. Containing his new theory about light and colors. *Philosophical Transactions of the Royal Society* 80: 3075–3087.

North, J.D. 2008. *Cosmos: an illustrated history of astronomy and cosmology*. Chicago, IL: University of Chicago Press.

Osterbrock, D.E. 1997. *Yerkes observatory, 1892–1950: the birth, near death, and resurrection of a scientific research institution*. London: University of Chicago Press.

Riess, A.G., et al. 2016a. A 2.4% determination of the local value of the Hubble constant. *Astrophysical Journal* 826: 56.

Vikram, V., et al. 2015. Wide-field lensing mass maps from DES science verification data: methodology and detailed analysis. *Physical Review D* 92(2): 022006.

Wright, H. 1966. *Explorer of the universe; a biography of George Ellery Hale*. New York: Dutton.

# Chapter 10
# Concluding Thoughts

*With the success of scientific theories in describing events, most people have come to believe that God allows the universe to evolve according to a set of laws and does not intervene in the universe to break these laws. However, the laws do not tell us what the universe should have looked like when it started – it would still be up to God…*

Steven Hawking (Hawking 1998)

*Superstrings and butterflies are examples illustrating two different aspects of the universe and two different notions of beauty. Superstrings come at the beginning and butterflies at the end because they are extreme examples… They mark the extreme limits of the territory over which science claims jurisdiction. Scientifically speaking, a butterfly is at least as mysterious as a superstring*

Freeman Dyson (Dyson 1989)

*There is a thing confusedly formed,*
*Born before heaven and earth.*
*Silent and void*
*It stands alone and does not change.*
*Goes round and does not weary.*
*It is capable of being the mother of the world*

(Lao Tzu, from Lau and Whalen (1969))

We have seen a dazzling array of human responses to the power of stars. Civilizations have invented creation stories, constellation tales and have attempted to connect the stars to their life on the Earth in many ways, ranging from stories to scientific investigations. In each culture, the response to the stars mirrored the environment, history, and belief system of the people and provided a means to bind them together.

It is interesting to compare our modern circumstances to those of ancient people. We can learn much from ancient tales, since human beings have not changed much. And yet our cosmology based on science is unique and offers us a perspective no other generation has shared. Our current picture of astrophysical cosmology is one in which observational constraints place increasingly precise limits on what were formerly the province of myth, legend, and folk tradition—the beginning and fate of the universe.

By observing galaxy clusters, motions within galaxies, and large census data on galaxies, astrophysics has placed very exact limits on how much of the universe is made of ordinary matter such as protons and neutrons (also known as $\Omega_b$ or the "baryonic" fraction), and $\Omega_m$, the total density of regular and dark matter in the universe.


B.E. Penprase, *The Power of Stars*, DOI 10.1007/978-3-319-52597-6_10

Additional constraints arise from observations of the microwave background radiation, in which small-scale variations give rise to minute differences in intensity from one part of the sky to the other. These small variations, or anisotropies, are typically only one part in 100,000 and require very precise telescopes to detect. By studying the sizes of variations on the sky, one can determine the values of many of the numbers that describe the mathematical model of our universe.

Cosmological parameters are something like the "holy grail" of modern astrophysics and typically include $H_o$ (Hubble's constant), $\Omega$ (the density of the universe), and $\Lambda$ (the fraction of dark energy). The last of these parameters is often written as $\Omega_\Lambda$ and is based on the recent discovery of an accelera-tion term in the Big Bang, popularly known as the "dark energy." One reason why the values of these parameters are so eagerly sought is that they hold the keys to understanding the ultimate fate of the universe—will the universe expand forever and eventually become cold and dark? Or will the expan-sion reverse and cause a "Big Crunch" in which matter collapses and heats up as gravity pulls the universe back into a singularity?

## The Concordance Cosmology

Recent research from the world's largest telescopes, a fleet of spacecraft poring over photons from the early universe, and a number of the best scientists has given us what is called a "concordance" cosmol-ogy—in which very different observational techniques are converging on answers to the size, age, and fate of the universe. Over decades, satellites of increasing sensitivity have been studying patterns in the cosmic microwave background radiation. These satellites, beginning with COBE, then WMAP, and now Planck have been converging on a spectrum of the cosmic microwaves and their pattern on the sky that locks in the reverberations in the early universe. These patterns are extremely sensitive to the size, den-sity, and composition of the early universe. Results from these recent astrophysics experiments enable a new era of "precision cosmology." These developments are so impressive that George Smoot, one of the designers of the COBE satellite, stated that "if you're religious, it's like seeing God" (LBL 1994).

Simultaneously, the Hubble Space Telescope, the largest ground-based telescopes and soon the James Webb Space Telescope scan the skies looking for supernovae—which are the brightest lights in the night sky and therefore visible at great distances. As astronomers collect more supernovae, and refine the cali-bration of their brightness with distances, they can also correlate these supernovae with the expansion rate of the universe with cosmic time—which is extremely sensitive to the density of the universe. These supernova measurements have been improving over decades. The new observations more accurately account for absorption from dust in the universe (which makes the supernovae slightly fainter) and also benefit from improved calibration of the "cosmic distance ladder" that allows astronomers to link measurements of nearby stars, nearby galaxies, and the far-flung galaxies containing supernovae.

The two techniques, cosmic microwaves and supernovae are complementary, and so the uncertain-ties have different levels when trying to measure the mix of dark matter and dark energy in the uni-verse. The overlap of their "error ellipses" gives a converging and smaller region of parameter space, which is locking in the fundamental numbers that dictate the fate of the universe—$\Omega_\Lambda$, the density of dark energy, and $\Omega_M$ the total density of matter (dark and ordinary gravitating particles).

A third complimentary technique is to use clusters of galaxies to probe the mass and distance from reconstructing arcs of light known as gravitational lenses. This technique allows astronomers to peer at clusters with thousands of galaxies and model the curved paths taken by the lensed light through vast swathes of the universe. These models can sense the distance and geometry of space over billions of light years and give a third ellipse in the space to give an even smaller convergence of uncertainty in values of $\Omega$.

Additional data from the BOSS Galaxy survey and other galaxies provides a fourth technique, which uses the three-dimensional arrangements of millions of galaxies to constrain models of galaxy clustering. These models are sensitive to the densities of dark matter and dark energy, provides a fourth complementary technique for constraining cosmological parameters. The overlap of the best estimates of $\Omega_\Lambda$ and $\Omega_M$ gives the concordance values that lock in fractions of dark energy and dark matter, which will determine the long-term fate of our universe. Figure 10.1 shows an example of the primary data used for each technique. Astronomers will steadily improve these values over the coming years and as the universe expands—our uncertainty in the values of its parameters continues to shrink!

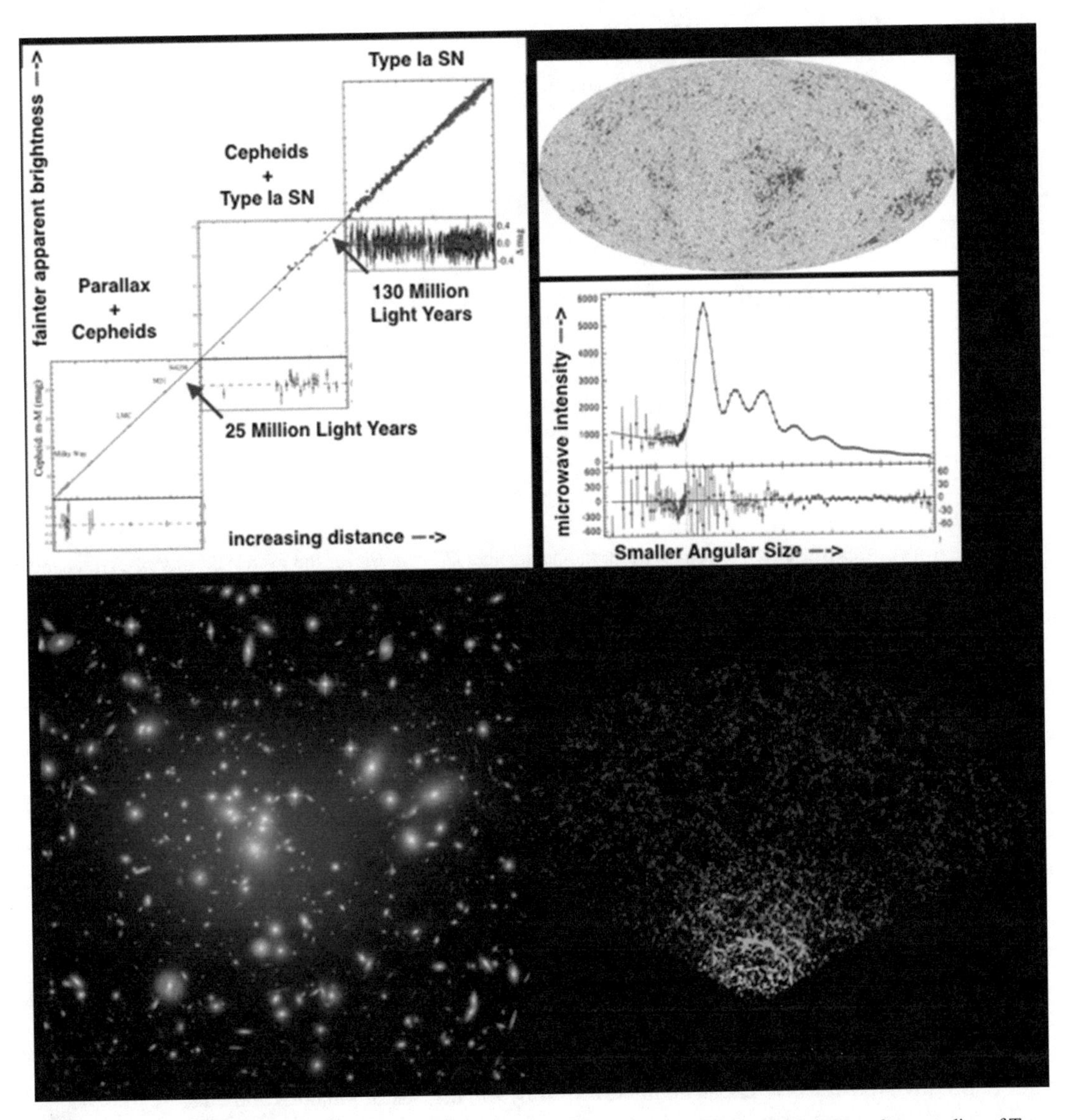


**Fig. 10.1** Representative data from four elements in the "concordance" cosmology. *Top left*: Data from studies of Type Ia supernovae using the Hubble Space Telescope. *Top right*: Data from the Cosmic Microwave Background (CMB) as measured by the Planck spacecraft (from Ade et al. 2014). *Bottom left*: The galaxy cluster Abell 1689, showing gravitational lensing from the massive galaxies and dark matter clustered near the center (from Jullo et al. 2010). *Bottom right*: Luminous red galaxies from the SDSS BOSS survey

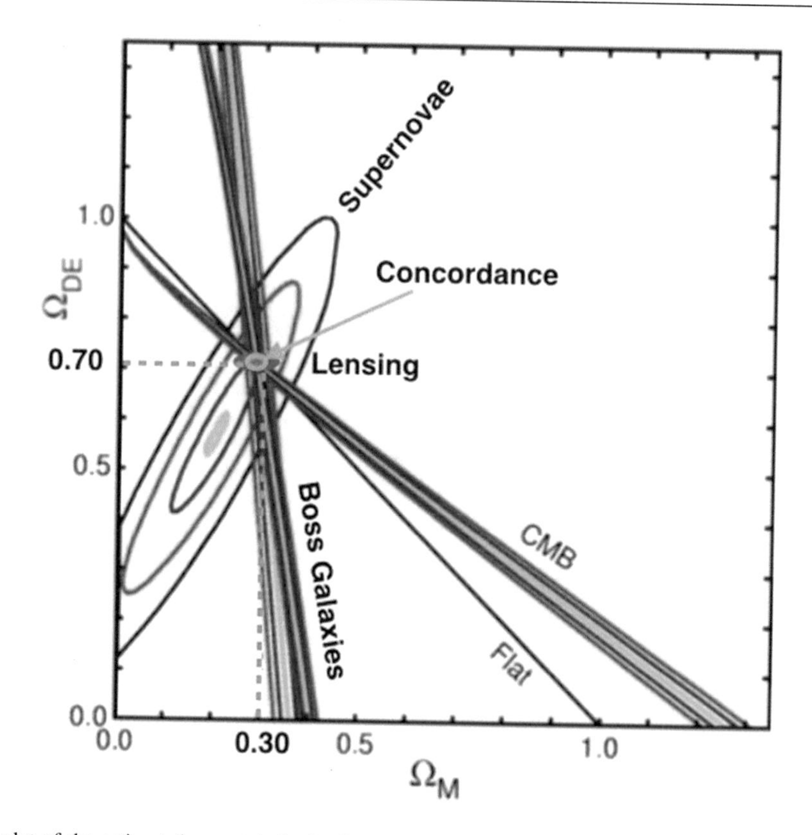


**Fig. 10.2** A plot of the estimated uncertainties in the parameters $\Omega_{DE}$ and $\Omega_M$ for the overall density of dark energy (*y*-axis) and the overall density of dark and ordinary matter (*x*-axis). The eclipsed and colored bands give the range of values within a standard deviation of uncertainties within each measurement for measurements of the Hubble constant from exploding stars (Supernovae), the Cosmic Microwave Background (CMB), Gravitational Lensing from clusters of galaxies (Lensing), and a large survey of redshifts from galaxies (BOSS Galaxies). The intersection of all of these measurements (which thankfully is consistent!) provides the basis of the "concordance" cosmology, which gives very precise constraints on the fundamental constants that determine the composition, and fate of our universe (figure by the author; adapted from Wright, 2014, with data from Dawson et al. 2013, Ade et al. 2014, and Jullo et al. 2010)

Figure 10.2 illustrates how all of these techniques come together to create what is called the "concordance" cosmology. Each of the four techniques provides an "error ellipse" in a plane where the *x* and *y* axes represent values of $\Omega_\Lambda$, the density of dark energy, and $\Omega_M$ the total density of matter. The overlap of these constraints is a very tight circle in the middle of the figure which is the joint set of points where all four techniques agree—the "concordance." The values of $\Omega_\Lambda$ and $\Omega_M$, developed by hundreds of astronomers across the earth, determine the ultimate fate of our physical universe. Our universe appears to be at a critical density—which is where the balance between the matter density and outward motion is poised in a delicate balance. This critical density is sometimes called a "flat universe" and it is the boundary between the "open" universe, where expansion continues forever, and the "closed" universe, where expansion is reversed by the masses of galaxies and dark matter.

Theoretical arguments have made the case for a critical density or "flat" universe for decades, and the latest observational data is consistent with this density. The irony is that even though the mass-energy density of the universe is sufficient to give a deceleration of the cosmic expansion, the unusual nature of dark energy causes it to retain its same intensity despite every larger volumes of space. The result is that Dark Energy will reverse what would have been a gentle deceleration and instead cause an exponential rate of expansion in timescales of hundreds of billions of years. The fate of the universe—according to the best cosmological models—is a slight deceleration of cosmic expansion over the next few tens of billions of years, and then a rising and exponential rate of expansion for the rest of the history of the universe. Our Milky Way in a few hundred billion years is likely to once again become a "universe" again—as nebulae were described by philosophers like Kant in the nineteenth century!

## The Answer to the Great Questions

Our cosmology has given us answers to the questions sought by the ancients. How old is the universe? The answer is well known—13.7 billion years old. Where is the universe going? This also is known—the Big Bang cosmology and measured parameters have determined that the universe is destined to expand outward to infinity, cooling, and darkening over billions of years, as the stars begin to exhaust their nuclear fuel.

The idea of measuring a single number that contains the fate of the universe, the ultimate answer may remind one of science fiction:

> Alright," said Deep Thought. "The Answer to the Great Question…"
> "Yes …!"
> "Of Life, the Universe, and Everything …" said Deep Thought.
> "Yes …!"
> "Is …" said Deep Thought, and paused.
> "Yes …!"
> "Is …"
> "Yes …!!!…?"
> "Forty-two," said Deep Thought, with infinite majesty and calm.
> [From Douglas Adam, "Hitchhiker's Guide to the Galaxy" (Adams 1979)]

In a like manner, modern cosmology can provide us with the answer to "the universe and everything." These answers, expressed as physical constants from the Concordance Cosmology, are listed below.

| | |
|---|---|
| $H_o = 70$ km s$^{-1}$ Mpc$^{-1}$ ± 3 | (Hubble's constant) |
| $\Omega_b = 0.045 \pm 0.005$ | (Baryonic density as a fraction of critical density) |
| $\Omega_\Lambda = 0.70 \pm 0.05$ | (Dark energy density as a fraction of critical density) |
| $\Omega_m = 0.30 \pm 0.05$ | (Total matter density, including dark matter) |

Unlike the number 42, however, these cosmological values are subject to constant refinement and adjustment as new telescopes, measurements, and mathematical models improve our understanding. It is also important to note that the apparent "concordance" is always subject to disruption, as seems to be the case from some papers reporting values of the Hubble constant based on new data that appear to be diverging. One group using CMB data from the most recent Planck satellite (Ade et al. 2014) reports a value of the Hubble constant of about 68 km s$^{-1}$ Mpc$^{-1}$ that is less the value obtained from the earlier WMAP satellite, and which is moving away from the latest values derived from studying supernovae. A recent publication (Riess et al. 2016) reported a value of the Hubble constant of 73 km s$^{-1}$ Mpc$^{-1}$, and the discrepancies between both groups are larger than their stated uncertainties. To preserve the "concordance"—most authors adopt values between 69 and 70 km s$^{-1}$ Mpc$^{-1}$ for the Hubble Constant, which is the value reported here. These divergences and refinements are how science works—and like the universe is subject to continuous evolution!

The numbers above imply that the matter that comprises our bodies, our planet, and all the visible material of galaxies observed with all of our telescopes makes a very small component of the universe, at a level of just 0.5% of the density of the universe. This visible "baryonic" matter is perhaps accurately described as a "contaminant" on the much large mass of dark matter and dark energy. The philosophical implications are obvious, since now it is clear that the observed matter which has been studied in physics and astronomy is the tip of a much larger "iceberg" of dark matter and dark energy.

Galaxies we observe have been likened to the "whitecaps" on top of much larger "waves" of matter and energy that pervade the universe. Paradoxically, along with the new exquisite precision of cosmological parameters comes the complete mystery of what substance comprises 98% of the mass and energy density of the universe!

## Telescopes for Mapping the Unknown

Could there be other completely unknown forces, particles, and entirely new types of telescopes that can help us study the universe? A history of science tells us that this is most likely the case. And science is rapidly deploying huge new detectors above, upon, and beneath the earth to study the universe in entirely new ways. The dark matter could be comprised of massive neutrinos, axions, supersymmetric particles (neutralinos, neutrinos, gravitinos), "scalar particles," magnetic monopoles, and other exotic particles predicted by new particle physics theories. Developing methods to detect these particles, and converting their signals into maps on the skies offer a chance to create entirely new types of telescopes that can study the universe in new ways.

One exciting example of a planetary scale particle telescope is the Pierre Auger Observatory, which has spread 1600 detectors across a 3000 km$^2$ site in Western Argentina. These detectors all work together as a telescope, and measure showers of electrons and muons that crash into earth after ultra-high energy cosmic rays are launched into Earth's atmosphere. These cosmic rays arrive to earth with an energy over a trillion times greater than the collisions in CERN's Large Hadron Collider. Some unknown cosmic accelerator has launched these high energy nuclei our way, and the physicists of over 15 countries are collaborating to find out more about the physics of these mysterious particles (Fig. 10.3).

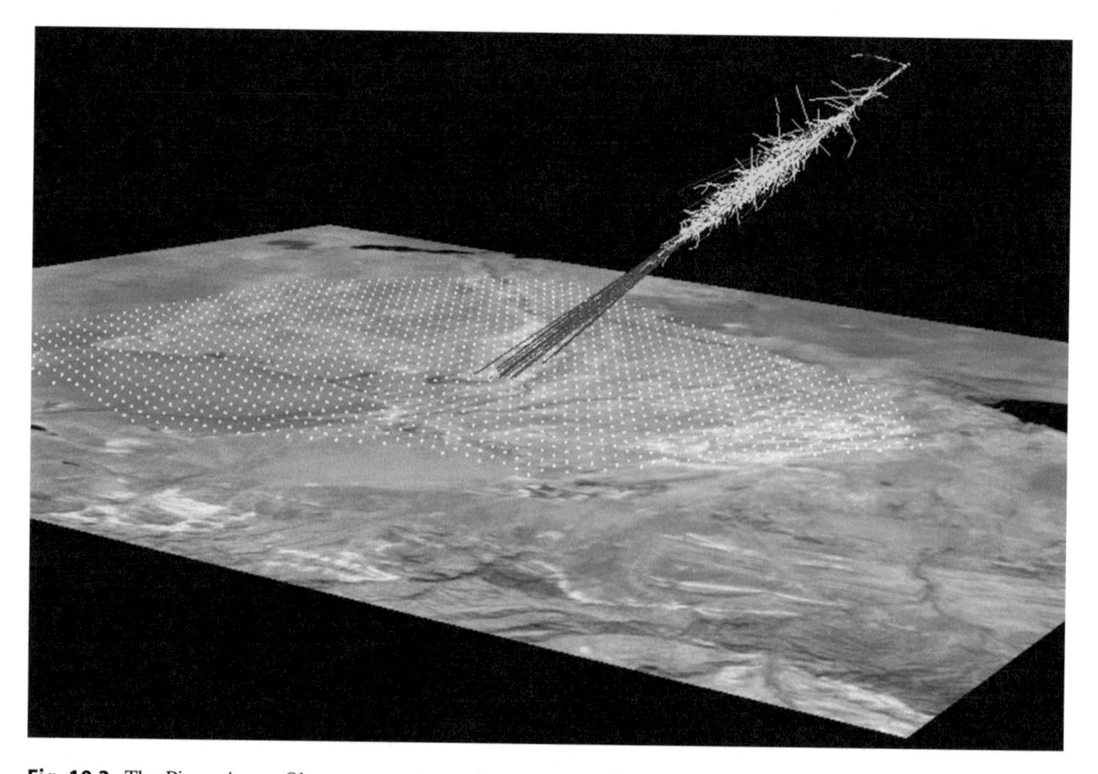

**Fig. 10.3** The Pierre Auger Observatory, a new telescope which maps the origins of ultra-high energy cosmic rays. The "telescope" consists of over 1600 detectors sprawling across 3000 km$^2$ in Western Argentina, and can detect cosmic rays with trillions of times the energy of the CERN accelerator

Another entirely new type of telescope would map the origins of cosmic neutrinos, which was a leading candidate for the dark matter particle. Neutrinos are created by the billions in the centers of stars—our sun alone is thought to produce a "neutrino flux" that bathes the surface of the Earth with about five million neutrinos per cubic centimeter each second! Neutrinos also react only vary rarely with ordinary matter. It is estimated that a neutrino of moderate energy can pass though over a light year of lead without a high probability of absorption. Neutrinos accumulate from stars, and from early emission in the Big Bang, and it is thought that each cubic meter of space in the universe is calculated to contain over 300 million neutrinos (Nave 2005). If the neutrino has even a very tiny mass, it could very nicely account for the dark matter. Laboratory experiments have limited the mass of the neutrino to $m_\nu < 0.2$ eV, which by comparison is less than one two-millionth of the mass of the tiny electron. These experiments limit neutrinos to be less than 7% of the mass of the universe, even if they have the 0.2 eV rest mass.

One of the leading neutrino experiments—which operates much like a colossal neutrino telescope—is the IceCube Neutrino Observatory, a vast array of light detectors imbedded in the Antarctic Ice Sheet built by 300 physicists from 48 institutions in 12 countries. The ultra-pure compressed Antarctic ice provides a free detector over 2 km thick—and physicists have suspended strings of light detectors within this ice sheet to detect the rare flashes of light that occur when cosmic neutrinos interact with the ice. Neutrinos rarely interact with matter, however, making them very difficult to detect. It takes about a light year of lead to stop half of the neutrinos, and so the largest earth-bound detectors only capture an infinitesimal fraction of the neutrinos for study. The IceCube detector uses the one cubic kilometer of ice to increase the odds of interactions with these neutrinos, which pass through the entire earth before emerging up through the ice cap to be detected by the array. The detector also can detect cosmic rays and each day detects 275 million cosmic rays and over 275 atmospheric neutrinos. These planetary-sized detectors are a different kind of telescope than those we discussed in the last chapter—and they are designed to detect new types of cosmic photons to help unravel some of the mysteries of dark matter, dark energy, and the environments surrounding black holes and galactic nuclei that create high energy particles (Fig. 10.4).

One of the most exciting developments of the past century is the direct detection of gravitational waves. The success of the LIGO experiment in detecting the first gravitational waves from outer space has inspired a race to build additional gravitational wave detectors in Italy and in India that together will form an earth-sized gravitational wave telescope. The LIGO detector involves a Michelson interferometer 4 miles long—that creates a pair of tunnels filled with laser light that bounces off incredibly stable mirrors on both ends. A gravitational wave causes all space and matter to expand and contract. LIGO is capable of sensing such distortions—which at the detection limit is one part in $10^{21}$. This means that over the 4-mile length of the interferometer, gravitational waves cause an expansion of only 1/1000 the width of a proton! Despite the astounding technical challenges, the LIGO group has prevailed and reported the first gravitational wave detections on Earth from the merging of a two 30-solar mass black holes at over a billion light years of distance! Soon a network of such facilities across the earth will detect dozens of distant black holes, and be able to peer deep inside the centers of galaxies—into regions where ordinary light cannot emerge. The historic opening of this new window into the workings of the universe cannot be understated in its importance—and like every other new window that has opened in the history of astronomy, gravitational wave astronomy will nearly certainly provide legions of new discoveries (Fig. 10.5).

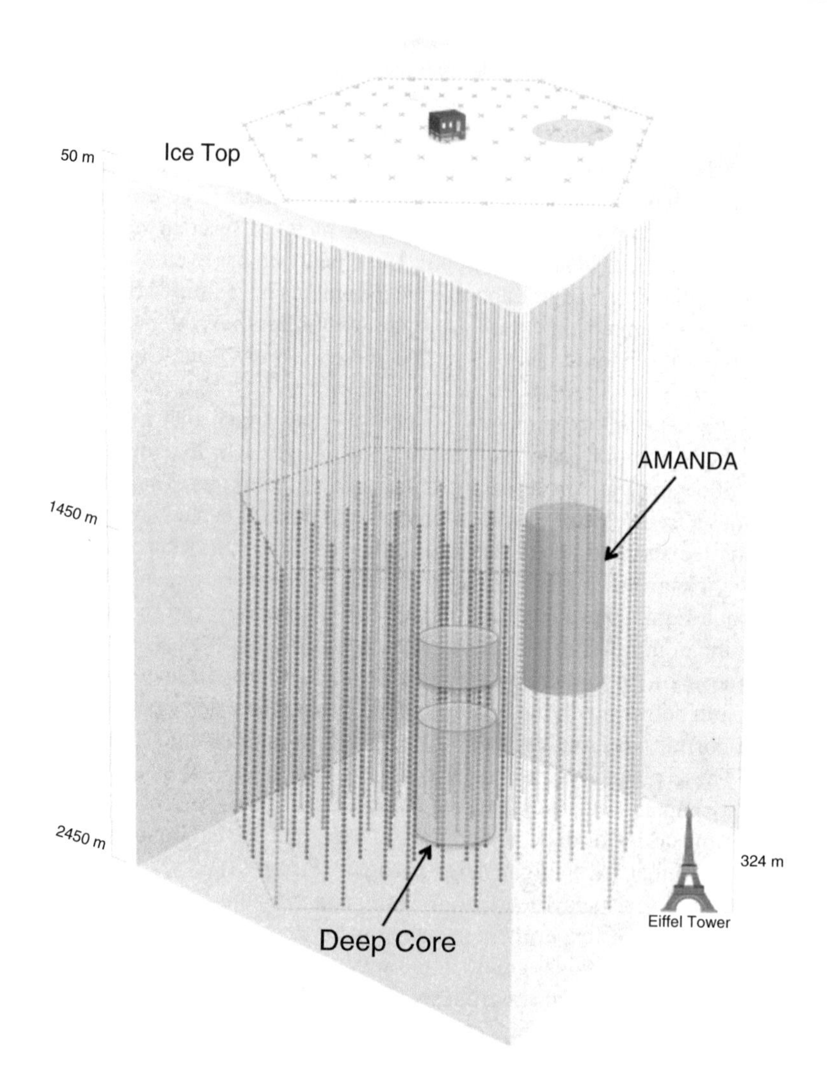


**Fig. 10.4** The IceCube Neutrino Observatory, which makes use of the 2 km thick Antarctic ice sheet to detect light from neutrinos interacting with the ice. The small object on lower right is the Eiffel Tower to scale! This type of planetary scale detector uses the other side of the earth as a filter for other particles, and then the Antarctic ice sheet as its detector medium

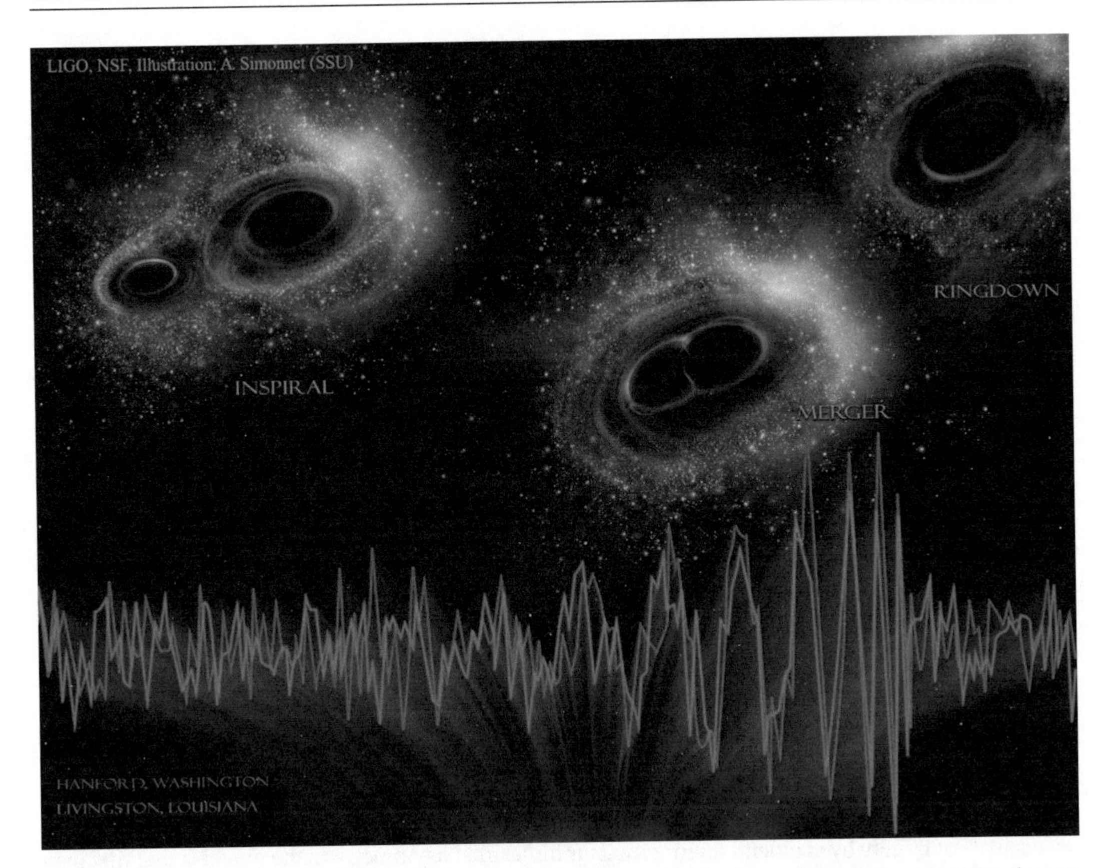


**Fig. 10.5** The LIGO detector has opened up an entirely new window to the universe, by detecting gravitational waves directly. This figure shows a schematic of the two 30 Mo black holes which merged over 1 billion years, giving a detectable signal on both LIGO sites on earth. Additional gravitational wave detectors are being built in Italy and India and will allow for a global gravitational wave telescope, capable of locating the sources of gravitational waves on the sky, and seeing through visible matter to peer into unseen regions of the universe (figure credit—from A. Simonnet, SSU, figure reprinted with permission, and available on http://apod.nasa.gov/apod/image/1602/BHmerger_LIGO_3600.jgp)

## The Fate of the Universe from Dark Energy

Dark energy is inferred from the fact that galaxies seem to be accelerating by an unknown process, based on very careful study of the Hubble law. Earlier cosmologists (such as Hubble) imagined that the universe would coast after the Big Bang and then slow down under its own gravity. New measurements are detecting the signature of a repulsive force, given the name "dark energy" that is expected to dominate the expansion of the universe at very long times into the future—perhaps in a trillion years. Physicists are only beginning to explain dark energy, which does not fit neatly into standard models of particles and forces. One conception of the dark energy is that it is the "vacuum zero point energy," a mysterious energy associated with empty space, and inferred from some particle physics and electromagnetic experiments. A phenomenon known as the Casimir effect describes how the vacuum energy density can be shown to exist through small and measurable forces due to fluctuations in the vacuum energy. Quantum mechanics predicts that virtual particle/anti-particle pairs can be produced within extremely short time intervals and the energy of these pairs can contribute to the vacuum energy.

Dark energy, despite our lack of knowledge about its nature, appears to have the deciding role in the long-term future of our universe. It can be shown that the universe obeys an exponential expansion in future times, since the dark energy is thought to be a "scalar field" with a fraction that increases in proportion while the matter of the universe is diluted through the expansion of space. This means that at large times in the future, the scale of the universe will expand exponentially, thereby diluting the density and decreasing the temperature of the universe toward absolute zero.

Philosophically, this presents our universe with a dismal future, and one from which we cannot escape, since the accuracy of modern observations makes it very difficult to exclude the near certain runaway expansion predicted by the dark matter and dark energy observed in the universe. Ancient people may have had a richer lore anchoring the future of their universe in their values and in a cosmology that celebrated their values, but our generation is the first to know precisely what the future fate of our physical universe may be hundreds of billions of years into the future.

Our modern cosmology also gives us a unique perspective on the place of humans and the Earth in the space and time of the physical universe. From the results of modern astrophysics and precision cosmology, it appears that we humans inhabit a location in the universe in which our (baryonic) composition appears to be a very minor constituent of the universe, during a time at which the density and temperature of the universe are suitable for life to exist, and where the universe is on the precipice of exponential expansion in which the density of matter and energy will drop suddenly, thereby isolating galaxies in the rapid cooling of the universe.

## The Accidental Universe

One conclusion from astrophysics worth deeply considering is that all of the matter in the universe seems to exist largely by accident—and from an infinitesimal asymmetry in the universe that tilts matter over anti-matter. Both matter and anti-matter were nearly in perfect balance in the early universe, at a level of one part in a billion. The presence of any residual matter at all, such as all the atoms in the universe—the galaxy, the sun, the planets, you, and me—appears due to this minute imperfection. This asymmetry illustrates the critical role that randomness and imperfection play in the existence of the universe.

In Japanese art, this intentional asymmetry is known as "*hacho.*" Asymmetry and imperfection is one of the key aesthetic principles in the architecture of temples and statuary to improve their beauty (Lowry 2002, p. 101). Japanese Zen aesthetics also refer to *fukinsei*, an asymmetry or irregularity that mirrors our imperfection, which is an intrinsic essence of the universe. Brush paintings of highest beauty, such as the *enso* (or Zen circle), are incomplete or asymmetric to embody this principle (http://www.presentationzen.com/).

These notions of imperfection and randomness subvert earlier Western aesthetics that guided much of European science—in which a deterministic or "causal" physics guided every observable phenomenon in the universe. As our understanding grows, we appear to approach a more sophisticated form of understanding of the universe—in which the linearity of a deterministic universe is replaced with probabilistic formulations that arise from quantum fluctuations—and necessarily impart imperfect and random imprints on the universe around us (Fig. 10.6).

**Fig. 10.6** Notions of aesthetics in many cultures include asymmetry as a principle of beauty, and also as a fundamental condition of the universe. Rocks within Zen rock gardens (*left*) are selected for their asymmetry and the *enso* symbol from Zen tradition (*right*) is drawn to be asymmetric and incomplete to embody this asymmetry principle, which is also confirmed from modern physics as a necessary condition for the existence of matter

As we improve our comprehension of the large-scale behavior of the universe, we must draw more deeply on our knowledge of the microscopic phenomena predicted by quantum physics and particle physics. The very large and very small are completely interconnected in our discovery—and in this way telescopes and particle accelerators—while working on opposite ends of size and time are able to converge and complement each other as they both race toward an understanding of the beginning of time. However, due to the breakdown in our existing physical laws at the highest energies and earliest times (the so-called Planck Era), our knowledge may be incomplete for decades to come in our description of the first instances of Big Bang.

Despite the amazing detail and predictive power of our astrophysics and the modern Big Bang cosmology, many would like to know—"what happened before the Big Bang?" and "what is the Big Bang expanding into?" Both of these questions are unanswerable by modern cosmology, however. The Big Bang explains how space, time, and all the forces of the universe are created but cannot explain what existed before space and time. In essence the Big Bang applies only to the observable universe we inhabit and we are limited in our understanding of any space or "multiverse" beyond this one much as hypothetical inhabitant of a black hole is limited in discovering anything of the universe outside of their event horizon.

More subtle questions can also be asked—is the universe the result of a random, stochastic process or instead a unique event? At the present time, we lack sufficient scientific knowledge to answer definitively, but we can state objectively that the universe displays set of properties more consistent with a random or "stochastic" process than the result of "fine tuning." The randomness of the universe is supported by observations which include the asymmetry of matter and anti-matter, the random fluctuations needed to produce the inflationary epoch, and the fact that our existence as a species relies on a long series of random events, culminating in self-aware life forms constructed of chemically altered and elementally anomalous debris in a small corner of a non-descript galaxy.

On the other hand, the nature of the initial singularity, the amazing possibility that the universe is poised between two infinities—one infinitesimal singularity and our later expansion into the void at exponential rates, suggests that in many ways we occupy a privileged position and one which begs the question of who and what could have started all of this?

## Our Origins from a Quantum Universe

As we discuss the beginning of our universe, we also begin to realize that this initial singularity belongs within the realm of quantum gravity, itself a paradox of modern physics. The quantum realm of physics challenges the natural "common sense" way of thinking of most people and goes against how our consciousness naturally has evolved. In the same way that creatures on the Earth have evolved vision to adapt to the particular frequencies of light that emerge from our Sun, so perhaps our minds have adapted to the particular natures which matter exhibits in our macroscopic universe. Many of our religious traditions reflect this "subject/object," "conscious identity" form of interaction with the universe, in which we attain identity or "soul" as we relate to the universe and in relation to the God that we acknowledge.

The subject/object construction and the emphasis on individuality is implicit in the familiar passage describing the creation of the universe from Genesis:

> And the Spirit of God moved upon the face of the waters.. And God said let there be light, and there was light…
> And God blessed them and God said unto them, Be fruitful and multiply, and replenish the earth and subdue it
> GeGenesis 1:3, 1:28

Islam likewise describes the creation as the construction of an objective reality by an active and conscious creator:

> Your Lord is God, who in six days created the heavens and the Earth and then mounted the throne: He throweth the veil of night over the day: it pursueth it swiftly: and he created the sun and the moon and the stars, subjected to laws by His behest: Is not all creation and its empire His? Blessed be God the Lord of the Worlds
> Sura VII, Al Araf: 52 (Leeming and Leeming 1995)

Modern physics, however, challenges some of these concepts of linear and deterministic reality. Quantum mechanics uses the "Schrödinger's cat" thought experiment to explain how a cat within a box cannot be determined to be alive or dead until the box is opened. This hypothetical describes how reality arises from quantum wave functions that collapse into a definite "eigenstate" once a measurement is performed. The larger philosophical implication is that the observer and observed are inseparable, that space and time are inseparable, and objective reality becomes split between that which is observed and that which has the potential to be realized. The future is determined by the quantum state which has not yet been measured—undermining classical notions of a deterministic universe.

The vastness of space, the subtle interplay between dark and light matter, and the quantum nature of reality challenge the human mind and give uniquely modern contributions to our understanding of the universe. It is important to recall that all of the conceptions we now consider ancient were once modern, and in coming centuries our current models of physical cosmology will take their place among these "ancient" models.

The idea of potentiality and probability, implicit in quantum mechanics, and the concepts of expanding space from the Big Bang cosmology have created a unique cultural legacy that complements more ancient ways of describing our universe. However, only by viewing both ancient and modern conceptions of the universe together can arrive at deeper understanding of the universe that can incorporate both the laws of modern physics and the unchanging nature of the human mind as it seeks meaning and morality. This kind of understanding allows us to fully appreciate the true power of the stars—as a means to more fully understand our humanity.

## References

Adams, D. 1979. *The hitch hiker's guide to the galaxy*. London: Pan Books.
Ade, P.A.R., et al. 2014. Planck 2013 results. XVI. Cosmological parameters. *Astronomy and Astrophysics* 571: A16.
Dawson, K.S. 2013. The Baryon Oscillation Spectroscopic Survey of SDSS-III. *Astronomical Journal* 145: 10.
Dyson, F.J. 1989. *Infinite in all directions: Gifford lectures given at Aberdeen, Scotland, April-November 1985*. New York: Harper & Row.
Hawking, S.W. 1998. *A brief history of time*. London/New York: Bantam Press.
Jullo, E., et al. 2010. Cosmological constraints from strong gravitational lensing in clusters of galaxies. *Science* 329(5994): 924.
Lao Tzu, D.C. 1969. *Tao TE Ching*. Trans. Lau, D.C., and Whalen, P. Baltimore: Penguin Books.
LBL. 1994. Professor George Smoot. From http://aether.lbl.gov/www/personnel/Smoot-bio.html.
Leeming, D.A., and M.A. Leeming. 1995. *A dictionary of creation myths*. New York: Oxford University Press.
Lowry, D. 2002. *Traditions: essays on the Japanese martial arts and ways*. Boston, MA: Tuttle.
Nave, C.R. 2005. Hyperphysics - the neutrino. From http://hyperphysics.phy-astr.gsu.edu/HBASE/particles/neutrino3.html.
Reynolds, G. Presentation Zen. http://www.presentationzen.com/presentationzen/2009/09/exposing-ourselves-to-traditional-japanese-aestheticideas-notions-that-may-seem-quite-foreign-to-most-of-us-is-a-goo.html. Accessed Sept 2016.
Riess, A.G., et al. 2016b. A 2.4% determination of the local value of the Hubble constant. *Astrophysical Journal* 826: 56.

# Index




B.E. Penprase, *The Power of Stars*, DOI 10.1007/978-3-319-52597-6